ALGORITHMICS
Theory and Practice

Gilles Brassard and Paul Bratley

Département d'informatique et de recherche opérationnelle

Université de Montréal

Prentice-Hall International, Inc.

© 1988 by Prentice-Hall, Inc.
A Division of Simon & Schuster
Englewood Cliffs, NJ 07632

Printed in the United States of America

10 9 8 7 6 5 4 3 2 1

ISBN 0-13-023169-X

Prentice-Hall International (UK) Limited, *London*
Prentice-Hall of Australia Pty. Limited, *Sydney*
Prentice-Hall Canada Inc., *Toronto*
Prentice-Hall Hispanoamericana, S.A., *Mexico*
Prentice-Hall of India Private Limited, *New Delhi*
Prentice-Hall of Japan, Inc., *Tokyo*
Simon & Schuster Asia Pte. Ltd., *Singapore*
Editora Prentice-Hall do Brasil, Ltda., *Rio de Janeiro*
Prentice-Hall, Inc., *Englewood Cliffs, New Jersey*

AL   ICS

for Isabelle and Pat

Contents

Preface

The explosion in computing we are witnessing is arousing extraordinary interest at every level of society. As the power of computing machinery grows, calculations once infeasible become routine. Another factor, however, has had an even more important effect in extending the frontiers of feasible computation: the use of efficient algorithms. For instance, today's typical medium-sized computers can easily sort 100,000 items in 30 seconds using a good algorithm, whereas such speed would be impossible, even on a machine a thousand times faster, using a more naive algorithm. There are other examples of tasks that can be completed in a small fraction of a second, but that would require millions of years with less efficient algorithms (read Section 1.7.3 for more detail).

The Oxford English Dictionary defines *algorithm* as an "erroneous refashioning of *algorism*" and says about *algorism* that it "passed through many pseudo-etymological perversions, including a recent *algorithm*". (This situation is not corrected in the OED Supplement.) Although the Concise Oxford Dictionary offers a more up-to-date definition for the word *algorithm*, quoted in the opening sentence of Chapter 1, we are aware of no dictionary of the English language that has an entry for *algorithmics*, the subject matter of this book.

We chose the word *algorithmics* to translate the more common French term *algorithmique*. (Although this word appears in some French dictionaries, the definition does not correspond to modern usage.) In a nutshell, algorithmics is the systematic study of the fundamental techniques used to design and analyse efficient algorithms. The same word was coined independently by several people, sometimes with slightly different meanings. For instance, Harel (1987) calls algorithmics "the spirit of computing", adopting the wider perspective that it is "the area of human study, knowledge and expertise that concerns algorithms".

Our book is neither a programming manual nor an account of the proper use of data structures. Still less is it a "cookbook" containing a long catalogue of programs ready to be used directly on a machine to solve certain specific problems, but giving at best a vague idea of the principles involved in their design. On the contrary, the aim of our book is to give the reader some basic tools needed to develop his or her own algorithms, in whatever field of application they may be required.

Thus we concentrate on the techniques used to design and analyse efficient algorithms. Each technique is first presented in full generality. Thereafter it is illustrated by concrete examples of algorithms taken from such different applications as optimization, linear algebra, cryptography, operations research, symbolic computation, artificial intelligence, numerical analysis, computing in the humanities, and so on. Although our approach is rigorous and theoretical, we do not neglect the needs of practitioners: besides illustrating the design techniques employed, most of the algorithms presented also have real-life applications.

To profit fully from this book, you should have some previous programming experience. However, we use no particular programming language, nor are the examples for any particular machine. This and the general, fundamental treatment of the material ensure that the ideas presented here will not lose their relevance. On the other hand, you should not expect to be able to use the algorithms we give directly: you will always be obliged to make the necessary effort to transcribe them into some appropriate programming language. The use of Pascal or similarly structured language will help reduce this effort to the minimum necessary.

Some basic mathematical knowledge is required to understand this book. Generally speaking, an introductory undergraduate course in algebra and another in calculus should provide sufficient background. A certain mathematical maturity is more important still. We take it for granted that the reader is familiar with such notions as mathematical induction, set notation, and the concept of a graph. From time to time a passage requires more advanced mathematical knowledge, but such passages can be skipped on the first reading with no loss of continuity.

Our book is intended as a textbook for an upper-level undergraduate or a lower-level graduate course in algorithmics. We have used preliminary versions at both the Université de Montréal and the University of California, Berkeley. If used as the basis for a course at the graduate level, we suggest that the material be supplemented by attacking some subjects in greater depth, perhaps using the excellent texts by Garey and Johnson (1979) or Tarjan (1983). Our book can also be used for independent study: anyone who needs to write better, more efficient algorithms can benefit from it. Some of the chapters, in particular the one concerned with probabilistic algorithms, contain original material.

It is unrealistic to hope to cover all the material in this book in an undergraduate course with 45 hours or so of classes. In making a choice of subjects, the teacher should bear in mind that the first two chapters are essential to understanding the rest of the book, although most of Chapter 1 can probably be assigned as independent reading. The other chapters are to a great extent independent of one another. An elementary course should certainly cover the first five chapters, without necessarily going over each and every example given there of how the techniques can be applied. The choice of the remaining material to be studied depends on the teacher's preferences and inclinations. The last three chapters, however, deal with more advanced topics; the teacher may find it interesting to discuss these briefly in an undergraduate class, perhaps to lay the ground before going into detail in a subsequent graduate class.

Each chapter ends with suggestions for further reading. The references from each chapter are combined at the end of the book in an extensive bibliography including well over 200 items. Although we give the origin of a number of algorithms and ideas, our primary aim is not historical. You should therefore not be surprised if information of this kind is sometimes omitted. Our goal is to suggest supplementary reading that can help you deepen your understanding of the ideas we introduce.

Almost 500 exercises are dispersed throughout the text. It is crucial to *read the problems*: their statements form an integral part of the text. Their level of difficulty is indicated as usual either by the absence of an asterisk (immediate to easy), or by the presence of one asterisk (takes a little thought) or two asterisks (difficult, maybe even a research project). The solutions to many of the difficult problems can be found in the references. No solutions are provided for the other problems, nor do we think it advisable to provide a solutions manual. We hope the serious teacher will be pleased to have available this extensive collection of unsolved problems from which homework assignments can be chosen. Several problems call for an algorithm to be implemented on a computer so that its efficiency may be measured experimentally and compared to the efficiency of alternative solutions. It would be a pity to study this material without carrying out at least one such experiment.

The first printing of this book by Prentice Hall is already in a sense a second edition. We originally wrote our book in French. In this form it was published by Masson, Paris. Although less than a year separates the first French and English printings, the experience gained in using the French version, in particular at an international summer school in Bayonne, was crucial in improving the presentation of some topics, and in spotting occasional errors. The numbering of problems and sections, however, is not always consistent between the French and English versions.

Writing this book would have been impossible without the help of many people. Our thanks go first to the students who have followed our courses in algorithmics over the years since 1979, both at the undergraduate and graduate levels. Particular thanks are due to those who kindly allowed us to copy their course notes: Denis Fortin, Laurent Langlois, and Sophie Monet in Montréal, and Luis Miguel and Dan Philip in Berkeley. We are also grateful to those people who used the preliminary versions of our book, whether they were our own students, or colleagues and students at other universities. The comments and suggestions we received were most valuable. Our warmest thanks, however, must go to those who carefully read and reread several chapters of the book and who suggested many improvements and corrections: Pierre

Beauchemin, André Chartier, Claude Crépeau, Bennett Fox, Claude Goutier, Pierre L'Ecuyer, Pierre McKenzie, Santiago Miro, Jean-Marc Robert, and Alan Sherman.

We are also grateful to those who made it possible for us to work intensively on our book during long periods spent away from Montréal. Paul Bratley thanks Georges Stamon and the Université de Franche-Comté. Gilles Brassard thanks Manuel Blum and the University of California, Berkeley, David Chaum and the CWI, Amsterdam, and Jean-Jacques Quisquater and Philips Research Laboratory, Bruxelles. He also thanks John Hopcroft, who taught him so much of the material included in this book, and Lise DuPlessis who so many times made her country house available; its sylvan serenity provided the setting and the inspiration for writing a number of chapters.

Denise St.-Michel deserves our special thanks. It was her misfortune to help us struggle with the text editing system through one translation and countless revisions. Annette Hall, of Editing, Design, and Production, Inc., was no less misfortuned to help us struggle with the last stages of production. The heads of the laboratories at the Université de Montréal's Département d'informatique et de recherche opérationnelle, Michel Maksud and Robert Gérin-Lajoie, provided unstinting support. We thank the entire team at Prentice Hall for their exemplary efficiency and friendliness; we particularly appreciate the help we received from James Fegen. We also thank Eugene L. Lawler for mentioning our French manuscript to Prentice Hall's representative in northern California, Dan Joraanstad, even before we plucked up the courage to work on an English version. The Natural Sciences and Engineering Research Council of Canada provided generous support.

Last but not least, we owe a considerable debt of gratitude to our wives, Isabelle and Pat, for their encouragement, understanding, and exemplary patience — in short, for putting up with us — while we were working on the French and English versions of this book.

Gilles Brassard
Paul Bratley

ALGORITHMICS

1

Preliminaries

1.1 WHAT IS AN ALGORITHM?

The Concise Oxford Dictionary defines an algorithm as a "process or rules for (esp. machine) calculation". The execution of an algorithm must not include any subjective decisions, nor must it require the use of intuition or creativity (although we shall see an important exception to this rule in Chapter 8). When we talk about algorithms, we shall mostly be thinking in terms of computers. Nonetheless, other systematic methods for solving problems could be included. For example, the methods we learn at school for multiplying and dividing integers are also algorithms. The most famous algorithm in history dates from the time of the Greeks: this is Euclid's algorithm for calculating the greatest common divisor of two integers. It is even possible to consider certain cooking recipes as algorithms, provided they do not include instructions like "Add salt to taste".

When we set out to solve a problem, it is important to decide which algorithm for its solution should be used. The answer can depend on many factors: the size of the instance to be solved, the way in which the problem is presented, the speed and memory size of the available computing equipment, and so on. Take elementary arithmetic as an example. Suppose you have to multiply two positive integers using only pencil and paper. If you were raised in North America, the chances are that you will multiply the multiplicand successively by each figure of the multiplier, taken from right to left, that you will write these intermediate results one beneath the other shifting each line one place left, and that finally you will add all these rows to obtain your answer. This is the "classic" multiplication algorithm.

However, here is quite a different algorithm for doing the same thing, sometimes called "multiplication *à la russe*". Write the multiplier and the multiplicand side by side. Make two columns, one under each operand, by repeating the following rule until the number under the multiplier is 1: divide the number under the multiplier by 2, ignoring any fractions, and double the number under the multiplicand by adding it to itself. Finally, cross out each row in which the number under the multiplier is even, and then add up the numbers that remain in the column under the multiplicand. For example, multiplying 19 by 45 proceeds as in Figure 1.1.1. In this example we get $19 + 76 + 152 + 608 = 855$. Although this algorithm may seem funny at first, it is essentially the method used in the hardware of many computers. To use it, there is no need to memorize any multiplication tables: all we need to know is how to add up, and how to double a number or divide it by 2.

45	19	19
22	38	---
11	76	76
5	152	152
2	304	----
1	608	608
		855

Figure 1.1.1.   Multiplication *à la russe*.

We shall see in Section 4.7 that there exist more efficient algorithms when the integers to be multiplied are very large. However, these more sophisticated algorithms are in fact slower than the simple ones when the operands are not sufficiently large.

At this point it is important to decide how we are going to *represent* our algorithms. If we try to describe them in English, we rapidly discover that natural languages are not at all suited to this kind of thing. Even our description of an algorithm as simple as multiplication *à la russe* is not completely clear. We did not so much as try to describe the classic multiplication algorithm in any detail. To avoid confusion, we shall in future specify our algorithms by giving a corresponding *program*. However, we shall not confine ourselves to the use of one particular programming language: in this way, the essential points of an algorithm will not be obscured by the relatively unimportant programming details.

We shall use phrases in English in our programs whenever this seems to make for simplicity and clarity. These phrases should not be confused with comments on the program, which will always be enclosed within braces. Declarations of scalar quantities (integer, real, or Boolean) are usually omitted. Scalar parameters of functions and procedures are passed by value unless a different specification is given explicitly, and arrays are passed by reference.

The notation used to specify that a function or a procedure has an array parameter varies from case to case. Sometimes we write, for instance

procedure *proc*1(*T* : **array**)

or even

    procedure $proc2(T)$

if the type and the dimensions of the array T are unimportant or if they are evident from the context. In such a case $\#T$ denotes the number of elements in the array T. If the bounds or the type of T are important, we write

    procedure $proc3(T[1..n])$

or more generally

    procedure $proc4(T[a..b] : integers)$.

In such cases n, a, and b should be considered as formal parameters, and their values are determined by the bounds of the actual parameter corresponding to T when the procedure is called. These bounds can be specified explicitly, or changed, by a procedure call of the form

    $proc3(T[1..m])$.

    To avoid proliferation of **begin** and **end** statements, the range of a statement such as **if**, **while**, or **for**, as well as that of a declaration such as **procedure**, **function**, or **record**, is shown by indenting the statements affected. The statement **return** marks the dynamic end of a procedure or a function, and in the latter case it also supplies the value of the function. The operators **div** and **mod** represent integer division (discarding any fractional result) and the remainder of a division, respectively. We assume that the reader is familiar with the concepts of recursion and of pointers. The latter are denoted by the symbol "↑". A reader who has some familiarity with Pascal, for example, will have no difficulty understanding the notation used to describe our algorithms. For instance, here is a formal description of multiplication *à la russe*.

```
function russe (A ,B )
  arrays X ,Y
  {initialization}
  X [1] ← A ;  Y [1] ← B
  i ← 1
  {make the two columns}
  while X [i ] > 1 do
    X [i +1] ← X [i ] div 2
    Y [i +1] ← Y [i ] + Y [i ]
    i ← i +1
  {add the appropriate entries}
  prod ← 0
  while i > 0 do
    if X [i ] is odd then prod ← prod + Y [i ]
    i ← i − 1
  return prod
```

If you are an experienced programmer, you will probably have noticed that the arrays X and Y are not really necessary, and that this program could easily be simplified. However, we preferred to follow blindly the preceding description of the algorithm, even if this is more suited to a calculation using pencil and paper than to computation on a machine. The following APL program describes exactly the same algorithm (although you might reasonably object to a program using logarithms, exponentiation, and multiplication by powers of 2 to describe an algorithm for multiplying two integers ...).

$$\nabla\ R \leftarrow A\ RUSAPL\ B;T$$
$$[1]\ \ R \leftarrow +/(2|\lfloor A \div T)/B \times T \leftarrow 1,2 * \iota \lfloor 2 * A\ \ \nabla$$

On the other hand, the following program, despite a superficial resemblance to the one given previously, describes quite a different algorithm.

function *not-russe* (A,B)
  arrays X,Y
  {initialization}
  $X[1] \leftarrow A;\ Y[1] \leftarrow B$
  $i \leftarrow 1$
  {make the two columns}
  while $X[i] > 1$ **do**
    $X[i+1] \leftarrow X[i] - 1$
    $Y[i+1] \leftarrow B$
    $i \leftarrow i + 1$
  {add the appropriate entries}
  $prod \leftarrow 0$
  while $i > 0$ **do**
    if $X[i] > 0$ **then** $prod \leftarrow prod + Y[i]$
    $i \leftarrow i - 1$
  return $prod$

We see that different algorithms can be used to solve the same problem, and that different programs can be used to describe the same algorithm. It is important not to lose sight of the fact that in this book we are interested in *algorithms,* not in the *programs* used to describe them.

1.2 PROBLEMS AND INSTANCES

Multiplication *à la russe* is not just a way to multiply 45 by 19. It gives a general solution to the *problem* of multiplying positive integers. We say that $(45, 19)$ is an *instance* of this problem. Most interesting problems include an infinite collection of instances. Nonetheless, we shall occasionally consider finite problems such as that of playing a perfect game of chess. An algorithm must work correctly on every instance of the problem it claims to solve. To show that an algorithm is incorrect, we need only find one instance of the problem for which it is unable to find a correct answer. On the other hand, it is usually more difficult to prove the correctness of an algorithm. When we specify a problem, it is important to define its *domain of definition*, that is, the set of instances to be considered. Although multiplication *à la russe* will not work if the first operand is negative, this does not invalidate the algorithm since (−45, 19) is *not* an instance of the problem being considered.

Any real computing device has a limit on the size of the instances it can handle. However, this limit cannot be attributed to the algorithm we choose to use. Once again we see that there is an essential difference between programs and algorithms.

1.3 THE EFFICIENCY OF ALGORITHMS

When we have a problem to solve, it is obviously of interest to find several algorithms that might be used, so we can choose the best. This raises the question of how to decide which of several algorithms is preferable. The *empirical* (or a posteriori) approach consists of programming the competing algorithms and trying them on different instances with the help of a computer. The *theoretical* (or a priori) approach, which we favour in this book, consists of determining mathematically the quantity of resources (execution time, memory space, etc.) needed by each algorithm *as a function of the size of the instances considered*.

The *size* of an instance x, denoted by $|x|$, corresponds formally to the number of bits needed to represent the instance on a computer, using some precisely defined and reasonably compact encoding. To make our analyses clearer, however, we often use the word "size" to mean any integer that in some way measures the number of components in an instance. For example, when we talk about sorting (see Section 1.7.1), an instance involving n items is generally considered to be of size n, even though each item would take more than one bit when represented on a computer. When we talk about numerical problems, we sometimes give the efficiency of our algorithms in terms of the *value* of the instance being considered, rather than its size (which is the number of bits needed to represent this value in binary).

The advantage of the theoretical approach is that it depends on neither the computer being used, nor the programming language, nor even the skill of the programmer. It saves both the time that would have been spent needlessly programming an inefficient algorithm and the machine time that would have been wasted testing it. It also allows us to study the efficiency of an algorithm when used on instances of any size. This is often not the case with the empirical approach, where practical considerations often force us to test our algorithms only on instances of moderate size. This last point is particularly important since often a newly discovered algorithm may only begin to perform better than its predecessor when both of them are used on large instances.

It is also possible to analyse algorithms using a *hybrid* approach, where the form of the function describing the algorithm's efficiency is determined theoretically, and then any required numerical parameters are determined empirically for a particular program and machine, usually by some form of regression. This approach allows predictions to be made about the time an actual implementation will take to solve an instance much larger than those used in the tests. If such an extrapolation is made solely on the basis of empirical tests, ignoring all theoretical considerations, it is likely to be less precise, if not plain wrong.

It is natural to ask at this point what *unit* should be used to express the theoretical efficiency of an algorithm. There can be no question of expressing this efficiency in seconds, say, since we do not have a standard computer to which all measurements might refer. An answer to this problem is given by the *principle of invariance*, according to which two different implementations of the same algorithm will not differ in efficiency by more than some multiplicative constant. More precisely, if two implementations take $t_1(n)$ and $t_2(n)$ seconds, respectively, to solve an instance of size n, then there always exists a positive constant c such that $t_1(n) \le ct_2(n)$ whenever n is sufficiently large. This principle remains true whatever the computer used (provided it is of a conventional design), regardless of the programming language employed and regardless of the skill of the programmer (provided that he or she does not actually modify the algorithm!). Thus, a change of machine may allow us to solve a problem 10 or 100 times faster, but only a change of algorithm will give us an improvement that gets more and more marked as the size of the instances being solved increases.

Coming back to the question of the unit to be used to express the theoretical efficiency of an algorithm, there will be no such unit: we shall only express this efficiency to within a multiplicative constant. We say that an algorithm takes a time *in the order of* $t(n)$, for a given function t, if there exist a positive constant c and an implementation of the algorithm capable of solving every instance of the problem in a time bounded above by $ct(n)$ seconds, where n is the size (or occasionally the value, for numerical problems) of the instance considered. The use of seconds in this definition is obviously quite arbitrary, since we only need change the constant to bound the time by $at(n)$ years or $bt(n)$ microseconds. By the principle of invariance any other implementation of the algorithm will have the same property, although the multiplicative constant may change from one implementation to another. In the next chapter we give a more rigorous treatment of this important concept known as the *asymptotic notation*. It will be clear from the formal definition why we say "*in* the order of" rather than the more usual "*of* the order of".

Certain orders occur so frequently that it is worth giving them a name. For example, if an algorithm takes a time in the order of n, where n is the size of the instance to be solved, we say that it takes *linear* time. In this case we also talk about a *linear algorithm*. Similarly, an algorithm is *quadratic*, *cubic*, *polynomial*, or *exponential* if it takes a time in the order of n^2, n^3, n^k, or c^n, respectively, where k and c are appropriate constants. Sections 1.6 and 1.7 illustrate the important differences between these orders of magnitude.

The *hidden multiplicative constant* used in these definitions gives rise to a certain danger of misinterpretation. Consider, for example, two algorithms whose implementations on a given machine take respectively n^2 days and n^3 seconds to solve an instance of size n. It is only on instances requiring more than 20 million years to solve that the quadratic algorithm outperforms the cubic algorithm! Nevertheless, from a theoretical point of view, the former is *asymptotically* better than the latter, that is to say, its performance is better on all sufficiently large instances.

The other resources needed to execute an algorithm, memory space in particular, can be estimated theoretically in a similar way. It may also be interesting to study the possibility of a trade-off between time and memory space: using more space sometimes allows us to reduce the computing time, and conversely. In this book, however, we concentrate on execution time.

Finally, note that logarithms to the base 2 are so frequently used in the analysis of algorithms that we give them their own special notation: thus "$\lg n$" is an abbreviation for $\log_2 n$. As is more usual, "ln" and "log" denote natural logarithms and logarithms to the base 10, respectively.

1.4 AVERAGE AND WORST-CASE ANALYSIS

The time taken by an algorithm can vary considerably between two different instances of the same size. To illustrate this, consider two elementary sorting algorithms: insertion and selection.

```
procedure insert (T [1 .. n ])
    for i ← 2 to n do
        x ← T [i];  j ← i − 1
        while j > 0 and x < T [ j ] do T [ j + 1] ← T [ j ]
                                        j ← j − 1
        T [ j + 1] ← x
```

and

```
procedure select (T [1 .. n ])
    for i ← 1 to n − 1 do
        minj ← i ;  minx ← T [i]
        for j ← i + 1 to n do
            if T [ j ] < minx then minj ← j
                                   minx ← T [ j ]
        T [minj ] ← T [i]
        T [i] ← minx
```

Problem 1.4.1. Simulate these two algorithms on the arrays

$$T = [3, 1, 4, 1, 5, 9, 2, 6, 5, 3], \ \ U = [1, 2, 3, 4, 5, 6], \ \text{and} \ \ V = [6, 5, 4, 3, 2, 1].$$

Make sure you understand how they work.   □

Let U and V be two arrays of n elements, such that U is already sorted in ascending order, whereas V is in descending order. Problem 1.4.1 shows that both these algorithms take more time on V than on U. In fact, V represents the worst case for these two algorithms: no array of n elements requires more work. Nonetheless, the time required by the selection sorting algorithm is not very sensitive to the original order of the array to be sorted: the test "**if** $T[j] < minx$" is executed exactly the same number of times in every case. The variation in execution time is only due to the number of times the assignments in the **then** part of this test are executed. To verify this, we programmed this algorithm in Pascal on a DEC VAX 780. We found that the time required to sort a given number of elements using selection sort does not vary by more than 15% whatever the initial order of the elements to be sorted. As Example 2.2.1 will show, the time required by $select(T)$ is quadratic, regardless of the initial order of the elements.

The situation is quite different if we compare the times taken by the insertion sort algorithm on the arrays U and V. On the one hand, $insert(U)$ is very fast, because the condition controlling the **while** loop is always false at the outset. The algorithm therefore performs in linear time. On the other hand, $insert(V)$ takes quadratic time, because the **while** loop is executed $i-1$ times for each value of i (see Example 2.2.3). The variation in time is therefore considerable, and moreover, it increases with the number of elements to be sorted. An implementation in Pascal on the DEC VAX 780 shows that $insert(U)$ takes less than one-fifth of a second if U is an array of 5,000 elements already in ascending order, whereas $insert(V)$ takes three and a half minutes when V is an array of 5,000 elements in descending order.

If such large variations can occur, how can we talk about the time taken by an algorithm solely in terms of the size of the instance to be solved? We usually consider the *worst case* of the algorithm, that is, for each size we only consider those instances of that size on which the algorithm requires the most time. Thus we say that insertion sorting takes quadratic time in the worst case.

Worst-case analysis is appropriate for an algorithm whose response time is critical. For example, if it is a question of controlling a nuclear power plant, it is crucial to know an upper limit on the system's response time, regardless of the particular instance to be solved. On the other hand, in a situation where an algorithm is to be used many times on many different instances, it may be more important to know the *average* execution time on instances of size n. We saw that the time taken by the insertion sort algorithm varies between the order of n and the order of n^2. If we can calculate the average time taken by the algorithm on the $n!$ different ways of initially ordering n elements (assuming they are all distinct), we shall have an idea of the likely time taken to sort an array initially in random order. We shall see in Example 2.2.3 that this average time is also in the order of n^2. The insertion sorting algorithm thus takes quadratic time both on the average and in the worst case, although in certain cases it can be much faster. In Section 4.5 we shall see another sorting algorithm that also takes quadratic time in the worst case, but that requires only a time in the order of $n \log n$ on the average. Even though this algorithm has a bad worst case, it is among the fastest algorithms known on the average.

It is usually harder to analyse the average behaviour of an algorithm than to analyse its behaviour in the worst case. Also, such an analysis of average behaviour can be misleading if in fact the instances to be solved are not chosen randomly when the algorithm is used in practice. For example, it could happen that a sorting algorithm might be used as an internal procedure in some more complex algorithm, and that for some reason it might mostly be asked to sort arrays whose elements are already nearly ordered. In this case, the hypothesis that each of the $n!$ ways of initially ordering n elements is equally likely fails. A useful analysis of the average behaviour of an algorithm therefore requires some a priori knowledge of the distribution of the instances to be solved, and this is normally an unrealistic requirement. In Chapter 8 we shall see how this difficulty can be circumvented for certain algorithms, and their behaviour made independent of the specific instances to be solved.

In what follows we shall only be concerned with worst-case analyses unless stated otherwise.

1.5 WHAT IS AN ELEMENTARY OPERATION?

An *elementary operation* is an operation whose execution time can be bounded above by a constant depending only on the particular implementation used (machine, programming language, and so on). Since we are only concerned with execution times of algorithms defined to within a multiplicative constant, it is only the number of elementary operations executed that matters in the analysis, not the exact time required by each of them. Equivalently, we say that elementary operations can be executed *at unit cost*. In the description of an algorithm it may happen that a line of program corresponds to a variable number of elementary operations. For example, if T is an array of n elements, the time required to compute

$$x \leftarrow \min\{ T[i] \mid 1 \le i \le n \}$$

increases with n, since it is an abbreviation for

$$x \leftarrow T[1]$$
for $i \leftarrow 2$ **to** n **do**
$\quad$ **if** $T[i] < x$ **then** $x \leftarrow T[i]$.

Similarly, some mathematical operations are too complex to be considered elementary. If we allowed ourselves to count the evaluation of a factorial and a test for divisibility at unit cost, regardless of the operand's size, Wilson's theorem would let us test an integer for primality with astonishing efficiency.

function *Wilson* (n)
$\quad$ {returns *true* if and only if n is prime}
$\quad$ **if** n divides $((n-1)! + 1)$ exactly **then return** *true*
$\qquad\qquad\qquad\qquad\qquad\qquad\qquad$ **else return** *false*

Can we consider addition and multiplication to be unit cost operations? In theory these operations are not elementary since the time needed to execute them increases with the length of the operands. In practice, however, it may be sensible to consider them as elementary operations so long as the operands concerned are of a reasonable size in the instances we expect to encounter. Two examples will illustrate what we mean.

> **function** *Not-Gauss* (*n*)
> {calculates the sum of the integers from 1 to *n* }
> *sum* ← 0
> **for** *i* ← 1 **to** *n* **do** *sum* ← *sum* + *i*
> **return** *sum*

and

> **function** *Fibonacci* (*n*)
> {calculates the *n* th term of the Fibonacci sequence (see section 1.7.5)}
> *i* ← 1; *j* ← 0
> **for** *k* ← 1 **to** *n* **do** *j* ← *i* + *j*
>   *i* ← *j* − *i*
> **return** *j*

In the algorithm called *Not-Gauss* the value of *sum* stays quite reasonable for all the instances that the algorithm can realistically be expected to meet in practice. If we are using a 32-bit machine, all the additions can be executed directly provided that *n* is no greater than 65,535. In theory, however, the algorithm should work for all possible values of *n*, so that no real machine can in fact execute these additions at unit cost if *n* is chosen sufficiently large. The analysis of the algorithm must therefore depend on its intended domain of application.

The situation is quite different in the case of *Fibonacci*. It suffices to take *n* = 47 to have the last addition "*j* ← *i* + *j* " cause arithmetic overflow on a 32-bit machine. As many as 45,496 bits are needed to hold the result corresponding to *n* = 65,535. It is therefore not realistic, as a practical matter, to consider that these additions can be carried out at unit cost; rather, we must attribute to them a cost proportional to the length of the operands concerned. In Example 2.2.8, this algorithm (there called *fib2*) is shown to take quadratic time, even though at first glance its execution time appears to be linear.

In the case of multiplication, although it may still be reasonable to consider this an elementary operation for sufficiently small operands, it is even more important to ensure that arithmetic operations do not overflow: it is easier to produce large operands by repeated multiplication than by addition. The following problem illustrates this danger.

** Problem 1.5.1.   Use Wilson's theorem (*n* is prime if and only if it is a divisor of (*n* − 1)! + 1), Newton's binomial theorem, and the divide-and-conquer technique discussed in Chapter 4 to design an algorithm capable of deciding in a time in the order of log *n*, given an integer *n*, whether or not *n* is prime. In your analysis you may assume that additions, multiplications, and tests of divisibility by an integer (but not calculations of factorials or exponentials) can be carried out in unit time, regardless of the size of the operands involved.   □

A similar problem can arise when we analyse algorithms involving real numbers if the required precision increases with the size of the instances to be solved. One typical example of this phenomenon is the use of De Moivre's formula to calculate values in the Fibonacci sequence (see Section 1.7.5). In most practical situations, however, the use of single precision floating point arithmetic proves satisfactory despite the inevitable loss of precision. When this is so, it is reasonable to count such arithmetic operations at unit cost.

To sum up, even deciding whether an instruction as apparently innocent as "$j \leftarrow i + j$" can be considered as elementary or not calls for the use of judgement. In what follows we count additions, subtractions, multiplications, divisions, modulo operations, Boolean operations, comparisons, and assignments at unit cost unless explicitly stated otherwise.

1.6 WHY DO WE NEED EFFICIENT ALGORITHMS ?

As computing equipment gets faster and faster, it may seem hardly worthwhile to spend our time trying to design more efficient algorithms. Would it not be easier simply to wait for the next generation of computers? The remarks made in the preceding sections show that this is not true. Suppose, to illustrate the argument, that to solve a particular problem you have available an exponential algorithm and a computer capable of running this algorithm on instances of size n in $10^{-4} \times 2^n$ seconds. Your program can thus solve an instance of size 10 in one-tenth of a second. Solving an instance of size 20 will take nearly two minutes. To solve an instance of size 30, even a whole day's computing will not be sufficient. Supposing you were able to run your computer without interruption for a year, you would only just be able to solve an instance of size 38.

Since you need to solve bigger instances than this, you buy a new computer one hundred times faster than the first. With the same algorithm you can now solve an instance of size n in only $10^{-6} \times 2^n$ seconds. You may feel you have wasted your money, however, when you figure out that now, when you run your new machine for a whole year, you cannot even solve an example of size 45. In general, if you were previously able to solve an instance of size n in some given time, your new machine will solve instances of size at best $n + 7$ in the same time.

Suppose you decide instead to invest in algorithmics. You find a cubic algorithm that can solve your problem. Imagine, for example, that using the original machine this new algorithm can solve an instance of size n in $10^{-2} \times n^3$ seconds. In one day you can now solve instances whose size is greater than 200; with one year's computation you can almost reach size 1,500. This is illustrated by Figure 1.6.1.

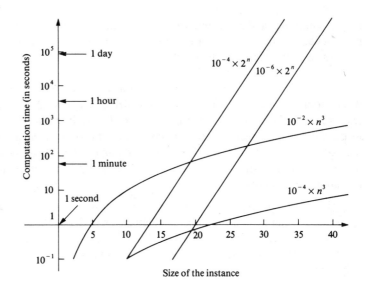

Figure 1.6.1. Algorithmics versus hardware.

Not only does the new algorithm offer a much greater improvement than the purchase of new machinery, it will also, supposing you are able to afford both, make such a purchase much more profitable. In fact, thanks to your new algorithm, a machine one hundred times faster than the old one will allow you to solve instances four or five times bigger in the same length of time. Nevertheless, the new algorithm should not be used uncritically on all instances of the problem, in particular on the rather small ones. On the original machine the new algorithm takes 10 seconds to solve an instance of size 10, which is one hundred times *slower* than the old algorithm. The new algorithm is faster only for instances of size 20 or greater. Naturally, it is possible to combine the two algorithms into a third one that looks at the size of the instance to be solved before deciding which method to use.

1.7 SOME PRACTICAL EXAMPLES

Maybe you are wondering whether it is really possible in practice to accelerate an algorithm to the extent suggested in the previous section. In fact, there have been cases where even more spectacular improvements have been made, even for well-established algorithms. Some of the following examples use large integers or real arithmetic. Unless we explicitly state the contrary, we shall simplify our presentation by ignoring the problems that may arise because of arithmetic overflow or loss of precision on a particular machine. Such problems can always be solved by using multiple-precision arithmetic (see Sections 1.7.2 and 4.7). Additions and multiplications are therefore generally taken to be elementary operations in the following paragraphs (except, of course, for Section 1.7.2).

1.7.1 Sorting

The sorting problem is of major importance in computer science, and in particular in algorithmics. We are required to arrange in ascending order a collection of n objects on which a total ordering is defined. Sorting problems are often found inside more complex algorithms. We have already seen two classic sorting algorithms in Section 1.4: insertion sorting and selection sorting. Both these algorithms, as we saw, take quadratic time both in the worst case and on the average.

Although these algorithms are excellent when n is small, other sorting algorithms are more efficient when n is large. Among others, we might use Williams's *heapsort* algorithm (see Example 2.2.4 and Problem 2.2.3), *mergesort* (see Section 4.4), or Hoare's *quicksort* algorithm (see Section 4.5). All these algorithms take a time in the order of $n \log n$ on the average; the first two take this same amount of time even in the worst case.

To have a clearer idea of the practical difference between a time in the order of n^2 and a time in the order of $n \log n$, we programmed insertion sort and *quicksort* in Pascal on a DEC VAX 780. The difference in efficiency between the two algorithms is marginal when the number of elements to be sorted is small. *Quicksort* is already almost twice as fast as insertion when sorting 50 elements, and three times as fast when sorting 100 elements. To sort 1,000 elements, insertion takes more than three seconds, whereas *quicksort* requires less than one-fifth of a second. When we have 5,000 elements to sort, the inefficiency of insertion sorting becomes still more pronounced: one and a half minutes are needed on average, compared to little more than one second for *quicksort*. In 30 seconds, *quicksort* can handle 100,000 elements; our estimate is that it would take nine and a half hours to carry out the same task using insertion sorting.

1.7.2 Multiplication of Large Integers

When a calculation requires very large integers to be manipulated, it can happen that the operands become too long to be held in a single word of the computer in use. Such operations thereupon cease to be elementary. When this occurs, we can use a representation such as FORTRAN's "double precision", or, more generally, multiple-precision arithmetic. In this case, we must ask ourselves how the time necessary to multiply two large integers increases with the size of the operands. We can measure this size by either the number of computer words needed to represent the operands on a machine or the length of their representation in decimal or binary. Since these measures differ only by a multiplicative constant, this choice does not alter our analysis of the order of efficiency of the algorithms in question. (This last remark would be false should we be considering exponential time algorithms—can you see why?)

Suppose two large integers, of sizes m and n, respectively, are to be multiplied. The classic algorithm of Section 1.1 can easily be transposed to this context. We see that it multiplies each word of one of the operands by each word of the other, and that it executes approximately one elementary addition for each of these multiplications. The time required is therefore in the order of mn. Multiplication *à la russe* also takes a time in the order of mn, provided we choose the smaller operand as the multiplier and the larger as the multiplicand. Thus, there is no reason for preferring it to the classic algorithm, particularly as the hidden constant is likely to be larger.

Problem 1.7.1. How much time does multiplication *à la russe* take if the multiplier is longer than the multiplicand?   □

As we mentioned in Section 1.1, more efficient algorithms exist to solve this problem. The simplest, which we shall study in Section 4.7, takes a time in the order of $nm^{\lg(3/2)}$, or approximately $nm^{0.59}$, where n is the size of the larger operand and m is the size of the smaller. If both operands are of size n, the algorithm thus takes a time in the order of $n^{1.59}$, which is preferable to the quadratic time taken by both the classic algorithm and multiplication *à la russe*.

The difference between the order of n^2 and the order of $n^{1.59}$ is less spectacular than that between the order of n^2 and the order of $n \log n$, which we saw in the case of sorting algorithms. To verify this, we programmed the classic algorithm and the algorithm of Section 4.7 in Pascal on a CDC CYBER 835 and tested them on operands of different sizes. To take account of the architecture of the machine, we carried out the calculations in base 2^{20} rather than in base 10. Integers of 20 bits are thus multiplied directly by the hardware of the machine, yet at the same time space is used quite efficiently (the machine has 60-bit words). Accordingly, the size of an operand is measured in terms of the number of 20-bit segments in its binary representation. The theoretically better algorithm of Section 4.7 gives little real improvement on operands of size 100 (equivalent to about 602 decimal digits): it takes about 300 milliseconds, whereas the classic algorithm takes about 400 milliseconds. For operands ten times this length, however, the fast algorithm is some three times more efficient than the classic algorithm: they take about 15 seconds and 40 seconds, respectively. The gain in efficiency continues to increase as the size of the operands goes up. As we shall see in Chapter 9, even more sophisticated algorithms exist for *much* larger operands.

1.7.3 Evaluating Determinants

Let

$$M = \begin{bmatrix} a_{1,1} & a_{1,2} & \cdots & a_{1,n} \\ a_{2,1} & a_{2,2} & \cdots & a_{2,n} \\ \cdot & \cdot & & \cdot \\ \cdot & \cdot & & \cdot \\ \cdot & \cdot & & \cdot \\ a_{n,1} & a_{n,2} & \cdots & a_{n,n} \end{bmatrix}$$

be an $n \times n$ matrix. The *determinant* of the matrix M, denoted by $\det(M)$, is often defined recursively: if $M[i,j]$ denotes the $(n-1) \times (n-1)$ submatrix obtained from M by deleting the ith row and the jth column, then

$$\det(M) = \sum_{j=1}^{n}(-1)^{j+1}a_{1,j}\ \det(M[1,j])\ .$$

If $n = 1$, the determinant is defined by $\det(M) = a_{1,1}$. Determinants are important in linear algebra, and we need to know how to calculate them efficiently.

If we use the recursive definition directly, we obtain an algorithm that takes a time in the order of $n!$ to calculate the determinant of an $n \times n$ matrix (see Example 2.2.5). This is even worse than exponential. On the other hand, another classic algorithm, Gauss-Jordan elimination, does the computation in cubic time. We programmed the two algorithms in Pascal on a CDC CYBER 835. The Gauss-Jordan algorithm finds the determinant of a 10×10 matrix in one-hundredth of a second; it takes about five and a half seconds on a 100×100 matrix. On the other hand, the recursive algorithm takes more than 20 seconds on a 5×5 matrix and 10 minutes on a 10×10 matrix; we estimate that it would take more than 10 million years to calculate the determinant of a 20×20 matrix, a task accomplished by the Gauss-Jordan algorithm in about one-twentieth of a second!

You should *not* conclude from this example that recursive algorithms are necessarily bad. On the contrary, Chapter 4 describes a technique where recursion plays a fundamental role in the design of efficient algorithms. In particular, Strassen discovered in 1969 a recursive algorithm that can calculate the determinant of an $n \times n$ matrix in a time in the order of $n^{\lg 7}$, or about $n^{2.81}$, thus proving that Gauss-Jordan elimination is not optimal.

1.7.4 Calculating the Greatest Common Divisor

Let m and n be two positive integers. The greatest common divisor of m and n, denoted by $\gcd(m,n)$, is the largest integer that divides both m and n exactly. When $\gcd(m,n) = 1$, we say that m and n are *coprime*. For example, $\gcd(6,15) = 3$ and $\gcd(10,21) = 1$. The obvious algorithm for calculating $\gcd(m,n)$ is obtained directly from the definition.

> **function** $gcd(m,n)$
> $\quad i \leftarrow \min(m,n) + 1$
> $\quad$ **repeat** $i \leftarrow i - 1$ **until** i divides both m and n exactly
> $\quad$ **return** i

The time taken by this algorithm is in the order of the difference between the smaller of the two arguments and their greatest common divisor. When m and n are of similar size and coprime, it therefore takes a time in the order of n.

A classic algorithm for calculating $\gcd(m,n)$ consists of first factorizing m and n, and then taking the product of the prime factors common to m and n, each prime factor being raised to the lower of its powers in the two arguments. For example, to calculate $\gcd(120,700)$ we first factorize $120 = 2^3 \times 3 \times 5$ and $700 = 2^2 \times 5^2 \times 7$. The common factors of 120 and 700 are therefore 2 and 5, and their lower powers are 2 and 1, respectively. The greatest common divisor of 120 and 700 is therefore $2^2 \times 5^1 = 20$.

Even though this algorithm is better than the one given previously, it requires us to factorize m and n, an operation we do not know how to do efficiently.

Nevertheless, there exists a much more efficient algorithm for calculating greatest common divisors. This is Euclid's famous algorithm.

> **function** *Euclid* (m, n)
>   **while** $m > 0$ **do**
> $t \leftarrow n \bmod m$
> $n \leftarrow m$
> $m \leftarrow t$
>   **return** n

Considering the arithmetic operations to have unit cost, this algorithm takes a time in the order of the logarithm of its arguments, even in the worst case (see Example 2.2.6), which is much faster than the preceding algorithms. To be historically exact, Euclid's original algorithm works using successive subtractions rather than by calculating a modulo.

1.7.5 Calculating the Fibonacci Sequence

The Fibonacci sequence is defined by the following recurrence:

$$\begin{cases} f_0 = 0; \ f_1 = 1 & \text{and} \\ f_n = f_{n-1} + f_{n-2} & \text{for } n \geq 2 . \end{cases}$$

The first ten terms of the sequence are therefore

$$0, 1, 1, 2, 3, 5, 8, 13, 21, 34.$$

This sequence has numerous applications in computer science, in mathematics, and in the theory of games. It is when applied to two consecutive terms of the Fibonacci sequence that Euclid's algorithm takes the longest time among all instances of comparable size. De Moivre proved the following formula (see Example 2.3.2):

$$f_n = \frac{1}{\sqrt{5}} [\phi^n - (-\phi)^{-n}] ,$$

where $\phi = (1 + \sqrt{5})/2$ is the *golden ratio*. Since $\phi^{-1} < 1$, the term $(-\phi)^{-n}$ can be neglected when n is large, which means that the value of f_n is in the order of ϕ^n. However, De Moivre's formula is of little immediate help in calculating f_n exactly, since the larger n becomes, the greater is the degree of precision required in the values of $\sqrt{5}$ and ϕ. On the CDC CYBER 835, a single-precision computation programmed in Pascal produces an error for the first time when calculating f_{66}.

The algorithm obtained directly from the definition of the Fibonacci sequence is the following.

> **function** *fib*1(n)
>   **if** $n < 2$ **then return** n
>   **else return** *fib*1$(n-1)$ + *fib*1$(n-2)$

This algorithm is very inefficient because it recalculates the same values many times. For instance, to calculate $fib1(5)$ we need the values of $fib1(4)$ and $fib1(3)$; but $fib1(4)$ also calls for the calculation of $fib1(3)$. We see that $fib1(3)$ will be calculated twice, $fib1(2)$ three times, $fib1(1)$ five times, and $fib1(0)$ three times. In fact, the time required to calculate f_n using this algorithm is in the order of the value of f_n itself, that is to say, in the order of ϕ^n (see Example 2.2.7).

To avoid wastefully recalculating the same values over and over, it is natural to proceed as in Section 1.5.

function $fib2(n)$
    $i \leftarrow 1; \ j \leftarrow 0$
    for $k \leftarrow 1$ **to** n **do** $j \leftarrow i + j$
                      $i \leftarrow j - i$
    return j

This second algorithm takes a time in the order of n, assuming we count each addition as an elementary operation (see Example 2.2.8). This is much better than the first algorithm. However, there exists a third algorithm that gives as great an improvement over the second algorithm as the second does over the first. This third algorithm, which at first sight appears quite mysterious, takes a time in the order of the logarithm of n (see Example 2.2.9). It will be explained in Chapter 4.

function $fib3(n)$
    $i \leftarrow 1; \ j \leftarrow 0; \ k \leftarrow 0; \ h \leftarrow 1$
    while $n > 0$ **do**
      if n is odd **then** $t \leftarrow jh$
                       $j \leftarrow ih + jk + t$
                       $i \leftarrow ik + t$
      $t \leftarrow h^2$
      $h \leftarrow 2kh + t$
      $k \leftarrow k^2 + t$
      $n \leftarrow n$ **div** 2
    return j

Once again, we programmed the three algorithms in Pascal on a CDC CYBER 835 in order to compare their execution times empirically. To avoid problems caused by arithmetic overflow (the Fibonacci sequence grows very rapidly: f_{100} is a number with 21 decimal digits), we carried out all the computations modulo 10^7, which is to say that we only obtained the seven least significant figures of the answer. Table 1.7.1 eloquently illustrates the difference that the choice of an algorithm can make. (All these times are approximate. Times greater than two minutes were estimated using the hybrid approach.) The time required by $fib1$ for $n > 50$ is so long that we did not bother to estimate it, with the exception of the case $n = 100$ on which $fib1$ would take well over 10^9 years! Note that $fib2$ is more efficient than $fib3$ on small instances.

TABLE 1.7.1 PERFORMANCE COMPARISON
BETWEEN MODULO 10^7 FIBONACCI ALGORITHMS

n	10	20	30	50
*fib*1	8 msec	1 sec	2 min	21 days
*fib*2	$\frac{1}{6}$ msec	$\frac{1}{3}$ msec	$\frac{1}{2}$ msec	$\frac{3}{4}$ msec
*fib*3	$\frac{1}{3}$ msec	$\frac{2}{5}$ msec	$\frac{1}{2}$ msec	$\frac{1}{2}$ msec

n	100	10,000	1,000,000	100,000,000
*fib*2	$1\frac{1}{2}$ msec	150 msec	15 sec	25 min
*fib*3	$\frac{1}{2}$ msec	1 msec	$1\frac{1}{2}$ msec	2 msec

Using the hybrid approach, we can estimate approximately the time taken by our implementations of these three algorithms. Writing $t_i(n)$ for the time taken by *fibi* on the instance n, we find

$$t_1(n) \approx \phi^{n-20} \text{ seconds,}$$

$$t_2(n) \approx 15n \text{ microseconds, and}$$

$$t_3(n) \approx \frac{1}{4} \log n \text{ milliseconds.}$$

It takes a value of n 10,000 times larger to make *fib*3 take one extra millisecond of computing time.

We could also have calculated all the figures in the answer using multiple-precision arithmetic. In this case the advantage that *fib*3 enjoys over *fib*2 is less marked, but their joint advantage over *fib*1 remains just as striking. Table 1.7.2 compares the times taken by these three algorithms when they are used in conjunction with an efficient implementation of the classic algorithm for multiplying large integers (see Problem 2.2.9, Example 2.2.8, and Problem 2.2.11).

TABLE 1.7.2. PERFORMANCE COMPARISON
BETWEEN EXACT FIBONACCI ALGORITHMS*

n	5	10	15	20	25
*fib*1	0.007	0.087	0.941	10.766	118.457
*fib*2	0.005	0.009	0.011	0.017	0.021
*fib*3	0.013	0.017	0.019	0.020	0.021

n	100	500	1,000	5,000	10,000
*fib*2	0.109	1.177	3.581	76.107	298.892
*fib*3	0.041	0.132	0.348	7.664	29.553

* All times are in seconds.

1.7.6 Fourier Transforms

The *Fast Fourier Transform* algorithm is perhaps the one algorithmic discovery that had the greatest practical impact in history. We shall come back to this subject in Chapter 9. For the moment let us only mention that Fourier transforms are of fundamental importance in such disparate applications as optics, acoustics, quantum physics, telecommunications, systems theory, and signal processing including speech recognition. For years progress in these areas was limited by the fact that the known algorithms for calculating Fourier transforms all took far too long.

The "discovery" by Cooley and Tukey in 1965 of a fast algorithm revolutionized the situation: problems previously considered to be infeasible could now at last be tackled. In one early test of the "new" algorithm the Fourier transform was used to analyse data from an earthquake that had taken place in Alaska in 1964. Although the classic algorithm took more than 26 minutes of computation, the "new" algorithm was able to perform the same task in less than two and a half seconds.

Ironically it turned out that an efficient algorithm had already been published in 1942 by Danielson and Lanczos. Thus the development of numerous applications had been hindered for no good reason for almost a quarter of a century. And if *that* were not sufficient, all the necessary theoretical groundwork for Danielson and Lanczos's algorithm had already been published by Runge and König in 1924!

1.8 WHEN IS AN ALGORITHM SPECIFIED?

At the beginning of this book we said that "the execution of an algorithm must not include any subjective decisions, nor must it require the use of intuition or creativity". In this case, can we reasonably maintain that *fib3* of Section 1.7.5 describes an algorithm? The problem arises because it is not realistic to consider that the multiplications in *fib3* are elementary operations. Any practical implementation must take this into account, probably by using a program package allowing arithmetic operations on very large integers. Since the exact way in which these multiplications are to be carried out is not specified in *fib3*, the choice may be considered a subjective decision, and hence *fib3* is not formally speaking an algorithm. That this distinction is not merely academic is illustrated by Problems 2.2.11 and 4.7.6, which show that indeed the order of time taken by *fib3* depends on the multiplication algorithm used. And what should we say about De Moivre's formula used as an algorithm?

Calculation of a determinant by the recursive method of Section 1.7.3 is another example of an incompletely presented algorithm. How are the recursive calls to be set up? The obvious approach requires a time in the order of n^2 to be used before each recursive call. We shall see in Problem 2.2.5 that it is possible to get by with a time in the order of n to set up not just one, but all the n recursive calls. However, this added subtlety does not alter the fact that the algorithm takes a time in the order of $n!$ to calculate the determinant of an $n \times n$ matrix.

To make life simple, we shall continue to use the word *algorithm* for certain incomplete descriptions of this kind. The details will be filled in later should our analyses require them.

1.9 DATA STRUCTURES

The use of well-chosen data structures is often a crucial factor in the design of efficient algorithms. Nevertheless, this book is *not* intended to be a manual on data structures. We suppose that the reader already has a good working knowledge of such basic notions as arrays, structures, pointers, and lists. We also suppose that he or she has already come across the mathematical concepts of directed and undirected graphs, and knows how to represent these objects efficiently on a computer. After a brief review of some important points, this section concentrates on the less elementary notions of heaps and disjoint sets. Chosen because they will be used in subsequent chapters, these two structures also offer interesting examples of the analysis of algorithms (see Example 2.2.4, Problem 2.2.3, and Example 2.2.10).

1.9.1 Lists

A *list* is a collection of nodes or elements of information arranged in a certain order. The corresponding data structure must allow us to determine efficiently, for example, which is the first node in the structure, which is the last, and which are the predecessor and the successor (if they exist) of any given node. Such a structure is frequently represented graphically by boxes and arrows, as in Figure 1.9.1. The information attached to a node is shown inside the corresponding box and the arrows show transitions from a node to its successor.

Such lists are subject to a number of operations: we might want to insert an additional node, to delete a node, to copy a list, to count the number of elements it contains, and so on. The different computer implementations that are commonly used differ in the quantity of memory required, and in the greater or less ease of carrying out certain operations. Here we content ourselves with mentioning the best-known techniques.

Implemented as an array by the declaration

type *tablist* = **record**
                counter : 0 .. *maxlength*
                value [1 .. *maxlength*] : *information*   ,

the elements of a list occupy the slots *value* [1] to *value* [*counter*], and the order of the elements is given by the order of their indices in the array. Using this implementation,

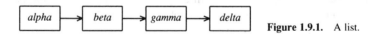

Figure 1.9.1. A list.

we can find the first and the last elements of the list rapidly, as we can the predecessor and the successor of a given node. On the other hand, inserting a new element or deleting one of the existing elements requires a worst-case number of operations in the order of the current size of the list.

This implementation is particularly efficient for the important structure known as the *stack*, which we obtain by restricting the permitted operations on a list : addition and deletion of elements are allowed only at one particular end of the list. However, it presents the major disadvantage of requiring that all the memory space potentially required be reserved from the outset of a program.

On the other hand, if pointers are used to implement a list structure, the nodes are usually represented by some such structure as

> **type** *node* = **record**
>   *value* : *information*
>   *next* : ↑*node*   ,

where each node includes an explicit pointer to its successor. In this case, provided a suitably powerful programming language is used, the space needed to represent the list can be allocated and recovered dynamically as the program proceeds.

Even if additional pointers are used to ensure rapid access to the first and last elements of the list, it is difficult when this representation is used to examine the kth element, for arbitrary k, without having to follow k pointers and thus to take a time in the order of k. However, once an element has been found, inserting new nodes or deleting an existing node can be done rapidly. In our example, a single pointer is used in each node to designate its successor : it is therefore easy to traverse the list in one direction, but not in the other. If a higher memory overhead is acceptable, it suffices to add a second pointer to each node to allow the list to be traversed rapidly in either direction.

1.9.2 Graphs

Intuitively speaking, a *graph* is a set of nodes joined by a set of lines or arrows. Consider Figure 1.9.2 for instance. We distinguish directed and undirected graphs. In the case of a directed graph the nodes are joined by arrows called *edges*. In the example of Figure 1.9.2 there exists an edge from *alpha* to *gamma* and another from *gamma* to *alpha*; *beta* and *delta*, however, are joined only in the direction indicated. In the case of an undirected graph, the nodes are joined by lines with no direction indicated, also called edges. In every case, the edges may form *paths* and *cycles*.

There are never more than two arrows joining any two given nodes of a directed graph (and if there *are* two arrows, then they must go in opposite directions), and there is never more than one line joining any two given nodes of an undirected graph. Formally speaking, a graph is therefore a pair $G = <N, A>$ where N is a set of nodes and $A \subseteq N \times N$ is a set of edges. An edge from node a to node b of a directed graph is denoted by the ordered pair (a, b), whereas an edge joining nodes a and b in an undirected graph is denoted by the set $\{a, b\}$.

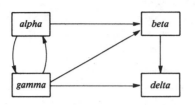

Figure 1.9.2. A directed graph.

There are at least two obvious ways to represent a graph on a computer. The first is illustrated by

type *adjgraph* = **record**
              value [1 .. *nbnodes*] : *information*
              adjacent [1 .. *nbnodes* , 1 .. *nbnodes*] : *Booleans*   .

If there exists an edge from node i of the graph to node j, then *adjacent* $[i,j]$ = *true* ; otherwise *adjacent* $[i,j]$ = *false*. In the case of an undirected graph, the matrix is necessarily symmetric.

With this representation it is easy to see whether or not two nodes are connected. On the other hand, should we wish to examine all the nodes connected to some given node, we have to scan a complete row in the matrix. This takes a time in the order of *nbnodes*, the number of nodes in the graph, independently of the number of edges that exist involving this particular node. The memory space required is quadratic in the number of nodes.

A second possible representation is as follows :

type *lisgraph* = **array**[1 .. *nbnodes*] **of**
              record
                 value : *information*
                 neighbours : *list*   .

Here we attach to each node i a list of its neighbours, that is to say of those nodes j such that an edge from i to j (in the case of a directed graph) or between i and j (in the case of an undirected graph) exists. If the number of edges in the graph is small, this representation is preferable from the point of view of the memory space used. It may also be possible in this case to examine all the neighbours of a given node in less than *nbnodes* operations on the average. On the other hand, to determine whether or not two given nodes i and j are connected directly, we have to scan the list of neighbours of node i (and possibly of node j, too), which is less efficient than looking up a Boolean value in an array.

A *tree* is an acyclic, connected, undirected graph. Equivalently, a tree may be defined as an undirected graph in which there exists exactly one path between any given pair of nodes. The same representations used to implement graphs can be used to implement trees.

1.9.3 Rooted Trees

Let G be a directed graph. If there exists in G a vertex r such that every other vertex can be reached from r by a unique path, then G is a *rooted tree* and r is its *root*. Any rooted tree with n nodes contains exactly $n - 1$ edges. It is usual to represent a rooted tree with the root at the top, like a family tree, as in Figure 1.9.3. In this example *alpha* is at the root of the tree. (When there is no danger of confusion, we shall use the simple term "tree" instead of the more correct "rooted tree".) Extending the analogy with a family tree, we say that *beta* is the *parent* of *delta* and the *child* of *alpha*, that *epsilon* and *zeta* are the *siblings* of *delta*, that *alpha* is an *ancestor* of *epsilon*, and so on.

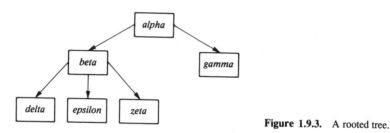

Figure 1.9.3.   A rooted tree.

A *leaf* of a rooted tree is a node with no children; the other nodes are called *internal* nodes. Although nothing in the definition indicates this, the branches of a rooted tree are often considered to be ordered: in the previous example *beta* is situated to the left of *gamma*, and (by analogy with a family tree once again) *delta* is the *eldest* sibling of *epsilon* and *zeta*. The two trees in Figure 1.9.4 may therefore be considered as different.

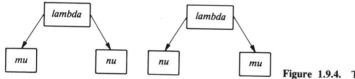

Figure 1.9.4.   Two distinct rooted trees.

On a computer, any rooted tree may be represented using nodes of the following type:

> **type** *treenode* = **record**
> value : information
> eldest-child, next-sibling : ↑treenode .

The rooted tree shown in Figure 1.9.3 would be represented as in Figure 1.9.5, where now the arrows no longer represent the edges of the rooted tree, but rather the pointers used in the computer representation. As in the case of lists, the use of additional pointers (for example, to the parent or the eldest sibling of a given node) may speed up certain operations at the price of an increase in the memory space needed.

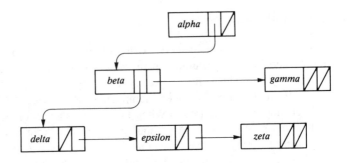

Figure 1.9.5.   Possible computer representation of a rooted tree.

The *depth* of a node in a rooted tree is the number of edges that need to be traversed to arrive at the node starting from the root. The *height of a node* is the number of edges in the longest path from the node in question to a leaf. The *height of a rooted tree* is the height of its root, and thus also the depth of its deepest leaf. Finally, the *level* of a node is equal to the height of the tree minus the depth of the node concerned. For example, *gamma* has depth 1, height 0, and level 1 in the tree of Figure 1.9.3.

If each node of a rooted tree can have up to *n* children, we say it is an *n-ary* tree. In this case, the positions occupied by the children are significant. For instance, the binary trees of Figure 1.9.6 are not the same: in the first case *b* is the elder child of *a* and the younger child is missing, whereas in the second case *b* is the younger child of *a* and the elder child is missing. In the important case of a binary tree, although the metaphor becomes somewhat strained, we naturally tend to talk about the *left-hand child* and the *right-hand child*.

There are several ways of representing an *n*-ary tree on a computer. One obvious representation uses nodes of the type

type *n-ary-node* = **record**
                value : *information*
                child [1 .. *n*] : ↑*n-ary-node*   .

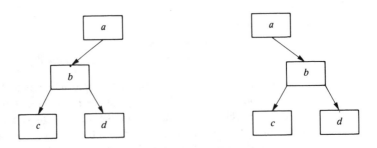

Figure 1.9.6.   Two distinct binary trees.

In the case of a binary tree we can also define

> **type** *binary-node* = **record**
>> *value : information*
>> *left-child , right-child* : ↑*binary-node* .

It is also sometimes possible, as we shall see in the following section, to represent a rooted tree using an array without any explicit pointers.

A binary tree is a *search tree* if the value contained in every internal node is larger than or equal to the values contained in its left-hand descendants, and less than or equal to the values contained in its right-hand descendants. An example of a search tree is given in Figure 5.5.1. This structure is interesting because it allows efficient searches for values in the tree.

Problem 1.9.1.   Suppose the value sought is held in a node at depth p in a search tree. Design an algorithm capable of finding this node starting at the root in a time in the order of p.   □

It is possible to update a search tree, that is, to delete nodes or to add new values, without destroying the search tree property. However, if this is done in an unconsidered fashion, it can happen that the resulting tree becomes badly unbalanced, in the sense that the height of the tree is in the order of the number of nodes it contains. More sophisticated methods, such as the use of AVL trees or 2-3 trees, allow such operations as searches and the addition or deletion of nodes in a time in the order of the logarithm of the number of nodes in the tree in the worst case. These structures also allow the efficient implementation of several additional operations. Since these concepts are not used in the rest of this book, here we only mention their existence.

1.9.4 Heaps

A heap is a special kind of rooted tree that can be implemented efficiently in an array without any explicit pointers. This interesting structure lends itself to numerous applications, including a remarkable sorting technique, called *heapsort* (see Problem 2.2.3), as well as the efficient implementation of certain dynamic priority lists.

A binary tree is *essentially complete* if each of its internal nodes possesses exactly two children, one on the left and one on the right, with the possible exception of a unique *special* node situated on level 1, which possesses only a left-hand child and no right-hand child. Moreover, all the leaves are either on level 0, or else they are on levels 0 and 1, and no leaf is found on level 1 to the left of an internal node at the same level. The unique special node, if it exists, is to the right of all the other level 1 internal nodes. This kind of tree can be represented using an array T by putting the nodes of depth k, from left to right, in the positions $T[2^k]$, $T[2^k+1]$, ..., $T[2^{k+1}-1]$ (with the possible exception of level 0, which may be incomplete). For instance, Figure 1.9.7 shows how to represent an essentially complete binary tree containing 10 nodes. The parent of the node represented in $T[i]$ is found in $T[i \text{ } \textbf{div} \text{ } 2]$ for $i > 1$, and

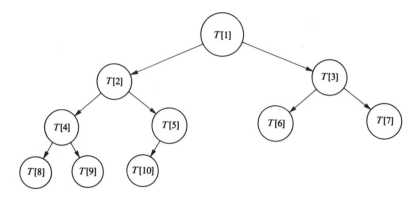

Figure 1.9.7.   An essentially complete binary tree.

the children of the node represented in $T[i]$ are found in $T[2i]$ and $T[2i+1]$, whenever they exist. The subtree whose root is in $T[i]$ is also easy to identify.

   A *heap* is an essentially complete binary tree, each of whose nodes includes an element of information called the *value* of the node. The *heap property* is that the value of each internal node is greater than or equal to the values of its children. Figure 1.9.8 gives an example of a heap. This same heap can be represented by the following array :

10	7	9	4	7	5	2	2	1	6

.

   The fundamental characteristic of this data structure is that the heap property can be restored efficiently after modification of the value of a node. If the value of the node increases, so that it becomes greater than the value of its parent, it suffices to exchange these two values and then to continue the same process upwards in the tree until the heap property is restored. We say that the modified value has been *percolated* up to its new position (one often encounters the rather strange term *sift-up* for this process). If, on the contrary, the value of a node is decreased so that it becomes less than the value of at least one of its children, it suffices to exchange the modified value with

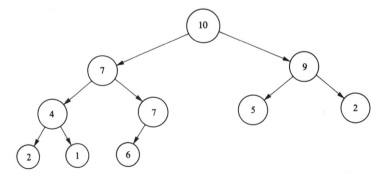

Figure 1.9.8.   A heap.

the larger of the values in the children, and then to continue this process downwards in the tree until the heap property is restored. We say that the modified value has been *sifted down* to its new position. The following procedures describe more formally the basic heap manipulation process. For the purpose of clarity, they are written so as to reflect as closely as possible the preceding discussion. If the reader wishes to make use of heaps for a "real" application, we encourage him or her to figure out how to avoid the inefficiency resulting from our use of the "exchange" instruction.

> **procedure** *alter-heap* $(T[1..n], i, v)$
> $\{T[1..n]$ is a heap; the value of $T[i]$ is set to v and the
>   heap property is re-established; we suppose that $1 \le i \le n\}$
> $x \leftarrow T[i]$
> $T[i] \leftarrow v$
>   **if** $v < x$ **then** *sift-down* (T, i)
>   **else** *percolate* (T, i)

> **procedure** *sift-down* $(T[1..n], i)$
> $\{$this procedure sifts node i down so as to re-establish the heap
>   property in $T[1..n]$; we suppose that T would be a heap if $T[i]$
>   were sufficiently large; we also suppose that $1 \le i \le n\}$
> $k \leftarrow i$
>   **repeat**
> $j \leftarrow k$
> $\{$find the larger child of node $j\}$
>   **if** $2j \le n$ **and** $T[2j] > T[k]$ **then** $k \leftarrow 2j$
>   **if** $2j < n$ **and** $T[2j+1] > T[k]$ **then** $k \leftarrow 2j+1$
>   exchange $T[j]$ and $T[k]$
> $\{$if $j = k$, then the node has arrived at its final position$\}$
>   **until** $j = k$

> **procedure** *percolate* $(T[1..n], i)$
> $\{$this procedure percolates node i so as to re-establish the
>   heap property in $T[1..n]$; we suppose that T would be a heap
>   if $T[i]$ were sufficiently small; we also suppose that
> $1 \le i \le n$; the parameter n is not used here$\}$
> $k \leftarrow i$
>   **repeat**
> $j \leftarrow k$
>   **if** $j > 1$ **and** $T[j \text{ div } 2] < T[k]$ **then** $k \leftarrow j \text{ div } 2$
>   exchange $T[j]$ and $T[k]$
>   **until** $j = k$

The heap is an ideal data structure for finding the largest element of a set, removing it, adding a new node, or modifying a node. These are exactly the operations we need to implement dynamic priority lists efficiently: the value of a node gives the priority of the corresponding event. The event with highest priority is always found at the root of the heap, and the priority of an event can be changed dynamically at all times. This is particularly useful in computer simulations.

function *find-max* $(T[1..n])$
    {returns the largest element of the heap $T[1..n]$}
    return $T[1]$

procedure *delete-max* $(T[1..n])$
    {removes the largest element of the heap $T[1..n]$
    and restores the heap property in $T[1..n-1]$}
    $T[1] \leftarrow T[n]$
    sift-down $(T[1..n-1],1)$

procedure *insert-node* $(T[1..n],v)$
    {adds an element whose value is v to the heap $T[1..n]$
    and restores the heap property in $T[1..n+1]$}
    $T[n+1] \leftarrow v$
    percolate $(T[1..n+1],n+1)$

It remains to be seen how we can create a heap starting from an array $T[1..n]$ of elements in an undefined order. The obvious solution is to start with an empty heap and to add elements one by one.

procedure *slow-make-heap* $(T[1..n])$
    {this procedure makes the array $T[1..n]$ into a heap,
    but rather inefficiently}
    for $i \leftarrow 2$ **to** n **do** *percolate* $(T[1..i],i)$

However, this approach is not particularly efficient (see Problem 2.2.2). There exists a cleverer algorithm for making a heap. Suppose, for example, that our starting point is the following array:

1	6	9	2	7	5	2	7	4	10

represented by the tree of Figure 1.9.9a. We begin by making each of the subtrees whose roots are at level 1 into a heap, by sifting down those roots, as illustrated in Figure 1.9.9b. The subtrees at the next higher level are then transformed into heaps, also by sifting down their roots. Figure 1.9.9c shows the process for the left-hand subtree. The other subtree at level 2 is already a heap. This results in an essentially complete binary tree corresponding to the array:

1	10	9	7	7	5	2	2	4	6

It only remains to sift down its root in order to obtain the desired heap. The final process thus goes as follows:

10	1	9	7	7	5	2	2	4	6
10	**7**	9	**1**	7	5	2	2	4	6
10	7	9	**4**	7	5	2	2	**1**	6

whose tree representation is shown in the previous example as Figure 1.9.8.

This algorithm can be described formally as follows.

procedure *make-heap* $(T[1..n])$
  {this procedure makes the array $T[1..n]$ into a heap}
  for $i \leftarrow (n \ \textbf{div} \ 2) \ \textbf{downto} \ 1 \ \textbf{do} \ \textit{sift-down} \ (T, i)$

We shall see in Example 2.2.4 that this algorithm allows the creation of a heap in linear time.

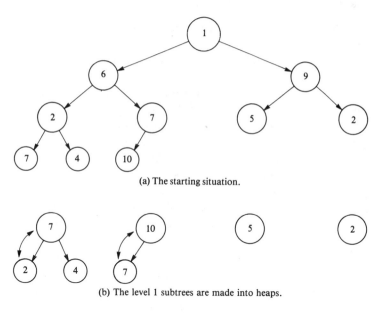

(a) The starting situation.

(b) The level 1 subtrees are made into heaps.

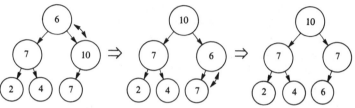

(c) One level 2 subtree is made into a heap (the other already is a heap).

Figure 1.9.9.   Making a heap.

Problem 1.9.2. Let $T[1..12]$ be an array such that $T[i]=i$ for each $i \leq 12$. Exhibit the state of the array after each of the following procedure calls:

> *make-heap* (T)
> *alter-heap* $(T,12,10)$
> *alter-heap* $(T,1,6)$
> *alter-heap* $(T,5,8)$.   □

Problem 1.9.3. Exhibit a heap $T[1..n]$ containing distinct values, such that the following sequence results in a different heap:

> $m \leftarrow$ *find-max* (T)
> *delete-max* (T)
> *insert-node* $(T[1..n-1],m)$.

Draw the heap after each operation.   □

The basic concept of a heap can be improved in several ways. For applications that need *percolate* more often than *sift-down* (Problems 3.2.12 and 3.2.13) it pays to have more than two children per internal node: this speeds up *percolate* (because the heap is shallower) at the cost of slowing down any operation that must consider every child at each level. It is still possible to represent such heaps in an array without explicit pointers, but care is needed to do it correctly.

For applications that have a tendency to sift down the (updated) root almost to the bottom level, it pays to ignore temporarily the new value stored at the root, treating this node as an empty location, and to sift it all the way down to a leaf. At this point, put back the relevant value into the empty leaf and percolate it to its proper position. The advantage of this procedure is that it requires only one comparison at each level in order to sift down the empty node, rather than two with the usual procedure (the children are compared to each other but not to their father). Experiments have shown this approach to yield an improvement in the classic heapsort algorithm (Problem 2.2.3).

Finally, the basic heap operations needed to implement a dynamic priority list can also be handled by data structures completely different from the heap we have considered so far. In particular, the *Fibonacci heap* (or *lazy binomial queue*) can process each *insert-node*, *find-max*, and *percolate* operation in constant time, and each *delete-max* in logarithmic time. As an application, this data structure allows the implementation of Dijkstra's algorithm in a time in the order of $a+n \log n$ for finding the length of the shortest path from a designated source node to each of the other nodes of a graph with n nodes and a edges (Section 3.2.2). The Fibonacci heap allows also the *merging* of priority lists in constant time, an operation beyond the (efficient) reach of classic heaps. (To be precise, the preceding times for Fibonacci heaps are correct in the *amortized* sense—a concept not discussed here.)

1.9.5 Disjoint Set Structures

Suppose we have N objects numbered from 1 to N. We wish to group these objects into disjoint sets, each object being in exactly one set at any given time. In each set we choose a *canonical* object, which will serve as a *label* for the set. Initially, the N objects are in N different sets, each containing exactly one object, which is necessarily the label for its set. Thereafter, we execute a series of operations of two kinds:

- for a given object, *find* which set contains it and return the label of this set; and
- given two distinct labels, *merge* the two corresponding sets.

How can we represent this situation efficiently on a computer?

One possible representation is obvious. Suppose we decide to use the smallest member of each set as the label: thus the set $\{7, 3, 16, 9\}$ will be called "set 3". If we now declare an array $set[1..N]$, it suffices to place the label of the set corresponding to each object in the appropriate array element. The two operations can be implemented by two procedures.

> **function** $find1(x)$
> {finds the label of the set containing object x}
> **return** $set[x]$

> **procedure** $merge1(a, b)$
> {merges the sets labelled a and b}
> $i \leftarrow a$; $j \leftarrow b$
> **if** $i > j$ **then** exchange i and j
> **for** $k \leftarrow 1$ **to** N **do**
>   **if** $set[k] = j$ **then** $set[k] \leftarrow i$

We wish to know the time required to execute an arbitrary series of n operations of the type *find* and *merge*, starting from the initial situation. If consulting or modifying one element of an array counts as an elementary operation, it is clear that *find1* takes constant time and that *merge1* takes a time in the order of N. A series of n operations therefore takes a time in the order of nN in the worst case.

We can do better than this. Still using a single array, we can represent each set as an "inverted" rooted tree. We adopt the following scheme: if $set[i]=i$, then i is both the label of a set and the root of the corresponding tree; if $set[i]=j\neq i$, then j is the parent of i in some tree. The array

1	2	3	2	1	3	4	3	3	4

therefore represents the trees given in Figure 1.9.10, which in turn represent the sets $\{1, 5\}$, $\{2, 4, 7, 10\}$ and $\{3, 6, 8, 9\}$. To merge two sets, we need now only change a single value in the array; on the other hand, it is harder to find the set to which an object belongs.

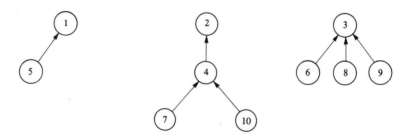

Figure 1.9.10.   Tree representation for disjoint sets.
(The figure shows the direction of the pointers in the array, not of the edges in the tree.)

function *find2(x)*
  {finds the label of the set containing object *x* }
  i ← *x*
  while *set* [*i*] ≠ *i* **do** *i* ← *set* [*i*]
  return *i*

procedure *merge2(a , b)*
  {merges the sets labelled *a* and *b* }
  if *a* < *b* **then** *set* [*b*] ← *a*
          else *set* [*a*] ← *b*

Problem 1.9.4.   If each consultation or modification of an array element
counts as an elementary operation, prove that the time needed to execute an arbitrary
sequence of *n* operations *find2* or *merge2* starting from the initial situation is in the
order of n^2 in the worst case.   □

In the case when *n* is comparable to *N*, we have not gained anything over the use
of *find*1 and *merge*1. The problem arises because after *k* calls on *merge2*, we may find
ourselves confronted by a tree of height *k*, so that each subsequent call on *find2* takes a
time in the order of *k*. Let us therefore try to limit the height of the trees produced.

So far, we have chosen arbitrarily to use the smallest member of a set as its label.
When we merge two trees whose heights are respectively h_1 and h_2, it would be better
to arrange matters so that it is always the root of the tree whose height is least that
becomes a child of the other root. In this way the height of the resulting merged tree
will be $\max(h_1, h_2)$ if $h_1 \neq h_2$, or h_1+1 if $h_1=h_2$. Using this technique, the trees do
not grow as rapidly.

Problem 1.9.5.   Prove by mathematical induction that if this tactic is used,
then after an arbitrary sequence of merge operations starting from the initial situation, a
tree containing *k* nodes will have a height at most $\lfloor \lg k \rfloor$.   □

The height of the trees can be maintained in an additional array *height* [1 .. *N*] so that *height* [*i*] gives the height of node *i* in its current tree. Whenever *a* is the label of a set, *height* [*a*] therefore gives the height of the corresponding tree. Initially, *height* [*i*] is set to zero for each *i*. The procedure *find2* is still relevant but we must modify *merge* accordingly.

> **procedure** *merge3*(*a* , *b*)
> {merges the sets labelled *a* and *b* ;
>   we suppose that *a* ≠ *b* }
> **if** *height* [*a*] = *height* [*b*]
> **then**
>   *height* [*a*] ← *height* [*a*] + 1
>   *set* [*b*] ← *a*
> **else**
>   **if** *height* [*a*] > *height* [*b*]
>   **then** *set* [*b*] ← *a*
>   **else** *set* [*a*] ← *b*

Problem 1.9.6.   Prove that the time needed to execute an arbitrary sequence of *n* operations *find2* and *merge3* starting from the initial situation is in the order of *n* log *n* in the worst case.   □

By modifying *find2*, we can make our operations faster still. When we are trying to determine the set that contains a certain object *x*, we first traverse the edges of the tree leading up from *x* to the root. Once we know the root, we can now traverse the same edges again, modifying each node encountered on the way to set its pointer directly to the root. This technique is called *path compression*. For example, when we execute the operation *find* (20) on the tree of Figure 1.9.11a, the result is the tree of

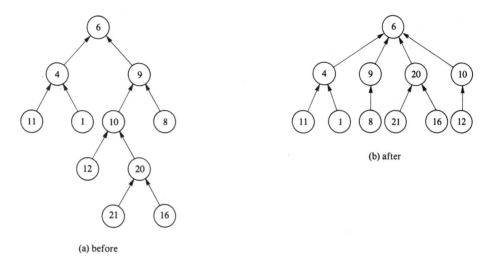

(a) before

(b) after

Figure 1.9.11.   Path compression.

Figure 1.9.11b: nodes 20, 10, and 9, which lay on the path from node 20 to the root, now point directly to the root. The pointers of the remaining nodes have not changed. This technique obviously tends to diminish the height of a tree and thus to accelerate subsequent *find* operations. On the other hand, the new *find* operation takes about twice as long as before. Is path compression a good idea? The answer is given when we analyse it in Example 2.2.10.

Using path compression, it is no longer true that the height of a tree whose root is a is given by *height* $[a]$. However, this remains an upper bound on the height. We call this value the *rank* of the tree, and change the name of the array accordingly. Our function becomes

```
function find3(x)
    {finds the label of the set containing object x }
    r ← x
    while set[r] ≠ r do r ← set[r]
    {r is the root of the tree}
    i ← x
    while i ≠ r do
        j ← set[i]
        set[i] ← r
        i ← j
    return r  .
```

From now on, when we use this combination of an array and of the procedures *find3* and *merge3* to deal with disjoint sets of objects, we say we are using a *disjoint set structure*.

Problem 1.9.7. A second possible tactic for merging two sets is to ensure that the root of the tree containing the smaller *number* of nodes becomes the child of the other root. Path compression does not change the number of nodes in a tree, so that it is easy to store this value exactly (whereas we could not efficiently keep track of the exact height of a tree after path compression).

Write a procedure *merge4* to implement this tactic, and give a result corresponding to the one in Problem 1.9.5. □

**Problem 1.9.8. Analyse the combined efficiency of *find3* together with your *merge4* from the previous problem. □

Problem 1.9.9. A canonical object has no parent, and we make no use of the rank of any object that is not canonical. Use this remark to implement a disjoint set structure that uses only one length N array rather than the two *set* and *rank*. (*Hint:* use negative values for the ranks.) □

1.10 REFERENCES AND FURTHER READING

We distinguish between three kinds of books on algorithm design. *Specific* books cover algorithms that are useful in a particular application area: sorting, searching, graph theory, Fourier transforms, and so on. *General* books cover several application areas: they give algorithms useful in each area. Finally, books on *algorithmics* concentrate on the techniques of algorithm design: they illustrate each technique by examples of algorithms taken from various applications areas. The excellent book of Harel (1987) takes a broader view at algorithmics and considers it as no less than "the spirit of computing".

The most complete collection of algorithms ever proposed is no doubt found in Knuth (1968, 1969, 1973), originally intended to consist of seven volumes. Several other well-known general books are worth mentioning: in chronological order Aho, Hopcroft, and Ullman (1974), Baase (1978), Dromey (1982), Sedgewick (1983), and Melhorn (1984a, 1984b, 1984c). Specific books will be referred to in the following chapters whenever they are relevant to our discussion; we may mention, however, Nilsson (1971), Brigham (1974), Borodin and Munro (1975), Christofides (1975), Lawler (1976), Reingold, Nievergelt, and Deo (1977), Gondran and Minoux (1979), Even (1980), Papadimitriou and Steiglitz (1982), and Tarjan (1983). Besides our own book and Harel (1987), we are aware of two more works on algorithmics: Horowitz and Sahni (1978) and Stinson (1985).

For a more popular account of algorithms, see Knuth (1977) and Lewis and Papadimitriou (1978). Multiplication *à la russe* is described in Warusfel (1961), a remarkable little French book of popular mathematics. Although we do not use any specific programming language in the present book, we suggest that a reader unfamiliar with Pascal would do well to look at one of the numerous books on this language, such as Jensen and Wirth (1985) or Lecarme and Nebut (1985).

Knuth (1973) contains a large number of sorting algorithms. Problem 1.5.1 comes from Shamir (1979). The algorithm capable of calculating the determinant of an $n \times n$ matrix in a time in the order of $n^{2.81}$ is given in Strassen (1969) and Bunch and Hopcroft (1974). The fast algorithm for calculating the Fibonacci sequence is explained in Gries and Levin (1980) and Urbanek (1980). For an introduction to Fourier transforms, read Demars (1981). We give more references on this subject in Chapter 9.

To reinforce our remarks in Sections 1.6 and 1.7, we encourage the reader to look at Bentley (1984), which offers experimental proof that intelligent use of algorithmics may allow a TRS-80 to run rings round a CRAY-1.

For more information about data structures, consult Knuth (1968, 1973), Stone (1972), Horowitz and Sahni (1976), Standish (1980), Aho, Hopcroft, and Ullman (1983), and Gonnet (1984). AVL trees come from Adel'son-Vel'skii and Landis (1962); they are described in detail in Knuth (1973); 2-3 trees come from Aho, Hopcroft, and Ullman (1974). Graphs and trees are presented from a mathematical standpoint in Berge (1958, 1970). The heap was introduced as a data structure for sorting in

Williams (1964). The improvements suggested at the end of the sub-section on heaps are described in Johnson (1975), Fredman and Tarjan (1984), Gonnet and Munro (1986), and Carlsson (1986, 1987). Carlsson (1986) also describes a data structure, which he calls the double-ended heap, or *deap*, that allows finding efficiently the largest and the smallest elements of a set. For ideas on building heaps faster, consult McDiarmid and Reed (1987). In this book, we give only some of the possible uses of disjoint set structures; for more applications see Hopcroft and Karp (1971) and Aho, Hopcroft, and Ullman (1974, 1976).

2

Analysing the Efficiency
of Algorithms

2.1 ASYMPTOTIC NOTATION

As we mentioned in Chapter 1, theoretical analyses of the efficiency of algorithms are carried out to within a multiplicative constant so as to take account of possible variations introduced by a change of implementation, of programming language, or of computer. To this end, we now introduce formally the asymptotic notation that will be used throughout the present book.

2.1.1 A Notation for "the order of "

Let $\mathbb{N}$ and $\mathbb{R}$ represent the set of natural numbers (positive or zero) and the set of real numbers, respectively. We denote the set of strictly positive natural numbers by $\mathbb{N}^+$, the set of strictly positive real numbers by $\mathbb{R}^+$, and the set of nonnegative real numbers by $\mathbb{R}^*$ (the latter being a nonstandard notation). The set {*true*, *false* } of Boolean constants is denoted by $\mathbb{B}$.

Let $f : \mathbb{N} \to \mathbb{R}^*$ be an arbitrary function. We define

$$O(f(n)) = \{ t : \mathbb{N} \to \mathbb{R}^* \mid (\exists c \in \mathbb{R}^+)(\exists n_0 \in \mathbb{N})(\forall n \geq n_0)[t(n) \leq cf(n)] \}.$$

In other words, $O(f(n))$ (read as "*the order of $f(n)$*") is the set of all functions $t(n)$ bounded above by a positive real multiple of $f(n)$, provided that n is sufficiently large (greater than some *threshold n_0*).

For convenience, we allow ourselves to misuse the notation from time to time. For instance, we say that $t(n)$ is in the order of $f(n)$ even if $t(n)$ is negative or undefined for some values $n < n_0$. Similarly, we talk about the order of $f(n)$ even when $f(n)$ is negative or undefined for a finite number of values of n ; in this case we must choose n_0 sufficiently large to be sure that such behaviour does not happen for $n \geq n_0$. For example, it is allowable to talk about the order of $n / \log n$, even though this function is not defined when $n = 0$ or $n = 1$, and it is correct to write $n^3 - 3n^2 - n - 8 \in O(n^3)$.

The principle of invariance mentioned in the previous chapter assures us that if some implementation of a given algorithm never takes more than $t(n)$ seconds to solve an instance of size n, then any other implementation of the same algorithm takes a time in the order of $t(n)$ seconds. We say that such an algorithm takes a time in the order of $f(n)$ for any function $f : \mathbb{N} \to \mathbb{R}^*$ such that $t(n) \in O(f(n))$. In particular, since $t(n) \in O(t(n))$, it takes a time in the order of $t(n)$ itself. In general, however, we try to express the order of the algorithm's running time using the simplest possible function f such that $t(n) \in O(f(n))$.

Problem 2.1.1.   Some implementation of a certain algorithm takes a time bounded above by

$$t(n) = 3 \text{ seconds} - 18n \text{ milliseconds} + 27n^2 \text{ microseconds}$$

to solve any instance of size n. (Such behaviour is unlikely since $t(n)$ decreases as n increases for $n \leq 333$.) Find the simplest possible function $f : \mathbb{N} \to \mathbb{R}^*$ such that the algorithm takes a time in the order of $f(n)$. Prove that $t(n) \in O(f(n))$.   □

Problem 2.1.2.   Which of the following statements are true ? Prove your answers.

i. $n^2 \in O(n^3)$

ii. $n^3 \in O(n^2)$

iii. $2^{n+1} \in O(2^n)$

iv. $(n+1)! \in O(n!)$

v. for any function $f : \mathbb{N} \to \mathbb{R}^*, f(n) \in O(n) \Rightarrow [f(n)]^2 \in O(n^2)$

vi. for any function $f : \mathbb{N} \to \mathbb{R}^*, f(n) \in O(n) \Rightarrow 2^{f(n)} \in O(2^n)$ □

Problem 2.1.3.   Prove that the following definition of $O(f(n))$ is equivalent to the preceding one, provided that $f(n)$ is strictly positive for all $n \in \mathbb{N}$:

$$O(f(n)) = \{ t : \mathbb{N} \to \mathbb{R}^* \mid (\exists c \in \mathbb{R}^+)(\forall n \in \mathbb{N})[t(n) \leq cf(n)] \}.$$

In other words, the threshold n_0 is not necessary in principle, even though it is often useful in practice.   □

Problem 2.1.4.   Prove that the relation " $\in O$ " is *transitive*: if $f(n) \in O(g(n))$ and $g(n) \in O(h(n))$, then $f(n) \in O(h(n))$. Conclude that if $g(n) \in O(h(n))$, then $O(g(n)) \subseteq O(h(n))$.   □

This asymptotic notation provides a way to define a partial order on functions and consequently on the relative efficiency of different algorithms to solve a given problem, as suggested by the following exercises.

Problem 2.1.5.   For arbitrary functions f and $g : \mathbb{N} \to \mathbb{R}^*$, prove that

i. $O(f(n)) = O(g(n))$ if and only if $f(n) \in O(g(n))$ and $g(n) \in O(f(n))$, and

ii. $O(f(n)) \subset O(g(n))$ if and only if $f(n) \in O(g(n))$ but $g(n) \notin O(f(n))$.   □

Problem 2.1.6.   Find two functions f and $g : \mathbb{N} \to \mathbb{N}^+$ such that $f(n) \notin O(g(n))$ and $g(n) \notin O(f(n))$. Prove your answer.   □

Problem 2.1.7.   For arbitrary functions f and $g : \mathbb{N} \to \mathbb{R}^*$, prove that $O(f(n) + g(n)) = O(\max(f(n), g(n)))$, where the sum and the maximum are to be taken pointwise.   □

The result of the preceding problem is useful for simplifying asymptotic calculations. For instance,

$$n^3 + 3n^2 + n + 8 \in O(n^3 + (3n^2 + n + 8))$$
$$= O(\max(n^3, 3n^2 + n + 8)) = O(n^3).$$

The last equality holds despite the fact that $\max(n^3, 3n^2 + n + 8) \neq n^3$ when $0 \leq n \leq 3$, because the asymptotic notation only applies when n is sufficiently large. However, we do have to ensure that $f(n)$ and $g(n)$ only take nonnegative values (possibly with a finite number of exceptions) to avoid false arguments like the following:

$$O(n^2) = O(n^3 + (n^2 - n^3))$$
$$= O(\max(n^3, n^2 - n^3)) = O(n^3).$$

A little manipulation is sufficient, however, to allow us to conclude that

$$n^3 - 3n^2 - n - 8 \in O(n^3)$$

because

$$n^3 - 3n^2 - n - 8 \in O(n^3 - 3n^2 - n - 8)$$
$$= O(\tfrac{1}{2}n^3 + (\tfrac{1}{2}n^3 - 3n^2 - n - 8))$$
$$= O(\max(\tfrac{1}{2}n^3, \tfrac{1}{2}n^3 - 3n^2 - n - 8))$$
$$= O(\tfrac{1}{2}n^3) = O(n^3).$$

Here again, the fact that $\tfrac{1}{2}n^3 - 3n^2 - n - 8$ is negative when $0 \le n \le 6$ is of no concern. False arguments of the kind illustrated in the following problem are also to be avoided.

Problem 2.1.8.   Find the error in the following argument:

$$\sum_{i=1}^{n} i \; = \; 1+2+\cdots+n \; \in \; O(1+2+\cdots+n) \; = \; O(\max(1, 2, \ldots, n)) \; = \; O(n). \;\; \square$$

Problem 2.1.9.   The notion of a limit is a powerful and versatile tool for comparing functions. Given f and $g : \mathbb{N} \to \mathbb{R}^+$, prove that

i. $\lim\limits_{n \to \infty} f(n)/g(n) \in \mathbb{R}^+ \;\Rightarrow\; O(f(n)) = O(g(n))$, and

ii. $\lim\limits_{n \to \infty} f(n)/g(n) = 0 \;\Rightarrow\; O(f(n)) \subset O(g(n)) = O(g(n) \pm f(n))$, but

iii. it can happen that $O(f(n)) = O(g(n))$ although the limit of $f(n)/g(n)$ does not exist as n tends to infinity, and

iv. it can happen that $O(f(n)) \subset O(g(n))$ when the limit of $f(n)/g(n)$ does not exist as n tends to infinity and when it is also not true that $O(g(n)) = O(g(n) - f(n))$. $\square$

De l'Hôpital's rule is often useful in order to apply the preceding problem. Recall that if $\lim\limits_{n \to \infty} f(n) = \lim\limits_{n \to \infty} g(n) = 0$, or if both these limits are infinite, then provided that the domains of f and g can be extended to some real interval $[n_0, +\infty)$ in such a way that the corresponding new functions $\hat{f}$ and $\hat{g}$ are differentiable on this interval and also that $\hat{g}'(x)$, the derivative of $\hat{g}(x)$, is never zero for $x \in [n_0, +\infty)$, then

$$\lim_{n \to \infty} f(n)/g(n) = \lim_{x \to \infty} \hat{f}'(x)/\hat{g}'(x),$$

provided that this last limit exists.

Problem 2.1.10.   Use de l'Hôpital's rule and Problems 2.1.5 and 2.1.9 to prove that $\log n \in O(\sqrt{n}\,)$ but that $\sqrt{n} \notin O(\log n)$. $\square$

Problem 2.1.11.   Let ε be an arbitrary real constant, $0 < \varepsilon < 1$. Use the relations "$\subset$" and "$=$" to put the orders of the following functions into a sequence:

$$n \log n, \ n^8, \ n^{1+\varepsilon}, \ (1+\varepsilon)^n, \ (n^2 + 8n + \log^3 n)^4, \ \text{and} \ n^2/\log n.$$

Prove your answer.   □

∗Problem 2.1.12.   Let f and $g : \mathbb{N} \to \mathbb{R}^+$ be two increasing functions, and let c be a strictly positive real constant such that for every integer n

i. $4g(n) \le g(2n) \le 8g(n)$ and

ii. $f(2n) \le 2f(n) + cg(n)$.

Prove that

iii. $f(n) \in O(g(n))$.   □

2.1.2 Other Asymptotic Notation

The notation we have just seen is useful for estimating an upper limit on the time that some algorithm will take on a given instance. It is also sometimes interesting to estimate a *lower* limit on this time. The following notation is proposed to this end:

$$\Omega(f(n)) = \{ \, t : \mathbb{N} \to \mathbb{R}^* \mid (\exists c \in \mathbb{R}^+)(\exists n_0 \in \mathbb{N})(\forall n \ge n_0)[t(n) \ge cf(n)] \, \}.$$

In other words, $\Omega(f(n))$, which we read unimaginatively as *omega of $f(n)$* is the set of all the functions $t(n)$ bounded below by a real positive multiple of $f(n)$, provided n is sufficiently large. The following exercise brings out the symmetry between the notation O and Ω.

Problem 2.1.13.   For arbitrary functions f and $g : \mathbb{N} \to \mathbb{R}^*$, prove that $f(n) \in O(g(n))$ if and only if $g(n) \in \Omega(f(n))$.   □

In a worst-case analysis there is, however, a fundamental asymmetry between the notation O and Ω. If an algorithm takes a time in $O(f(n))$ in the worst case, there exists a real positive constant c such that a time of $cf(n)$ is sufficient for the algorithm to solve the worst instance of size n, for each sufficiently large n. This time is obviously also sufficient to solve all the other instances of size n, since they cannot take more time than the worst case. Assuming only a finite number of instances of each given size exists, there can thus be only a finite number of instances, all of size less than the threshold, on which the algorithm takes a time greater than $cf(n)$. These instances can all be taken care of as suggested by Problem 2.1.3, by using a bigger constant. On the other hand, if an algorithm takes a time in $\Omega(f(n))$ in the worst case, there exists a real positive constant d such that the algorithm takes a time longer than

$df(n)$ to solve the worst instance of size n, for each sufficiently large n. This in no way rules out the possibility that a much shorter time might suffice to solve some other instances of size n. Thus there can exist an infinity of instances for which the algorithm takes a time less than $df(n)$. Insertion sort, which we saw in Section 1.4, provides a typical example of such behaviour: it takes a time in $\Omega(n^2)$ in the worst case, despite the fact that a time in the order of n is sufficient to solve arbitrarily large instances in which the items are already sorted.

We shall be happiest if, when we analyse the asymptotic behaviour of an algorithm, its execution time is bounded simultaneously both above and below by positive real multiples (possibly different) of the same function. For this reason we introduce a final notation

$$\Theta(f(n)) = O(f(n)) \cap \Omega(f(n)),$$

called *the exact order of $f(n)$*.

Problem 2.1.14. Prove that $f(n) \in \Theta(g(n))$ if and only if

$$(\exists c, d \in \mathbb{R}^+)(\exists n_0 \in \mathbb{N})(\forall n \geq n_0)[cg(n) \leq f(n) \leq dg(n)]. \qquad \Box$$

The following problem shows that the Θ notation is no more powerful than the O notation for comparing the respective orders of two functions.

Problem 2.1.15. For arbitrary functions f and $g : \mathbb{N} \rightarrow \mathbb{R}^*$, prove that the following statements are equivalent:

   i. $O(f(n)) = O(g(n))$,

  ii. $\Theta(f(n)) = \Theta(g(n))$, and

 iii. $f(n) \in \Theta(g(n))$. $\qquad \Box$

Problem 2.1.16. Continuing Problem 2.1.9, prove that if f and $g : \mathbb{N} \rightarrow \mathbb{R}^+$ are two arbitrary functions, then

   i. $\displaystyle\lim_{n \to \infty} f(n)/g(n) \in \mathbb{R}^+ \implies f(n) \in \Theta(g(n))$,

  ii. $\displaystyle\lim_{n \to \infty} f(n)/g(n) = 0 \implies f(n) \in O(g(n))$ but $f(n) \notin \Theta(g(n))$, and

 iii. $\displaystyle\lim_{n \to \infty} f(n)/g(n) = +\infty \implies f(n) \in \Omega(g(n))$ but $f(n) \notin \Theta(g(n))$. $\qquad \Box$

Problem 2.1.17.   Prove the following assertions :

i. $\log_a n \in \Theta(\log_b n)$ whatever the values of $a, b > 1$ (so that we generally do not bother to specify the base of a logarithm in an asymptotic expression), but

ii. $2^{\log_a n} \notin \Theta(2^{\log_b n})$ if $a \neq b$,

iii. $\sum_{i=1}^{n} i^k \in \Theta(n^{k+1})$ for any given integer $k \geq 0$ (this works even for *real* $k > -1$; the hidden constant in the Θ notation may depend on the value of k),

iv. $\log(n!) \in \Theta(n \log n)$, and

v. $\sum_{i=1}^{n} i^{-1} \in \Theta(\log n)$.   □

2.1.3 Asymptotic Notation with Several Parameters

It may happen when we analyse an algorithm that its execution time depends simultaneously on more than one parameter of the instance in question. This situation is typical of certain algorithms for problems involving graphs, where the time depends on both the number of vertices and the number of edges. In such cases the notion of the "size of the instance" that we have used so far may lose much of its meaning. For this reason the asymptotic notation is generalized in a natural way to functions of several variables.

Let $f : \mathbb{N} \times \mathbb{N} \to \mathbb{R}^*$ be an arbitrary function. We define

$$O(f(m,n)) = \{ t : \mathbb{N} \times \mathbb{N} \to \mathbb{R}^* \mid (\exists c \in \mathbb{R}^+)(\exists m_0, n_0 \in \mathbb{N})$$
$$(\forall n \geq n_0)(\forall m \geq m_0)[t(m,n) \leq cf(m,n)] \}.$$

Other generalizations are defined similarly.

There is nevertheless an essential difference between an asymptotic notation with only one parameter and one with several : unlike the result obtained in Problem 2.1.3, it can happen that the thresholds m_0 and n_0 are indispensable. This is explained by the fact that while there are never more than a finite number of values of $n \geq 0$ such that $n \geq n_0$ is not true, there are in general an infinite number of pairs $<m,n>$ such that $m \geq 0$ and $n \geq 0$ yet such that $m \geq m_0$ and $n \geq n_0$ are not both true.

*Problem 2.1.18.   Give an example of a function $f : \mathbb{N} \times \mathbb{N} \to \mathbb{R}^+$ such that

$$O(f(m,n)) \neq \{ t : \mathbb{N} \times \mathbb{N} \to \mathbb{R}^* \mid (\exists c \in \mathbb{R}^+)(\forall m, n \in \mathbb{N})[t(m,n) \leq cf(m,n)] \}. \quad □$$

2.1.4 Operations on Asymptotic Notation

To simplify some calculations, we can manipulate the asymptotic notation using arithmetic operators. For instance, $O(f(n)) + O(g(n))$ represents the set of functions obtained by adding pointwise any function in $O(f(n))$ to any function in $O(g(n))$. Intuitively this represents the order of the time taken by an algorithm composed of a first stage taking a time in the order of $f(n)$ followed by a second stage taking a time in the order of $g(n)$. The hidden constants that multiply $f(n)$ and $g(n)$ may well be different, but this is of no consequence (see Problem 2.1.19(i)).

More formally, if X and Y are sets of functions from $\mathbb{N}$ into $\mathbb{R}^*$ and if **op** is any binary operator, then "X **op** Y " denotes

$$\{ t : \mathbb{N} \to \mathbb{R}^* \mid (\exists f \in X)(\exists g \in Y)(\exists n_0 \in \mathbb{N})(\forall n \ge n_0)[t(n) = f(n) \text{ op } g(n)] \}.$$

If g is a function from $\mathbb{N}$ into $\mathbb{R}^*$, we stretch the notation by writing X **op** g to denote X **op** $\{g\}$. Furthermore, if $a \in \mathbb{R}^*$, we use X **op** a to denote X **op** Id_a, where $Id_a : \mathbb{N} \to \mathbb{R}^*$ is the constant function $Id_a(n) = a$ for every integer n. We also use the symmetrical notation g **op** X and a **op** X, and all this theory of operations on sets is extended in the obvious way to operators other than binary.

This notation occasionally conflicts with the usual mathematical conventions. For instance, $[O(f(n))]^2$ does *not* denote the set of pairs of functions chosen from the set $O(f(n))$. Similarly, $O(f(n)) \times O(g(n))$ does *not* denote the Cartesian product of $O(f(n))$ and $O(g(n))$. On the other hand, if N is the set of vertices of a directed graph, then $N \times N$ denotes as usual the set of possible edges between these vertices. In every case but one the context removes any potential ambiguity. Still, there is one case that must be treated cautiously. If the symbol "$-$" were used to denote the difference of two sets, a genuinely ambiguous situation would arise: what would $O(n^3) - O(n^2)$ mean, for example? To avoid this ambiguity, we use "$-$" only to denote arithmetic subtraction, including pointwise subtraction of functions, reserving the symbol "$\backslash$" to denote set difference: $A \backslash B = \{x \in A \mid x \notin B\}$.

Notice the subtle difference between $[\Theta(f(n))]^2$ and $\Theta(f(n)) \times \Theta(f(n))$. Although they both denote the same set as $\Theta([f(n)]^2)$, this requires a proof (Problem 2.1.19(ii)). To belong to $[\Theta(f(n))]^2$, a function $g(n)$ must be the pointwise square of some function in $\Theta(f(n))$; to belong to $\Theta(f(n)) \times \Theta(f(n))$, however, it suffices for $g(n)$ to be the pointwise product of two *possibly different* functions, each a member of $\Theta(f(n))$. To understand the first notation, think of it as $\Theta(f(n)) \exp \{Id_2\}$, where "**exp**" denotes the binary exponentiation operator and "Id_2" is the constant function $Id_2(n) = 2$ for all n. Similarly, $n \times \Theta(f(n))$ denotes

$$\{ t : \mathbb{N} \to \mathbb{R}^* \mid (\exists g(n) \in \Theta(f(n)))(\exists n_0 \in \mathbb{N})(\forall n \ge n_0)[t(n) = n \times g(n)] \},$$

which is not at all the same as $\sum_{i=1}^{n} \Theta(f(n)) = \Theta(f(n)) + \Theta(f(n)) + \cdots + \Theta(f(n))$.

Problem 2.1.19.   Let f and g be arbitrary functions from $\mathbb{N}$ into $\mathbb{R}^*$. Prove the following identities:

i. $\Theta(f(n)) + \Theta(g(n)) = \Theta(f(n) + g(n)) = \Theta(\max(f(n), g(n)))$
   $= \max(\Theta(f(n)), \Theta(g(n)))$;

ii. $\Theta([f(n)]^2) = [\Theta(f(n))]^2 = \Theta(f(n)) \times \Theta(f(n))$;

iii. $[1+\Theta(1)]^n = 2^{\Theta(n)}$, but $[\Theta(1)]^n \neq 2^{\Theta(n)}$; and

iv. $f(n) \in \prod_{i=0}^{n} \Theta(1)$ $\square$

Another kind of operation on the asymptotic notation allows it to be nested. Let X be a set of functions from $\mathbb{N}$ into $\mathbb{R}^*$, possibly defined by some asymptotic notation. Now $O(X)$ denotes

$$\bigcup_{f \in X} O(f(n)) = \{ t : \mathbb{N} \to \mathbb{R}^* \mid (\exists f \in X) [t \in O(f(n))] \}.$$

$\Omega(X)$ and $\Theta(X)$ are defined similarly.

Example 2.1.1.   Although this expression can be simplified, the natural way to express the execution time required by Dixon's integer factorization algorithm (Section 8.5.3) is

$$O(e^{O(\sqrt{\ln n \ln \ln n}\,)}),$$

where n is the value of the integer to be factorized. $\square$

2.1.5 Conditional Asymptotic Notation

Many algorithms are easier to analyse if initially we only consider instances whose size satisfies a certain condition, such as being a power of 2. Conditional asymptotic notation handles this situation. Let $f : \mathbb{N} \to \mathbb{R}^*$ be any function and let $P : \mathbb{N} \to \mathbb{B}$ be a predicate. We define

$$O(f(n) \mid P(n)) = \{ t : \mathbb{N} \to \mathbb{R}^* \mid (\exists c \in \mathbb{R}^+)(\exists n_0 \in \mathbb{N})(\forall n \geq n_0)$$

$$[P(n) \Rightarrow t(n) \leq cf(n)] \}.$$

In other words, $O(f(n) \mid P(n))$, which we read as *the order of $f(n)$ when $P(n)$*, is the set of all functions $t(n)$ bounded above by a real positive multiple of $f(n)$ whenever n is sufficiently large and provided the condition $P(n)$ holds. The notation $O(f(n))$ defined previously is thus equivalent to $O(f(n) \mid P(n))$ where $P(n)$ is the predicate whose value is always *true*. The notation $\Omega(f(n) \mid P(n))$ and $\Theta(f(n) \mid P(n))$ is defined similarly, as is the notation with several parameters.

The principal reason for using this conditional notation is that it can generally be eliminated once it has been used to make the analysis of an algorithm easier. You probably used this idea for solving Problem 2.1.12. A function $f : \mathbb{N} \to \mathbb{R}^*$ is *eventually nondecreasing* if $(\exists n_0 \in \mathbb{N})(\forall n \geq n_0)[f(n) \leq f(n+1)]$, which implies by mathematical induction that $(\exists n_0 \in \mathbb{N})(\forall n \geq n_0)(\forall m \geq n)[f(n) \leq f(m)]$. Let $b \geq 2$ be any integer. Such a function is *b-smooth* if, as well as being eventually nondecreasing, it satisfies the condition $f(bn) \in O(f(n))$. It turns out that any function that is b-smooth for some integer $b \geq 2$ is also c-smooth for every integer $c \geq 2$ (prove it!); we shall therefore in future simply refer to such functions as being *smooth*. The following problem assembles these ideas.

∗ Problem 2.1.20.   Let $b \geq 2$ be any integer, let $f : \mathbb{N} \to \mathbb{R}^*$ be a smooth function, and let $t : \mathbb{N} \to \mathbb{R}^*$ be an eventually nondecreasing function such that $t(n) \in X(f(n) \mid n$ is a power of $b)$, where X stands for one of O, Ω, or Θ. Prove that $t(n) \in X(f(n))$. Furthermore, if $t(n) \in \Theta(f(n))$, prove that $t(n)$ is also smooth. Give two specific examples to illustrate that the conditions "$t(n)$ is eventually nondecreasing" and "$f(bn) \in O(f(n))$" are both necessary to obtain these results.   ☐

We illustrate this principle using an example suggested by the algorithm for merge sorting given in Section 4.4.

Example 2.1.2.   Let $t(n)$ be defined by the following equation:

$$t(n) = \begin{cases} a & \text{if } n = 1 \\ t(\lfloor n/2 \rfloor) + t(\lceil n/2 \rceil) + bn & \text{otherwise,} \end{cases}$$

where a and b are arbitrary real positive constants. The presence of floors and ceilings makes this equation hard to analyse exactly. However, if we only consider the cases when n is a power of 2, the equation becomes

$$t(n) = \begin{cases} a & \text{if } n = 1 \\ 2t(n/2) + bn & \text{if } n > 1 \text{ is a power of 2.} \end{cases}$$

The techniques discussed in Section 2.3, in particular Problem 2.3.6, allow us to infer immediately that $t(n) \in \Theta(n \log n \mid n$ is a power of 2). In order to apply the result of the previous problem to conclude that $t(n) \in \Theta(n \log n)$, we need only show that $t(n)$ is an eventually nondecreasing function and that $n \log n$ is smooth.

The proof that $(\forall n \geq 1)[t(n) \leq t(n+1)]$ is by mathematical induction. First, note that $t(1) = a \leq 2(a+b) = t(2)$. Let n be greater than 1. By the induction hypothesis, assume that $(\forall m < n)[t(m) \leq t(m+1)]$. In particular, $t(\lfloor n/2 \rfloor) \leq t(\lfloor (n+1)/2 \rfloor)$ and $t(\lceil n/2 \rceil) \leq t(\lceil (n+1)/2 \rceil)$. Therefore

$$t(n) = t(\lfloor n/2 \rfloor) + t(\lceil n/2 \rceil) + bn \leq t(\lfloor (n+1)/2 \rfloor) + t(\lceil (n+1)/2 \rceil) + b(n+1) = t(n+1).$$

A word of caution is important here. One might be tempted to claim that $t(n)$ is eventually nondecreasing because such is obviously the case with $n \log n$. This argumentation is irrelevant and fallacious because the relation between $t(n)$ and $n \log n$ has *only* been demonstrated thus far when n is a power of 2. The proof that $t(n)$ is nondecreasing must use its recursive definition.

Finally, $n \log n$ is smooth because it is clearly eventually nondecreasing and

$$2n \log(2n) = 2n(\log 2 + \log n) = (2\log 2)n + 2n \log n$$
$$\in O(n + n \log n) = O(\max(n, n \log n)) = O(n \log n). \qquad \square$$

2.1.6 Asymptotic Recurrences

When analysing algorithms, we do not always find ourselves faced with equations as precise as those in Example 2.1.2 in the preceding section. More often we have to deal with *inequalities* such as

$$t(n) \le \begin{cases} t_1(n) & \text{if } n \le n_0 \\ t(\lfloor n/2 \rfloor) + t(\lceil n/2 \rceil) + cn & \text{otherwise} \end{cases}$$

and simultaneously

$$t(n) \ge \begin{cases} t_2(n) & \text{if } n \le n_0 \\ t(\lfloor n/2 \rfloor) + t(\lceil n/2 \rceil) + dn & \text{otherwise,} \end{cases}$$

for some constants $c, d \in \mathbb{R}^+$, $n_0 \in \mathbb{N}$, and for appropriate initial functions $t_1, t_2 : \mathbb{N} \to \mathbb{R}^+$. Our asymptotic notation allows these constraints to be expressed succinctly as

$$t(n) \in t(\lfloor n/2 \rfloor) + t(\lceil n/2 \rceil) + \Theta(n).$$

To solve such inequalities, it is convenient to convert them first to equalities. To this end, define $f : \mathbb{N} \to \mathbb{R}$ by

$$f(n) = \begin{cases} 1 & \text{if } n = 1 \\ f(\lfloor n/2 \rfloor) + f(\lceil n/2 \rceil) + n & \text{otherwise.} \end{cases}$$

We saw in the previous section that $f(n) \in \Theta(n \log n)$.

Coming back now to the function $t(n)$ satisfying the preceding inequalities, let $u = \max(c, \max\{t_1(n) \mid n \le n_0\})$ and $v = \min(d, \min\{t_2(n)/f(n) \mid n \le n_0\})$. It is easy to prove by mathematical induction that $v \le t(n)/f(n) \le u$ for every integer n. We immediately conclude that $t(n) \in \Theta(f(n)) = \Theta(n \log n)$.

This change from the original inequalities to a parametrized equation is useful from two points of view. Obviously it saves having to prove independently both $t(n) \in O(n \log n)$ and $t(n) \in \Omega(n \log n)$. More importantly, however, it allows us to confine our analysis in the initial stages to the easier case where n is a power of 2. It is then possible, using the conditional asymptotic notation and the technique explained in Problem 2.1.20, to generalize our results automatically to the case where n is an arbitrary integer. This could not have been done directly with the original inequalities, since they do not allow us to conclude that $t(n)$ is eventually nondecreasing, which in turn prevents us from applying Problem 2.1.20.

2.1.7 Constructive Induction

Mathematical induction is used primarily as a proof technique. Too often it is employed to prove assertions that seem to have been produced from nowhere like a rabbit out of a hat. While the truth of these assertions is thus established, their origin remains mysterious. However, mathematical induction is a tool sufficiently powerful to allow us to discover not merely the proof of a theorem, but also its exact statement. By applying this technique, we can simultaneously prove the truth of a partially specified assertion and discover the missing specifications thanks to which the assertion is correct. As we shall see in Examples 2.1.4 and 2.2.5, this technique of *constructive induction* is especially useful for solving certain recurrences that occur in the context of the analysis of algorithms. We begin with a simple example.

Example 2.1.3.   Let the function $f : \mathbb{N} \to \mathbb{N}$ be defined by the following recurrence:

$$f(n) = \begin{cases} 0 & \text{if } n = 0 \\ n + f(n-1) & \text{otherwise.} \end{cases}$$

It is clear that $f(n) = \sum_{i=0}^{n} i$. Pretend for a moment that you do not know that $f(n) = n(n+1)/2$, but that you are looking for some such formula. Obviously $f(n) = \sum_{i=0}^{n} i \le \sum_{i=0}^{n} n = n^2$, and so $f(n) \in O(n^2)$. This suggests that we formulate a hypothesis that $f(n)$ might be a quadratic polynomial. We therefore try the *partially specified induction hypothesis HI(n)* according to which $f(n) = an^2 + bn + c$. This hypothesis is partial in the sense that a, b, and c are not yet known. The technique of constructive induction consists of trying to prove this incomplete hypothesis by mathematical induction. Along the way we hope to pick up sufficient information about the constants to determine their values.

Supposing that $HI(n-1)$ is true for some $n \ge 1$, we know that

$$f(n) = n + f(n-1) = n + a(n-1)^2 + b(n-1) + c = an^2 + (1+b-2a)n + (a-b+c).$$

If we are to conclude $HI(n)$, it must be the case that $f(n) = an^2 + bn + c$. By equating the coefficients of each power of n, we obtain two nontrivial equations for our three unknowns: $1 + b - 2a = b$ and $a - b + c = c$. From these it follows that $a = b = \frac{1}{2}$, the value of c being as yet unconstrained. We now have therefore a new, more complete hypothesis, which we continue to call $HI(n)$: $f(n) = n^2/2 + n/2 + c$. We have just shown that if $HI(n-1)$ is true for some $n \ge 1$, then so is $HI(n)$. It remains to establish the truth of $HI(0)$ in order to conclude by mathematical induction that $HI(n)$ is true for every integer n. Now $HI(0)$ says precisely that $f(0) = a0^2 + b0 + c = c$. Knowing that $f(0) = 0$, we conclude that $c = 0$ and that $f(n) = n^2/2 + n/2$ is true for every integer n.   □

In the case of the preceding example it would have been simpler to determine the values of a, b, and c by constructing three linear equations using the values of $HI(n)$ for $n = 0$, 1, and 2.

$$c = 0$$
$$a + b + c = 1$$
$$4a + 2b + c = 3$$

Solving this system gives us immediately $a = \frac{1}{2}$, $b = \frac{1}{2}$, and $c = 0$. However, using this approach does not *prove* that $f(n) = n^2/2 + n/2$, since nothing allows us to assert a priori that $f(n)$ is in fact given by a quadratic polynomial. Thus once the constants are determined we must in any case follow this with a proof by mathematical induction.

Some recurrences are more difficult to solve than the one given in Example 2.1.3. Even the techniques we shall see in Section 2.3 will prove insufficient on occasion. However, in the context of asymptotic notation, an exact solution of the recurrence equations is generally unnecessary, since we are only interested in establishing an upper bound on the quantity of interest. In this setting constructive induction can be exploited to the hilt.

Example 2.1.4.   Let the function $t : \mathbb{N}^+ \to \mathbb{R}^+$ be given by the recurrence

$$t(n) = \begin{cases} a & \text{if } n = 1 \\ bn^2 + nt(n-1) & \text{otherwise,} \end{cases}$$

where a and b are arbitrary real positive constants. Although this equation is not easy to solve exactly, it is sufficiently similar to the recurrence that characterizes the factorial ($n! = n \times (n-1)!$) that it is natural to conjecture that $t(n) \in \Theta(n!)$. To establish this, we shall prove independently that $t(n) \in O(n!)$ and that $t(n) \in \Omega(n!)$. The technique of constructive induction is useful in both cases. For simplicity, we begin by proving that $t(n) \in \Omega(n!)$, that is, there exists a real positive constant u such that $t(n) \geq un!$ for every positive integer n. Suppose by the partially specified induction hypothesis that $t(n-1) \geq u(n-1)!$ for some $n > 1$. By definition of $t(n)$, we know that $t(n) = bn^2 + nt(n-1) \geq bn^2 + nu(n-1)! = bn^2 + un! \geq un!$. Thus we see that $t(n) \geq un!$ is always true, regardless of the value of u, provided that $t(n-1) \geq u(n-1)!$. In order to conclude that $t(n) \geq un!$ for every positive integer n, it suffices to show that this is true for $n = 1$; that is, $t(1) \geq u$. Since $t(1) = a$, this is the same as saying that $u \leq a$. Taking $u = a$, we have established that $t(n) \geq an!$ for every positive integer n, and thus that $t(n) \in \Omega(n!)$.

Encouraged by this success, we now try to show that $t(n) \in O(n!)$ by proving the existence of a real positive constant v such that $t(n) \leq vn!$ for every positive integer n. Suppose by the partially specified induction hypothesis that $t(n-1) \leq v(n-1)!$ for some $n > 1$. As usual, this allows us to affirm that $t(n) = bn^2 + nt(n-1) \leq bn^2 + vn!$. However our aim is to show that $t(n) \leq vn!$. Unfortunately no positive value of v allows us to conclude that $t(n) \leq vn!$ given that $t(n) \leq bn^2 + vn!$. It seems then that constructive induction has nothing to offer in this context, or perhaps even that the hypothesis to the effect that $t(n) \in O(n!)$ is false.

In fact, it is possible to obtain the result we hoped for. Rather than trying to prove directly that $t(n) \leq vn!$, we use constructive induction to determine real positive constants v and w such that $t(n) \leq vn! - wn$ for any positive integer n. This idea may seem odd, since $t(n) \leq vn! - wn$ is a stronger statement than $t(n) \leq vn!$, which we were unable to prove. We may hope for success, however, on the grounds that if the statement to be proved is stronger, then so too is the induction hypothesis it allows us to use.

Suppose then by the partially specified induction hypothesis that $t(n-1) \leq v(n-1)! - w(n-1)$ for some $n > 1$. Using the definition of $t(n)$, we conclude that

$$t(n) = bn^2 + nt(n-1) \leq bn^2 + n(v(n-1)! - w(n-1)) = vn! + ((b-w)n + w)n.$$

To conclude that $t(n) \leq vn! - wn$, it is necessary and sufficient that $(b-w)n + w \leq -w$. This inequality holds if and only if $n \geq 3$ and $w \geq bn/(n-2)$. Since $n/(n-2) \leq 3$ for every $n \geq 3$, we may in particular choose $w = 3b$ to ensure that $t(n) \leq vn! - wn$ is a consequence of the hypothesis $t(n-1) \leq v(n-1)! - w(n-1)$, independently of the value of v, provided that $n \geq 3$.

All that remains is to adjust the constant v to take care of the cases $n \leq 2$. When $n = 1$, we know that $t(1) = a$. If we are to conclude that $t(n) \leq v - 3b$, it is necessary and sufficient that $v \geq a + 3b$. When $n = 2$, we can apply the recurrence definition of $t(n)$ to find $t(2) = 4b + 2t(1) = 4b + 2a$. If we are to conclude that $t(n) \leq 2v - 6b$, it is necessary and sufficient that $v \geq a + 5b$, which is stronger than the previous condition. In particular, we may choose $v = a + 5b$.

The conclusion from all this is that $t(n) \in \Theta(n!)$ since

$$an! \leq t(n) \leq (a + 5b)n! - 3bn$$

for every positive integer n. If you got lost in the preceding argument, you may wish to prove this assertion, which is now completely specified, by straightforward mathematical induction. □

The following problem is not so easy; it illustrates well the advantage obtained by using constructive induction, thanks to which we were able to prove that $t(n) \in \Theta(n!)$ without ever finding an exact expression for $t(n)$.

*Problem 2.1.21. To complete Example 2.1.4, solve exactly the recurrence defining $t(n)$. Determine in terms of a and b the real positive constant c such that

$$\lim_{n \to \infty} \frac{t(n)}{n!} = c.$$

Verify that $a \leq c \leq a + 5b$. □

Problem 2.1.22. Let $k \in \mathbb{N}$ and $a, b \in \mathbb{R}^+$ be arbitrary constants. Let $g : \mathbb{N}^+ \to \mathbb{R}^+$ be the function defined by the recurrence

$$g(n) = \begin{cases} a & \text{if } n = 1 \\ bn^k + ng\,(n-1) & \text{otherwise.} \end{cases}$$

Prove that $g(n) \in \Theta(n!)$.   □

2.1.8 For Further Reading

The notation used in this chapter is not universally accepted. You may encounter three major differences in other books. The most striking is the widespread use of statements such as $n^2 = O(n^3)$ where we would write $n^2 \in O(n^3)$. Use of such "one-way equalities" (for one would not write $O(n^3) = n^2$) is hard to defend except on historical grounds. With this definition we say that the execution time of some algorithm is *of the order of* $f(n)$ (or *is* $O(f(n))$) rather than saying it is *in* the order of $f(n)$.

The second difference is less striking but more important, since it can lead an incautious reader astray. Some authors define

$$\Omega(f(n)) = \{\, t : \mathbb{N} \to \mathbb{R}^* \mid (\exists c \in \mathbb{R}^+)\,(\forall n_0 \in \mathbb{N})\,(\exists n \geq n_0)\,[t(n) \geq cf(n)]\, \}.$$

Notice the quantifier reversal. With this definition it suffices that there exist an infinite number of instances x that force some algorithm to take at least $cf(|x|)$ steps in order to conclude that this algorithm takes a time in $\Omega(f(n))$. This corresponds more closely to our intuitive idea of what a lower bound on the performance of an algorithm should look like. Furthermore, with this definition, the asymmetry between O and Ω noted after Problem 2.1.13 is neatly avoided. Unfortunately, the notation becomes difficult to handle, in particular because Ω thus defined is not transitive, and because it makes Θ asymmetric.

The third difference concerns the definition of O. We often find

$$O(f(n)) = \{\, t : \mathbb{N} \to \mathbb{R} \mid (\exists c \in \mathbb{R}^+)\,(\exists n_0 \in \mathbb{N})\,(\forall n \geq n_0)\,[\,|t(n)| \leq cf(n)]\, \}$$

where $|t(n)|$ denotes (here only) the absolute value of $t(n)$. Using this definition, one would write $n^3 - n^2 \in n^3 + O(n^2)$. Of course, the meaning of "such-and-such an algorithm takes a time in $O(n^2)$" does not change since algorithms cannot take negative time. On the other hand, a statement such as $\Theta(f(n)) + \Theta(g(n)) = \Theta(\max(f(n), g(n)))$ is no longer true.

Problem 2.1.23.   Why?   □

When we want the equivalent of this definition of $O(f(n))$ we write $\pm O(f(n))$.

2.2 ANALYSIS OF ALGORITHMS

There is no magic formula for analysing the efficiency of an algorithm. It is largely a question of judgement, intuition, and experience. Often a first analysis gives rise to a complicated-looking function, involving summations or recurrences. The next step is to simplify this function using the asymptotic notation as well as the techniques explained in Section 2.3. Here are some examples.

Example 2.2.1. Selection sort.   Consider the selection sorting algorithm given in Section 1.4. Most of the execution time is spent carrying out the instructions in the inner loop, including the implicit control statements for this loop. The time taken by each trip round the inner loop can be bounded above by a constant a. The complete execution of the inner loop for a given value of i therefore takes at most a time $b + a(n-i)$, where b is a second constant introduced to take account of the time spent initializing the loop. One trip round the outer loop is therefore bounded above by $c + b + a(n-i)$, where c is a third constant, and finally, the complete algorithm takes a time not greater than $d + \sum_{i=1}^{n-1}[c + b + a(n-i)]$, for a fourth constant d. We can simplify this expression to $\frac{a}{2}n^2 + (b+c-a/2)n + (d-c-b)$, from which we conclude that the algorithm takes a time in $O(n^2)$. A similar analysis for the lower bound shows that in fact it takes a time in $\Theta(n^2)$.   □

In this first example we gave all the details of our argument. Details like the initialization of the loops are rarely considered explicitly. It is often sufficient to choose some instruction in the algorithm as a *barometer* and to count how many times this instruction is executed. This figure gives us the exact order of the execution time of the complete algorithm, provided that the time taken to execute the chosen instruction can itself be bounded above by a constant. In the selection sort example, one possible barometer is the test in the inner loop, which is executed exactly $n(n-1)/2$ times when n items are sorted. The following example shows, however, that such simplifications should not be made incautiously.

Example 2.2.2. Choosing a barometer.   When an algorithm includes several nested loops, as is the case with selection sort, any instruction of the inner loop can usually be used as a barometer. However, there are cases where it is necessary to take account of the implicit control of the loops. Consider the following algorithm (which is reminiscent of the *countsort* algorithm discussed in Section 10.1):

```
k ← 0
  for i ← 1 to n do
    for j ← 1 to T[i] do
      k ← k + T[j] ,
```

where T is an array of n integers such that $0 \le T[i] \le i$ for every $i \le n$. Let s be the sum of the elements of T. How much time does the algorithm take?

For each value of i the instruction "$k \leftarrow k + T[j]$" is executed $T[i]$ times. The total number of times it is executed is therefore $\sum_{i=1}^{n} T[i] = s$ times. If indeed this instruction could serve as a barometer, we would conclude that the algorithm takes a time in the exact order of s. A simple example is sufficient to convince us that this is not the case. Suppose that $T[i] = 1$ whenever i is a perfect square and that $T[i] = 0$ otherwise. In this case $s = \lfloor \sqrt{n} \rfloor$. However, the algorithm clearly takes a time in $\Omega(n)$ since each element of T is considered at least once. The problem arises because we can only neglect the time spent initializing and controlling the loops provided we include *something* each time the loop is executed.

The detailed analysis of this algorithm is as follows. Let a be the time taken by one trip round the inner loop, including the time spent on loop control. To execute the inner loop completely for a given value of i therefore takes a time $b + aT[i]$, where the constant b represents the initialization time for the loop. This time is *not* zero when $T[i] = 0$. Next, the time taken to execute one trip round the outer loop is $c + b + aT[i]$, where c is a new constant. Finally, the complete algorithm takes a time $d + \sum_{i=1}^{n}(c + b + aT[i])$, for yet another constant d. When simplified, this expression yields $(c+b)n + as + d$. The time therefore depends on two independent parameters n and s and cannot be expressed as a function of just one of these variables. This situation is typical of algorithms for handling graphs, where the execution time often depends on both the number of vertices and the number of edges. Coming back to our algorithm, Problem 2.1.7 says that we can express its execution time in asymptotic notation in two ways: $\Theta(n+s)$ or $\Theta(\max(n,s))$.

With a little experience the same algorithm can be analysed more succinctly. The instruction in the inner loop is executed exactly s times. To this we must add n to take account of the control of the outer loop and of the fact that the inner loop is initialized n times. The total time taken by the algorithm is therefore in $\Theta(n+s)$.   □

We now give an example of analysis of the average behaviour of an algorithm. As Section 1.4 points out, analyses of average behaviour are usually harder than analyses of the worst case, and they presuppose that we know a priori the probability distribution of the instances to be solved.

Example 2.2.3. Insertion sort.   Consider the insertion sorting algorithm given in Section 1.4. The time that this algorithm takes to sort n elements depends on their original order. We use the comparison "$x < T[j]$" as our barometer. The number of comparisons carried out *between elements* is a good measure of the complexity of most sorting algorithms, as we shall see in Chapter 10. (Here we do not count the implicit comparisons involved in control of the **for** loop, nor the comparison "$j > 0$".)

Suppose for a moment that i is fixed. Let $x = T[i]$, as in the algorithm. The worst case arises when x is less than $T[j]$ for every j between 1 and $i-1$, since in this case we have to compare x to $T[i-1], T[i-2], \ldots, T[1]$ before we leave the **while** loop because $j = 0$. Thus the algorithm makes $i-1$ comparisons. This can happen for every value of i from 2 to n when the array is initially sorted in descending order. The total number of comparisons is therefore $\sum_{i=2}^{n}(i-1) = n(n-1)/2 \in \Theta(n^2)$. In the worst case insertion sorting thus takes a time in $\Theta(n^2)$. Notice that selection sorting systematically makes the same number of comparisons between elements that insertion sorting makes in the worst case.

To determine the time needed by the insertion sorting algorithm on the average, suppose that the n items to be sorted are all distinct and that each permutation of these items has the same probability of occurrence. If i and k are such that $1 \le k \le i$, the probability that $T[i]$ is the kth largest element among $T[1], T[2], \ldots, T[i]$ is $1/i$ because this happens for $\binom{n}{i}(i-1)!(n-i)! = n!/i$ of the $n!$ possible permutations of n elements. For a given value of i, $T[i]$ can therefore be situated with equal probability in any position relative to the items $T[1], T[2], \ldots, T[i-1]$. With probability $1/i$, $T[i] < T[i-1]$ is false at the outset, and the first comparison $x < T[j]$ gets us out of the **while** loop. The same probability applies to any given number of comparisons up to $i-2$ included. On the other hand, the probability is $2/i$ that $i-1$ comparisons will be carried out, since this happens both when $x < T[1]$ and when $T[1] \le x < T[2]$. The average number of comparisons made for a given value of i is therefore

$$ c_i = \frac{1}{i} \left[2(i-1) + \sum_{k=1}^{i-2} k \right] $$

$$ = \frac{(i-1)(i+2)}{2i} = \frac{i+1}{2} - \frac{1}{i} . $$

These events are independent for different values of i. The average number of comparisons made by the algorithm when sorting n items is therefore

$$ \sum_{i=2}^{n} c_i = \sum_{i=2}^{n} \left[\frac{i+1}{2} - \frac{1}{i} \right] $$

$$ = \frac{n^2 + 3n}{4} - H_n $$

$$ \in \Theta(n^2). $$

Here $H_n = \sum_{i=1}^{n} i^{-1}$, the nth term of the *harmonic series*, is negligible compared to the dominant term $n^2/4$ because $H_n \in \Theta(\log n)$, as shown in Problem 2.1.17.

The insertion sorting algorithm makes on the average about half the number of comparisons that it makes in the worst case, but this number is still in $\Theta(n^2)$. Although the algorithm takes a time in $\Omega(n^2)$ both on the average and in the worst case, a time in $O(n)$ is sufficient for an infinite number of instances.   □

When we analyse an algorithm, we often have to evaluate the sum of arithmetic, geometric, and other series.

Problem 2.2.1.   Prove that for any positive integers n and d

$$\sum_{k=0}^{d} 2^k \lg(n/2^k) = 2^{d+1} \lg \frac{n}{2^{d-1}} - 2 - \lg n .$$

Rather than simply proving this formula by mathematical induction, try to see how you might have discovered it for yourself.   □

Example 2.2.4. Making a heap.   Consider the "*make-heap*" algorithm given at the end of Section 1.9.4: this algorithm constructs a heap starting from an array T of n items. As a barometer we use the instructions in the **repeat** loop of the algorithm used to sift down a node. Let m be the largest number of trips round the loop that can be caused by calling *sift-down* (T, n, i). Denote by j_t the value of j after execution of the assignment "$j \leftarrow k$" on the t th trip round the loop. Obviously $j_1 = i$. Moreover, if $1 < t \leq m$, then at the end of the $(t-1)$st trip round the loop we had $j \neq k$; therefore $k \geq 2j$. This shows that $j_t \geq 2j_{t-1}$ for $1 < t \leq m$. But it is impossible for k (and thus j) to exceed n. Consequently

$$n \geq j_m \geq 2j_{m-1} \geq 4j_{m-2} \geq \cdots \geq 2^{m-1}i .$$

Thus $2^{m-1} \leq n/i$, which implies that $m \leq 1 + \lg(n/i)$.

The total number of trips round the **repeat** loop when constructing a heap can now be bounded above by

$$\sum_{i=1}^{\lfloor n/2 \rfloor} (1 + \lg(n/i)) . \qquad (*)$$

To simplify this expression, notice first that for any $k \geq 0$

$$\sum_{i=2^k}^{2^{k+1}-1} \lg(n/i) \leq 2^k \lg(n/2^k) .$$

The interesting part of the sum $(*)$ can therefore be decomposed into sections corresponding to powers of 2. Let $d = \lfloor \lg(n/2) \rfloor$.

$$\sum_{i=1}^{\lfloor n/2 \rfloor} \lg(n/i) \leq \sum_{k=0}^{d} 2^k \lg(n/2^k) \leq 2^{d+1} \lg(n/2^{d-1})$$

(by Problem 2.2.1). But $d = \lfloor \lg(n/2) \rfloor$ implies that $d+1 \leq \lg n$ and $d-1 > \lg(n/8)$. Hence

$$\sum_{i=1}^{\lfloor n/2 \rfloor} \lg(n/i) \leq 3n .$$

From $(*)$ we thus conclude that $\lfloor n/2 \rfloor + 3n$ trips round the **repeat** loop are enough to construct a heap, so that this can be done in a time in $O(n)$. Since any algorithm for constructing a heap must look at each element of the array at least once, we obtain our final result that the construction of a heap of size n can be carried out in a time in $\Theta(n)$.

A different approach yields the same result. Let $t(k)$ stand for the time needed to build a heap of height at most k in the worst case. Assume $k \geq 2$. In order to construct the heap, the algorithm first transforms each of the two subtrees attached to the root into heaps of height at most $k-1$ (the right hand subtree could be of height $k-2$). The algorithm then sifts the root down a path whose length is at most k, which takes a time in the order of k in the worst case. We thus obtain the asymptotic recurrence $t(k) \in 2t(k-1) + O(k)$. The techniques of Section 2.3, in particular Example 2.3.5, can be used to conclude that $t(k) \in O(2^k)$. But a heap containing n elements is of height $\lfloor \lg n \rfloor$, hence it can be built in at most $t(\lfloor \lg n \rfloor)$ steps, which is in $O(n)$ since $2^{\lfloor \lg n \rfloor} \leq n$. □

Problem 2.2.2. In Section 1.9.4 we saw another algorithm for making a heap (*slow-make-heap*). Analyse the worst case for this algorithm and compare it to the algorithm analysed in Example 2.2.4. □

Problem 2.2.3. Analysis of heapsort. Williams invented the heap to serve as the underlying data structure for the following sorting algorithm.

```
procedure heapsort (T [1 .. n ])
    {T is an array to be sorted}
    make-heap (T )
    for i ← n step −1 to 2 do
        exchange T [1] and T [i]
        sift-down (T [1 .. i − 1], 1)
    {T is sorted}
```

What is the order of the execution time required by this algorithm in the worst case? □

*Problem 2.2.4. Find the exact order of the execution time for Williams's *heapsort*, both in the worst case and on the average. For a given number of elements, what are the best and the worst ways to arrange the elements initially insofar as the execution time of the algorithm is concerned? □

Example 2.2.5. Recursive calculation of determinants. We now analyse the algorithm derived from the recursive definition of a determinant (Section 1.7.3). In our analysis, additions and multiplications are considered to be elementary operations. We therefore ignore for the time being the problems posed by the fact that the size of the operands can become very large during the execution of the algorithm.

Let $t(n)$ be the time taken by some implementation of this algorithm working on an $n \times n$ matrix. When n is greater than 1, most of the work done by the algorithm consists of calling itself recursively n times to work on $(n-1) \times (n-1)$ matrices.

Besides this, the matrices for the recursive calls have to be set up and some other housekeeping done, which takes a time in $\Theta(n^2)$ for each of the n recursive calls if we do this without thinking too much about it (but see Problem 2.2.5). This gives us the following asymptotic recurrence: $t(n) \in nt(n-1) + \Theta(n^3)$. By Problem 2.1.22 the algorithm therefore takes a time in $\Theta(n!)$ to calculate the determinant of an $n \times n$ matrix. $\qquad\square$

Problem 2.2.5.   Example 2.2.5 supposes that the time needed to compute a determinant, excluding the time taken by the recursive calls, is in $\Theta(n^3)$. Show that this time can be reduced to $\Theta(n)$. By Problem 2.1.22, however, this does not affect the fact that the complete algorithm takes a time in $\Theta(n!)$. $\qquad\square$

*Problem 2.2.6.   Analyse the algorithm again, taking account this time of the fact that the operands may become very large during execution of the algorithm. Assume that you know how to add two integers of size n in a time in $\Theta(n)$ and that you can multiply an integer of size m by an integer of size n in a time in $\Theta(mn)$. $\qquad\square$

Example 2.2.6. Analysis of Euclid's algorithm.   Recall that Euclid's algorithm calculates the greatest common divisor of two integers (Section 1.7.4).

> **function** *Euclid* (m, n)
> **while** $m > 0$ **do**
> $\quad t \leftarrow n \bmod m$
> $\quad n \leftarrow m$
> $\quad m \leftarrow t$
> **return** n

We first show that for any two integers m and n such that $n \geq m$, it is always true that $n \bmod m < n/2$.

- If $m > n/2$, then $1 \leq n/m < 2$, and so $\lfloor n/m \rfloor = 1$, which means that $n \bmod m = n - m < n - n/2 = n/2$.

- If $m \leq n/2$, then $(n \bmod m) < m \leq n/2$.

Let k be the number of trips round the loop made by the algorithm working on the instance $<m, n>$. For each integer $i \leq k$, let n_i and m_i be the values of n and m at the end of the ith trip round the loop. In particular, $m_k = 0$ causes the algorithm to terminate and $m_i \geq 1$ for every $i < k$. The values of m_i and n_i are defined by the following equations for $1 \leq i \leq k$, where m_0 and n_0 are the initial values of m and n:

$$n_i = m_{i-1}$$
$$m_i = n_{i-1} \bmod m_{i-1}.$$

Clearly $n_i > m_i$ for each $i \geq 1$. Using the preceding observation, we have

$$m_i = n_{i-1} \bmod m_{i-1} < n_{i-1}/2 = m_{i-2}/2$$

for every $i \geq 2$. Suppose for the moment that k is odd. Define d by $k = 2d + 1$. Then

$$m_{k-1} < m_{k-3}/2 < m_{k-5}/4 < \cdots < m_0/2^d.$$

But recall that $m_{k-1} \geq 1$, and so $m_0 \geq 2^d$. Finally, $k = 2d + 1 \leq 1 + 2\lg m_0$. The case when k is even is handled similarly, remembering that $m_1 = n_0 \bmod m_0 < m_0$.

In conclusion, the number of trips round the loop made by Euclid's algorithm working on the integers m and n, and therefore the time required by the algorithm, are in $O(\log m)$, provided the instructions in the loop can be considered as elementary. □

*Problem 2.2.7.   Prove that the worst case for Euclid's algorithm arises when we calculate the greatest common divisor of two consecutive numbers from the Fibonacci sequence. □

Example 2.2.7. Analysis of the algorithm *fib*1.   We now analyse the algorithm *fib*1 of Section 1.7.5, still not taking account of the large size of the operands involved. Let $t(n)$ be the time taken by some implementation of this algorithm working on the integer n. We give without explanation the corresponding asymptotic recurrence: $t(n) \in t(n-1) + t(n-2) + \Theta(1)$.

Once again, the recurrence looks so like the one used to define the Fibonacci sequence that it is tempting to suppose that $t(n)$ must be in $\Theta(f_n)$. However, as in Example 2.1.4, constructive induction cannot be used directly to find a constant d such that $t(n) \leq df_n$. On the other hand, it is easy to use this technique to find three real positive constants a, b, and c such that $af_n \leq t(n) \leq bf_n - c$ for any positive integer n. The algorithm *fib*1 therefore takes a time in $\Theta(f_n) = \Theta(\phi^n)$ to calculate the nth term of the Fibonacci sequence, where $\phi = (1+\sqrt{5})/2$. □

Problem 2.2.8.   Using constructive induction, prove that $af_n \leq t(n) \leq bf_n - c$ for appropriate constants a, b, and c, and give values for these constants. □

Problem 2.2.9.   Prove that the algorithm *fib*1 takes a time in $\Theta(f_n)$ even if we take into account that we need a time in $\Theta(n)$ to add two integers of size n. (Since the value of f_n is in $\Theta(\phi^n)$, its size is in $\Theta(n \lg \phi) = \Theta(n)$.) □

Example 2.2.8. Analysis of the algorithm *fib*2.   It is clear that the algorithm *fib*2 takes a time equal to $a + bn$ on any instance n, for appropriate constants a and b. This time is therefore in $\Theta(n)$.

What happens, however, if we take account of the size of the operands involved? Let a be a constant such that the time to add two numbers of size n is bounded above by an, and let b be a constant such that the size of f_n is bounded above by bn for every integer $n \geq 2$. Notice first that the values of i and j at the beginning of the kth trip round the **for** loop are respectively f_{k-2} and f_{k-1} (where we take $f_{-1} = 1$). The kth trip round the loop therefore consists of calculating $f_{k-2} + f_{k-1}$ and $f_k - f_{k-2}$, which takes a time bounded above by $ab(2k-1)$ for $k \geq 3$, plus some constant time c to carry out the assignments and the loop control. For each of the first two trips round the loop, the time is bounded above by $c + 2a$. Let d be an appropriate constant to account for necessary initializations. Then the time taken by $fib2$ on an integer $n \geq 2$ is bounded above by

$$d + 2(c + 2a) + \sum_{k=3}^{n} ab(2k-1) = abn^2 + (d + 2c + 4a - 4ab),$$

which is in $O(n^2)$. It is easy to see by symmetry that the algorithm takes a time in $\Theta(n^2)$.   □

Example 2.2.9. Analysis of the algorithm *fib3*.   The analysis of *fib3* is relatively easy if we do not take account of the size of the operands. To see this, take the instructions in the **while** loop as our barometer. To evaluate the number of trips round the loop, let n_t be the value of n at the end of the tth trip; in particular $n_1 = \lfloor n/2 \rfloor$. It is obvious that $n_t = \lfloor n_{t-1}/2 \rfloor \leq n_{t-1}/2$ for every $2 \leq t \leq m$. Consequently

$$n_t \leq n_{t-1}/2 \leq n_{t-2}/4 \leq \cdots \leq n_1/2^{t-1} \leq n/2^t .$$

Let $m = 1 + \lfloor \lg n \rfloor$. The preceding equation shows that $n_m \leq n/2^m < 1$. But n_m is a nonnegative integer, and so $n_m = 0$, which is the condition for ending the loop. We conclude that the loop is executed at most m times, which implies that the algorithm *fib3* takes a time in $O(\log n)$.   □

Problem 2.2.10.   Prove that the execution time of the algorithm *fib3* on an integer n is in $\Theta(\log n)$ if no account is taken of the size of the operands.   □

**Problem 2.2.11.   Determine the exact order of the execution time of the algorithm *fib3* used on an integer n. Assume that addition of two integers of size n takes a time in $\Theta(n)$ and that multiplication of an integer of size n by an integer of size m takes a time in $\Theta(mn)$. Compare your result to that obtained in Example 2.2.8. If you find the result disappointing, look back at the table at the end of Section 1.7.5 and remember that the hidden constants can have practical importance! In Section 4.7 we shall see a multiplication algorithm that can be used to improve the performance of the algorithm *fib3* (Problem 4.7.5), but not, of course, that of *fib2* (why not?).   □

Problem 2.2.12.   Consider the following algorithm :

for $i \leftarrow 0$ **to** n **do**
    $j \leftarrow i$
    while $j \neq 0$ **do** $j \leftarrow j$ **div** 2 .

Supposing that integer division by 2, assignments, and loop control can all be carried out at unit cost, it is clear that this algorithm takes a time in $\Omega(n) \cap O(n \log n)$. Find the exact order of its execution time. Prove your answer.   □

* **Problem 2.2.13.**   Answer the same question as in the preceding problem, this time for the algorithm

for $i \leftarrow 0$ **to** n **do**
    $j \leftarrow i$
    while j is odd **do** $j \leftarrow j$ **div** 2 .

Show a relationship between this algorithm and the act of counting from 0 to $n+1$ in binary.   □

Example 2.2.10. Analysis of disjoint set structures.   It can happen that the analysis of an algorithm is facilitated by the addition of extra instructions and counters that have nothing to do with the execution of the algorithm proper. For instance, this is so when we look at the algorithms *find3* and *merge3* used to handle the disjoint set structures introduced in Section 1.9.5. The analysis of these algorithms is the most complicated case we shall see in this book. We begin by introducing a counter called *global* and a new array *cost* $[1..N]$. Their purpose will be explained later. The array *set* $[1..N]$ keeps the meaning given to it in algorithms *find3* and *merge3* : *set* $[i]$ gives the parent of node i in its tree, except when *set* $[i]=i$, which indicates that i is the root of its tree. The array *rank* $[1..N]$ plays the role of *height* $[1..N]$ in algorithm *merge3* : *rank* $[i]$ denotes the rank of node i (see Section 1.9.5). We also introduce a strictly increasing function $F : \mathbb{N} \rightarrow \mathbb{N}$ (specified later) and its "inverse" $G : \mathbb{N} \rightarrow \mathbb{N}$ defined by $G(n) = \min\{m \in \mathbb{N} \mid F(m) \geq n\}$. Finally, define the *group* of an element of rank r as $G(r)$. The algorithms become

procedure *init*
    {initializes the trees}
    global $\leftarrow 0$
    for $i \leftarrow 1$ **to** N **do** *set* $[i] \leftarrow i$
                        rank $[i] \leftarrow 0$
                        cost $[i] \leftarrow 0$

function *find* (x)
    {finds the label of the set containing object x}
    $r \leftarrow x$
    while $set[r] \neq r$ **do** $r \leftarrow set[r]$
    {r is the root of the tree}
    $i \leftarrow x$
    while $i \neq r$ **do**
      if $G(rank[i]) < G(rank[set[i]])$ **or** $r = set[i]$
      then $global \leftarrow global + 1$
      else $cost[i] \leftarrow cost[i] + 1$
      $j \leftarrow set[i]$
      $set[i] \leftarrow r$
      $i \leftarrow j$
    return r

procedure *merge* (a, b)
    {merges the sets labelled a and b;
     we suppose that $a \neq b$}
    if $rank[a] = rank[b]$
    then
      $rank[a] \leftarrow rank[a] + 1$
      $set[b] \leftarrow a$
    else
      if $rank[a] > rank[b]$
      then $set[b] \leftarrow a$
      else $set[a] \leftarrow b$.

With these modifications the time taken by a call on the procedure *find* can be reckoned to be in the order of 1 plus the increase of $global + \sum_{i=1}^{N} cost[i]$ occasioned by the call. The time required for a call on the procedure *merge* can be bounded above by a constant. Therefore the total time required to execute an arbitrary sequence of n calls on *find* and *merge*, including initialization, is in

$$O(N + n + global + \sum_{i=1}^{N} cost[i]),$$

where *global* and $cost[i]$ refer to the final values of these variables after execution of the sequence. In order to obtain an upper bound on these values, the following remarks are relevant:

1. once an element ceases to be the root of a tree, it never becomes a root thereafter and its rank no longer changes;

2. the rank of a node that is not a root is always strictly less than the rank of its parent;

3. the rank of an element never exceeds the logarithm (to the base 2) of the number of elements in the corresponding tree;

4. at every moment and for every value of k, there are not more than $N/2^k$ elements of rank k; and

5. at no time does the rank of an element exceed $\lfloor \lg N \rfloor$, nor does its group ever exceed $G(\lfloor \lg N \rfloor)$.

Remarks (1) and (2) are obvious if one simply looks at the algorithms. Remark (3) has a simple proof by mathematical induction, which we leave to the reader. Remark (5) derives directly from remark (4). To prove the latter, define $sub_k(i)$ for each element i and rank k : if node i never attains rank k, $sub_k(i)$ is the empty set; otherwise $sub_k(i)$ is the set of nodes that are in the tree whose root is i at that precise moment when the rank of i becomes k. (Note that i is necessarily a root at that moment, by remark (1).) By remark (3), $sub_k(i) \neq \varnothing \Rightarrow \# sub_k(i) \geq 2^k$. By remark (2), $i \neq j \Rightarrow sub_k(i) \cap sub_k(j) = \varnothing$. Hence, if there were more than $N/2^k$ elements i such that $sub_k(i) \neq \varnothing$, there would have to be more than N elements in all, which proves remark (4).

The fact that G is nondecreasing allows us to conclude, using remarks (2) and (5), that the increase in the value of *global* caused by a call on the procedure *find* cannot exceed $1 + G(\lfloor \lg N \rfloor)$. Consequently, after the execution of a sequence of n operations, the final value of this variable is in $O(1 + nG(\lfloor \lg N \rfloor))$. It only remains to find an upper bound on the final value of $cost[i]$ for each element i in terms of its final rank.

Note first that $cost[i]$ remains at zero while i is a root. What is more, the value of $cost[i]$ only increases when a path compression causes the parent of node i to be changed. In this case the rank of the new parent is necessarily greater than the rank of the old parent by remark (2). But the increase in $cost[i]$ stops as soon as i becomes the child of a node whose group is greater than its own. Let r be the rank of i at the instant when i stops being a root, should this occur. By remark (1) this rank does not change subsequently. Using all the preceding observations, we see that $cost[i]$ cannot increase more than $F(G(r)) - F(G(r) - 1) - 1$ times. We conclude from this that the final value of $cost[i]$ is less than $F(G(r))$ for every node $i \in final(r)$, where $final(r)$ denotes the set of elements that cease to be a root when they have rank $r \geq 1$ (while, on the other hand, $cost[i]$ remains at zero for those elements that never cease to be a root or that do so when they have rank zero). Let $K = G(\lfloor \lg N \rfloor) - 1$. The rest is merely manipulation.

$$\sum_{i=1}^{N} cost[i] = \sum_{g=0}^{K} \sum_{r=F(g)+1}^{F(g+1)} \sum_{i \in final(r)} cost[i]$$

$$\leq \sum_{g=0}^{K} \sum_{r=F(g)+1}^{F(g+1)} \sum_{i \in final(r)} F(G(r))$$

$$\leq \sum_{g=0}^{K} \sum_{r=F(g)+1}^{F(g+1)} (N/2^r)F(g+1)$$

$$\leq N \sum_{g=0}^{K} F(g+1)/2^{F(g)}$$

It suffices therefore to put $F(g+1) = 2^{F(g)}$ to balance *global* and $\sum_{i=1}^{N} cost[i]$ and so to obtain $\sum_{i=1}^{N} cost[i] \leq NG(\lfloor \lg N \rfloor)$. The time taken by the sequence of n calls on *find* and *merge* with a universe of N elements, including the initialization time, is therefore in

$$O(N + n + global + \sum_{i=1}^{N} cost[i]) \subseteq O(N + n + nG(\lfloor \lg N \rfloor) + NG(\lfloor \lg N \rfloor))$$

$$= O(\max(N,n)(1 + G(\lfloor \lg N \rfloor))) .$$

Now that we have decided that $F(g+1) = 2^{F(g)}$, with the initial condition $F(0) = 0$, what can we say about the function G? This function, which is often denoted by $\lg^*$, can be defined by

$$G(N) = \lg^* N = \min\{k \mid \underbrace{\lg \lg \cdots \lg}_{k \text{ times}} N \leq 0\} .$$

The function $\lg^*$ increases very slowly : $\lg^* N \leq 5$ for every $N \leq 65{,}536$ and $\lg^* N \leq 6$ for every $N \leq 2^{65{,}536}$. Notice also that $\lg^* N - \lg^*(\lfloor \lg N \rfloor) \leq 2$, so that $\lg^*(\lfloor \lg N \rfloor) \in \Theta(\lg^* N)$. The algorithms that we have just analysed can therefore execute a sequence of n calls on *find* and *merge* with a universe of N elements in a time in $O(n \lg^* N)$, provided $n \geq N$, which is to most intents and purposes linear.

This bound can be improved by refining the argument in a way too complex to give here. We content ourselves with mentioning that the exact analysis involves the use of Ackermann's function (Problem 5.8.7) and that the time taken by the algorithm is not linear in the worst case.   □

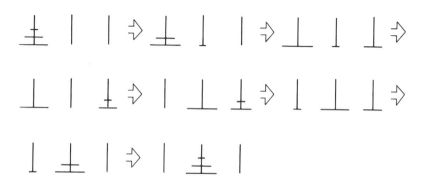

Figure 2.2.1   The towers of Hanoi.

Example 2.2.11. The towers of Hanoi.   It is said that after creating the world, God set on Earth three rods made of diamond and 64 rings of gold. These rings are all different in size. At the creation they were threaded on one of the rods in order of size, the largest at the bottom and the smallest at the top. God also created a monastery close by the rods. The monks' task in life is to transfer all the rings onto another rod. The only operation permitted consists of moving a single ring from one rod to another, in such a way that no ring is ever placed on top of another smaller one. When the monks have finished their task, according to the legend, the world will come to an end. This is probably the most reassuring prophecy ever made concerning the end of the world, for if the monks manage to move one ring per second, working night and day without ever resting nor ever making a mistake, their work will still not be finished 500,000 million years after they began!

The problem can obviously be generalized to an arbitrary number of rings. For example, with $n = 3$, we obtain the solution given in Figure 2.2.1. To solve the general problem, we need only realize that to transfer the m smallest rings from rod i to rod j (where $1 \leq i \leq 3, 1 \leq j \leq 3, i \neq j$, and $m \geq 1$), we can first transfer the smallest $m-1$ rings from rod i to rod $6-i-j$, next transfer the mth ring from rod i to rod j, and finally retransfer the $m-1$ smallest rings from rod $6-i-j$ to rod j. Here is a formal description of this algorithm; to solve the original instance, all you have to do (!) is to call it with the arguments (64, 1, 2).

> **procedure** $Hanoi(m, i, j)$
>   {moves the m smallest rings from rod i to rod j }
>   **if** $m > 0$ **then** $Hanoi(m-1, i, 6-i-j)$
>   **write** i "$\rightarrow$" j
> $Hanoi(m-1, 6-i-j, j)$

To analyse the execution time of this algorithm, let us see how often the instruction **write**, which we use as a barometer, is executed. The answer is a function of m, which we denote $e(m)$. We obtain the following recurrence:

$$e(m) = \begin{cases} 1 & \text{if } m = 1 \\ 2e(m-1) + 1 & \text{if } m > 1, \end{cases}$$

from which we find that $e(m) = 2^m - 1$ (see Example 2.3.4). The algorithm therefore takes a time in the exact order of 2^n to solve the problem with n rings. □

Problem 2.2.14.   Prove that the algorithm of Example 2.2.11 is optimal in the sense that it is impossible with the given constraints to move n rings from one rod to another in less than $2^n - 1$ operations. □

*Problem 2.2.15.   Give a nonrecursive algorithm to solve this problem. (It is cheating simply to rewrite the above algorithm using an explicit stack to simulate the recursive calls.) □

2.3 SOLVING RECURRENCES USING
THE CHARACTERISTIC EQUATION

We have seen that the indispensable last step when analysing an algorithm is often to solve a system of recurrences. With a little experience and intuition such recurrences can often be solved by intelligent guesswork. This approach, which we do not illustrate here, generally proceeds in four stages: calculate the first few values of the recurrence, look for regularity, guess a suitable general form, and finally, prove by mathematical induction that this form is correct. Fortunately there exists a technique that can be used to solve certain classes of recurrence almost automatically.

2.3.1 Homogeneous Recurrences

Our starting point is the resolution of homogeneous linear recurrences with constant coefficients, that is, recurrences of the form

$$a_0 t_n + a_1 t_{n-1} + \cdots + a_k t_{n-k} = 0 \qquad (*)$$

where

    i. the t_i are the values we are looking for. The recurrence is linear because it does not contain terms of the form $t_i t_{i+j}$, t_i^2, and so on;

    ii. the coefficients a_i are constants; and

    iii. the recurrence is homogeneous because the linear combination of the t_i is equal to zero.

After a while intuition may suggest we look for a solution of the form

$$t_n = x^n$$

where x is a constant as yet unknown. If we try this solution in (*), we obtain

$$a_0 x^n + a_1 x^{n-1} + \cdots + a_k x^{n-k} = 0.$$

This equation is satisfied if $x = 0$, a trivial solution of no interest, or else if

$$a_0 x^k + a_1 x^{k-1} + \cdots + a_k = 0.$$

This equation of degree k in x is called the *characteristic equation* of the recurrence (*).

Suppose for the time being that the k roots $r_1, r_2, \ldots, r_k$ of this characteristic equation are all distinct (they could be complex numbers). It is then easy to verify that any linear combination

$$t_n = \sum_{i=1}^{k} c_i r_i^n$$

of terms r_i^n is a solution of the recurrence (*), where the k constants $c_1, c_2, \ldots, c_k$ are determined by the initial conditions. (We need exactly k initial conditions to determine the values of these k constants.) The remarkable fact, which we do not prove here, is that (*) has *only* solutions of this form.

      Example 2.3.1.   Consider the recurrence

$$t_n - 3t_{n-1} - 4t_{n-2} = 0 \quad n \geq 2$$

subject to $t_0 = 0$, $t_1 = 1$.

      The characteristic equation of the recurrence is

$$x^2 - 3x - 4 = 0$$

whose roots are -1 and 4. The general solution therefore has the form

$$t_n = c_1(-1)^n + c_2 4^n.$$

The initial conditions give

$$\begin{aligned} c_1 + c_2 &= 0 \quad & n = 0 \\ -c_1 + 4c_2 &= 1 \quad & n = 1 \end{aligned}$$

that is, $c_1 = -\frac{1}{5}$, $c_2 = \frac{1}{5}$.

      We finally obtain

$$t_n = \frac{1}{5}[4^n - (-1)^n].$$ □

      Example 2.3.2. Fibonacci.   Consider the recurrence

$$t_n = t_{n-1} + t_{n-2} \quad n \geq 2$$

subject to $t_0 = 0$, $t_1 = 1$.

(This is the definition of the Fibonacci sequence; see Section 1.7.5.)

      The recurrence can be rewritten in the form $t_n - t_{n-1} - t_{n-2} = 0$, so the characteristic equation is

$$x^2 - x - 1 = 0$$

whose roots are

$$r_1 = \frac{1+\sqrt{5}}{2} \quad \text{and} \quad r_2 = \frac{1-\sqrt{5}}{2}.$$

The general solution is therefore of the form

$$t_n = c_1 r_1^n + c_2 r_2^n.$$

The initial conditions give

$$c_1 + \quad c_2 = 0 \qquad n = 0$$
$$r_1 c_1 + r_2 c_2 = 1 \qquad n = 1$$

from which it is easy to obtain

$$c_1 = \frac{1}{\sqrt{5}}, \qquad c_2 = -\frac{1}{\sqrt{5}}.$$

Thus $t_n = \frac{1}{\sqrt{5}}(r_1^n - r_2^n)$. To show that this is the same as the result obtained by De Moivre mentioned in Section 1.7.5, we need to note only that $r_1 = \phi$ and $r_2 = -\phi^{-1}$.   □

* Problem 2.3.1.   Consider the recurrence

$$t_n = 2t_{n-1} - 2t_{n-2} \qquad n \geq 2$$

subject to $t_0 = 0$, $t_1 = 1$.

Prove that $t_n = 2^{n/2}\sin(n\pi/4)$, not by mathematical induction but by using the characteristic equation.   □

Now suppose that the roots of the characteristic equation are not all distinct. Let

$$p(x) = a_0 x^k + a_1 x^{k-1} + \cdots + a_k$$

be the polynomial in the characteristic equation, and let r be a multiple root. For every $n \geq k$, consider the nth degree polynomial defined by

$$h(x) = x[x^{n-k} p(x)]' = a_0 n x^n + a_1(n-1)x^{n-1} + \cdots + a_k(n-k)x^{n-k}.$$

Let $q(x)$ be the polynomial such that $p(x) = (x-r)^2 q(x)$. We have that

$$h(x) = x[(x-r)^2 x^{n-k} q(x)]' = x[2(x-r)x^{n-k} q(x) + (x-r)^2[x^{n-k} q(x)]'].$$

In particular, $h(r) = 0$. This shows that

$$a_0 n r^n + a_1(n-1)r^{n-1} + \cdots + a_k(n-k)r^{n-k} = 0,$$

that is, $t_n = nr^n$ is also a solution of (*). More generally, if m is the multiplicity of the root r, then $t_n = r^n$, $t_n = nr^n$, $t_n = n^2 r^n$,...., $t_n = n^{m-1}r^n$ are all possible solutions of (*). The general solution is a linear combination of these terms and of the terms contributed by the other roots of the characteristic equation. Once again there are k constants to be determined by the initial conditions.

Example 2.3.3.   Consider the recurrence

$$t_n = 5t_{n-1} - 8t_{n-2} + 4t_{n-3} \qquad n \geq 3$$

subject to $t_0 = 0$, $t_1 = 1$, $t_2 = 2$.

The recurrence can be written

$$t_n - 5t_{n-1} + 8t_{n-2} - 4t_{n-3} = 0$$

and so the characteristic equation is

$$x^3 - 5x^2 + 8x - 4 = 0$$

or $(x-1)(x-2)^2 = 0$.

The roots are 1 (of multiplicity 1) and 2 (of multiplicity 2). The general solution is therefore

$$t_n = c_1 1^n + c_2 2^n + c_3 n 2^n.$$

The initial conditions give

$$
\begin{array}{ll}
c_1 + c_2 = 0 & n = 0 \\
c_1 + 2c_2 + 2c_3 = 1 & n = 1 \\
c_1 + 4c_2 + 8c_3 = 2 & n = 2
\end{array}
$$

from which we find $c_1 = -2$, $c_2 = 2$, $c_3 = -\frac{1}{2}$. Therefore

$$t_n = 2^{n+1} - n 2^{n-1} - 2.$$ □

2.3.2 Inhomogeneous Recurrences

We now consider recurrences of a slightly more general form.

$$a_0 t_n + a_1 t_{n-1} + \cdots + a_k t_{n-k} = b^n p(n) \qquad (**)$$

The left-hand side is the same as (*), but on the right-hand side we have $b^n p(n)$, where

   i. b is a constant; and
   ii. $p(n)$ is a polynomial in n of degree d.

For example, the recurrence might be

$$t_n - 2t_{n-1} = 3^n.$$

In this case $b = 3$ and $p(n) = 1$, a polynomial of degree 0. A little manipulation allows us to reduce this example to the form (*). To see this, we first multiply the recurrence by 3, obtaining

$$3t_n - 6t_{n-1} = 3^{n+1}.$$

If we replace n by $n+1$ in the original recurrence, we get

$$t_{n+1} - 2t_n = 3^{n+1}.$$

Finally, subtracting these two equations, we have

$$t_{n+1} - 5t_n + 6t_{n-1} = 0,$$

which can be solved by the method of Section 2.3.1. The characteristic equation is

$$x^2 - 5x + 6 = 0$$

that is, $(x-2)(x-3) = 0$.

Intuitively we can see that the factor $(x-2)$ corresponds to the left-hand side of the original recurrence, whereas the factor $(x-3)$ has appeared as a result of our manipulation to get rid of the right-hand side.

Here is a second example.

$$t_n - 2t_{n-1} = (n+5)3^n$$

The necessary manipulation is a little more complicated: we must

a. multiply the recurrence by 9
b. replace n in the recurrence by $n+2$, and
c. replace n in the recurrence by $n+1$ and then multiply by -6,

obtaining respectively

$$9t_n \quad - 18t_{n-1} = \quad (n+5)3^{n+2}$$
$$t_{n+2} - \quad 2t_{n+1} = \quad (n+7)3^{n+2}$$
$$-6t_{n+1} + 12t_n \quad = -6(n+6)3^{n+1}.$$

Adding these three equations, we obtain

$$t_{n+2} - 8t_{n+1} + 21t_n - 18t_{n-1} = 0.$$

The characteristic equation of this new recurrence is

$$x^3 - 8x^2 + 21x - 18 = 0$$

that is, $(x-2)(x-3)^2 = 0$.

Once again, we can see that the factor $(x-2)$ comes from the left-hand side of the original recurrence, whereas the factor $(x-3)^2$ is the result of our manipulation.

Generalizing this approach, we can show that to solve (**) it is sufficient to take the following characteristic equation:

$$(a_0 x^k + a_1 x^{k-1} + \cdots + a_k)(x-b)^{d+1} = 0.$$

Once this equation is obtained, proceed as in the homogeneous case.

Example 2.3.4.   The number of movements of a ring required in the Towers of Hanoi problem (see Example 2.2.11) is given by

$$t_n = 2t_{n-1} + 1 \qquad n \geq 1$$

subject to $t_0 = 0$.

The recurrence can be written

$$t_n - 2t_{n-1} = 1,$$

which is of the form (**) with $b = 1$ and $p(n) = 1$, a polynomial of degree 0. The characteristic equation is therefore

$$(x-2)(x-1) = 0$$

where the factor $(x-2)$ comes from the left-hand side and the factor $(x-1)$ comes from the right-hand side. The roots of this equation are 1 and 2, so the general solution of the recurrence is

$$t_n = c_1 1^n + c_2 2^n.$$

We need two initial conditions. We know that $t_0 = 0$; to find a second initial condition we use the recurrence itself to calculate

$$t_1 = 2t_0 + 1 = 1.$$

We finally have

$$c_1 + c_2 = 0 \qquad n = 0$$
$$c_1 + 2c_2 = 1 \qquad n = 1$$

from which we obtain the solution

$$t_n = 2^n - 1. \qquad\qquad \square$$

If all we want is the order of t_n, there is no need to calculate the constants in the general solution. In the previous example, once we know that

$$t_n = c_1 1^n + c_2 2^n$$

we can already conclude that $t_n \in \Theta(2^n)$. For this it is sufficient to notice that t_n, the number of movements of a ring required, is certainly neither negative nor a constant, since clearly $t_n \geq n$. Therefore $c_2 > 0$, and the conclusion follows.

In fact we can obtain a little more. Substituting the general solution back into the original recurrence, we find

$$1 = t_n - 2t_{n-1}$$
$$= c_1 + c_2 2^n - 2(c_1 + c_2 2^{n-1})$$
$$= -c_1.$$

Whatever the initial condition, it is therefore always the case that c_1 must be equal to -1.

Problem 2.3.2.   There is nothing surprising in the fact that we can determine one of the constants in the general solution without looking at the initial condition; on the contrary! Why?   □

Example 2.3.5.   Consider the recurrence

$$t_n = 2t_{n-1} + n.$$

This can be written

$$t_n - 2t_{n-1} = n,$$

which is of the form (**) with $b = 1$ and $p(n) = n$, a polynomial of degree 1. The characteristic equation is therefore

$$(x-2)(x-1)^2 = 0$$

with roots 2 (multiplicity 1) and 1 (multiplicity 2). The general solution is

$$t_n = c_1 2^n + c_2 1^n + c_3 n 1^n.$$

In the problems that interest us, we are always looking for a solution where $t_n \geq 0$ for every n. If this is so, we can conclude immediately that t_n must be in $O(2^n)$.   □

Problem 2.3.3.   By substituting the general solution back into the recurrence, prove that in the preceding example $c_2 = -2$ and $c_3 = -1$ whatever the initial condition. Conclude that all the interesting solutions of the recurrence must have $c_1 > 0$, and hence that they are all in $\Theta(2^n)$.   □

A further generalization of the same type of argument allows us finally to solve recurrences of the form

$$a_0 t_n + a_1 t_{n-1} + \cdots + a_k t_{n-k} = b_1^n p_1(n) + b_2^n p_2(n) + \cdots \qquad (***)$$

where the b_i are distinct constants and the $p_i(n)$ are polynomials in n respectively of degree d_i. It suffices to write the characteristic equation

$$(a_0 x^k + a_1 x^{k-1} + \cdots + a_k)(x-b_1)^{d_1+1}(x-b_2)^{d_2+1} \cdots = 0,$$

which contains one factor corresponding to the left-hand side and one factor corresponding to each term on the right-hand side, and to solve the problem as before.

Example 2.3.6.   Solve

$$t_n = 2t_{n-1} + n + 2^n \qquad n \geq 1$$

subject to $t_0 = 0$.

The recurrence can be written

$$t_n - 2t_{n-1} = n + 2^n,$$

which is of the form (***) with $b_1 = 1$, $p_1(n) = n$, $b_2 = 2$, $p_2(n) = 1$. The degree of $p_1(n)$ is 1, and $p_2(n)$ is of degree 0. The characteristic equation is

$$(x-2)(x-1)^2(x-2) = 0,$$

which has roots 1 and 2, both of multiplicity 2. The general solution of the recurrence is therefore of the form

$$t_n = c_1 1^n + c_2 n 1^n + c_3 2^n + c_4 n 2^n.$$

Using the recurrence, we can calculate $t_1 = 3$, $t_2 = 12$, $t_3 = 35$. We can now determine c_1, c_2, c_3 and c_4 from

$$
\begin{array}{lll}
c_1 \quad\quad + c_3 \quad\quad\quad\quad = 0 & \quad n = 0 \\
c_1 + c_2 + 2c_3 + \quad 2c_4 = 3 & \quad n = 1 \\
c_1 + 2c_2 + 4c_3 + \quad 8c_4 = 12 & \quad n = 2 \\
c_1 + 3c_2 + 8c_3 + 24c_4 = 35 & \quad n = 3
\end{array}
$$

arriving finally at

$$t_n = -2 - n + 2^{n+1} + n 2^n.$$

We could obviously have concluded that $t_n \in O(n 2^n)$ without calculating the constants.   ☐

Problem 2.3.4.   Prove that all the solutions of this recurrence are in fact in $\Theta(n 2^n)$, regardless of the initial condition.   ☐

Problem 2.3.5.   If the characteristic equation of the recurrence (***) is of degree

$$m = k + (d_1+1) + (d_2+1) + \cdots,$$

then the general solution contains m constants $c_1, c_2, \ldots, c_m$. How many constraints on these constants can be obtained without using the initial conditions? (See Problems 2.3.3 and 2.3.4.)   ☐

2.3.3 Change of Variable

It is sometimes possible to solve more complicated recurrences by making a change of variable. In the following examples we write $T(n)$ for the term of a general recurrence, and t_k for the term of a new recurrence obtained by a change of variable.

Example 2.3.7.   Here is how we can find the order of $T(n)$ if n is a power of 2 and if

$$T(n) = 4T(n/2) + n \quad\quad n > 1.$$

Replace n by 2^k (so that $k = \lg n$) to obtain $T(2^k) = 4T(2^{k-1}) + 2^k$. This can be written

$$t_k = 4t_{k-1} + 2^k$$

if $t_k = T(2^k) = T(n)$. We know how to solve this new recurrence: the characteristic equation is

$$(x-4)(x-2) = 0$$

and hence $t_k = c_1 4^k + c_2 2^k$.

Putting n back instead of k, we find

$$T(n) = c_1 n^2 + c_2 n.$$

$T(n)$ is therefore in $O(n^2 \mid n$ is a power of 2). $\qquad\qquad\qquad\qquad$ □

Example 2.3.8.   Here is how to find the order of $T(n)$ if n is a power of 2 and if

$$T(n) = 4T(n/2) + n^2 \qquad n > 1.$$

Proceeding in the same way, we obtain successively

$$T(2^k) = 4T(2^{k-1}) + 4^k$$

$$t_k = 4t_{k-1} + 4^k.$$

The characteristic equation is $(x-4)^2 = 0$, and so

$$t_k = c_1 4^k + c_2 k 4^k$$

$$T(n) = c_1 n^2 + c_2 n^2 \lg n.$$

Thus $T(n) \in O(n^2 \log n \mid n$ is a power of 2). $\qquad\qquad\qquad\qquad$ □

Example 2.3.9.   Here is how to find the order of $T(n)$ if n is a power of 2 and if

$$T(n) = 2T(n/2) + n \lg n \qquad n > 1.$$

As before, we obtain

$$T(2^k) = 2T(2^{k-1}) + k \, 2^k$$

$$t_k = 2t_{k-1} + k \, 2^k .$$

The characteristic equation is $(x-2)^3 = 0$, and so

$$t_k = c_1 2^k + c_2 k \, 2^k + c_3 k^2 2^k$$

$$T(n) = c_1 n + c_2 n \lg n + c_3 n \lg^2 n.$$

Hence, $T(n) \in O(n \log^2 n \mid n$ is a power of 2). $\qquad\qquad\qquad\qquad$ □

Example 2.3.10.   We want to find the order of $T(n)$ if n is a power of 2 and if

$$T(n) = 3T(n/2) + cn \qquad (c \text{ is constant}, n = 2^k > 1).$$

We obtain successively

$$T(2^k) = 3T(2^{k-1}) + c\,2^k$$

$$t_k = 3t_{k-1} + c\,2^k.$$

The characteristic equation is $(x-3)(x-2) = 0$, and so

$$t_k = c_1 3^k + c_2 2^k$$

$$T(n) = c_1 3^{\lg n} + c_2 n$$

and hence since $a^{\lg b} = b^{\lg a}$

$$T(n) = c_1 n^{\lg 3} + c_2 n.$$

Finally, $T(n) \in O(n^{\lg 3} \mid n$ is a power of 2). $\qquad\qquad \square$

Remark. In Examples 2.3.7 to 2.3.10 the recurrence given for $T(n)$ only applies when n is a power of 2. It is therefore inevitable that the solution obtained should be in conditional asymptotic notation. In each of these four cases, however, it is sufficient to add the condition that $T(n)$ is eventually nondecreasing to be able to conclude that the asymptotic results obtained apply unconditionally for all values of n. This follows from problem 2.1.20 since the functions n^2, $n^2 \log n$, $n \log^2 n$ and $n^{\lg 3}$ are smooth.

* Problem 2.3.6.   The constants $n_0 \geq 1$, $b \geq 2$ and $k \geq 0$ are integers, whereas a and c are positive real numbers. Let $T : \mathbb{N} \to \mathbb{R}^+$ be an eventually nondecreasing function such that

$$T(n) = aT(n/b) + cn^k \qquad n > n_0$$

when n/n_0 is a power of b. Show that the exact order of $T(n)$ is given by

$$T(n) \in \begin{cases} \Theta(n^k) & \text{if } a < b^k \\ \Theta(n^k \log n) & \text{if } a = b^k \\ \Theta(n^{\log_b a}) & \text{if } a > b^k. \end{cases}$$

Rather than proving this result by constructive induction, obtain it using the techniques of the characteristic equation and change of variable. This result is generalized in Problem 2.3.13. $\qquad\qquad \square$

Problem 2.3.7.   Solve the following recurrence exactly for n a power of 2:

$$T(n) = 2T(n/2) + \lg n \qquad n \geq 2$$

subject to $T(1) = 1$.

Express your solution as simply as possible using the Θ notation.   □

Problem 2.3.8.   Solve the following recurrence exactly for n of the form 2^{2^k}:

$$T(n) = 2T(\sqrt{n}) + \lg n \qquad n \geq 4$$

subject to $T(2) = 1$.

Express your solution as simply as possible using the Θ notation.   □

2.3.4 Range Transformations

When we make a change of variable, we transform the domain of the recurrence. It is sometimes useful to transform the range instead in order to obtain something of the form (***). We give just one example of this approach. We want to solve

$$T(n) = n\, T^2(n/2) \qquad n > 1$$

subject to $T(1) = 6$ for the case when n is a power of 2. The first step is a change of variable: put $t_k = T(2^k)$, which gives

$$t_k = 2^k\, t_{k-1}^2 \qquad k > 0$$

subject to $t_0 = 6$.

At first glance, none of the techniques we have seen applies to this recurrence since it is not linear, and furthermore, one of the coefficients is not constant. To transform the range, we create a new recurrence by putting $V_k = \lg t_k$, which yields

$$V_k = k + 2V_{k-1} \qquad k > 0$$

subject to $V_0 = \lg 6$.

The characteristic equation is $(x-2)(x-1)^2 = 0$, and so

$$V_k = c_1 2^k + c_2 1^k + c_3 k\, 1^k.$$

From $V_0 = 1 + \lg 3$, $V_1 = 3 + 2\lg 3$, and $V_2 = 8 + 4\lg 3$ we obtain $c_1 = 3 + \lg 3$, $c_2 = -2$, and $c_3 = -1$, and hence

$$V_k = (3 + \lg 3)2^k - k - 2.$$

Finally, using $t_k = 2^{V_k}$ and $T(n) = t_{\lg n}$, we obtain

$$T(n) = \frac{2^{3n-2}3^n}{n}.$$

2.3.5 Supplementary Problems

Problem 2.3.9.   Solve the following recurrence exactly:

$$t_n = t_{n-1} + t_{n-3} - t_{n-4} \qquad n \geq 4$$

subject to $t_n = n$ for $0 \leq n \leq 3$. Express your answer as simply as possible using the Θ notation. $\square$

Problem 2.3.10.   Solve the following recurrence exactly for n a power of 2:

$$T(n) = 5T(n/2) + (n \lg n)^2 \qquad n \geq 2$$

subject to $T(1) = 1$. Express your answer as simply as possible using the Θ notation. $\square$

Problem 2.3.11.   (Multiplication of large integers: see Sections 1.7.2, 4.1, and 4.7). Consider any constants $c \in \mathbb{R}^+$ and $n_0 \in \mathbb{N}$. Let $T : \mathbb{N} \to \mathbb{R}^*$ be an eventually nondecreasing function such that

$$T(n) \leq T(\lfloor n/2 \rfloor) + T(\lceil n/2 \rceil) + T(1 + \lceil n/2 \rceil) + cn \qquad n > n_0.$$

Prove that $T(n) \in O(n^{\lg 3})$. *Hint*: observe that $T(n) \leq 3T(1 + \lceil n/2 \rceil) + cn$ for $n > n_0$, make the change of variable $T'(n) = T(n+2)$, use Example 2.3.10 to solve for $T'(n)$ when n is a power of 2, and use problem 2.1.20 to conclude for $T(n)$. $\square$

Problem 2.3.12.   Solve the following recurrence exactly:

$$t_n = t_{n-1} + 2t_{n-2} - 2t_{n-3} \qquad n \geq 3$$

subject to $t_n = 9n^2 - 15n + 106$ for $0 \leq n \leq 2$. Express your answer as simply as possible using the Θ notation. $\square$

*Problem 2.3.13.   Recurrences arising from the analysis of divide-and-conquer algorithms (Chapter 4) can usually be handled by Problem 2.3.6. In some cases, however, a more general result is required (and the technique of the characteristic equation does not always apply).

The constants $n_0 \geq 1$ and $b \geq 2$ are integers, whereas a and d are real positive constants. Define

$$X = \{ n \in \mathbb{N} \mid \log_b (n/n_0) \in \mathbb{N} \} = \{ n \in \mathbb{N} \mid (\exists i \in \mathbb{N})[n = n_0 b^i] \}.$$

Let $f : X \to \mathbb{R}^*$ be an arbitrary function. Define the function $T : X \to \mathbb{R}^*$ by the recurrence

$$T(n) = \begin{cases} d & \text{if } n = n_0 \\ a T(n/b) + f(n) & \text{if } n \in X, n > n_0. \end{cases}$$

Let $p = \log_b a$. It turns out that the simplest way to express $T(n)$ in asymptotic notation depends on how $f(n)$ compares to n^p. In what follows, all asymptotic notation is implicitly conditional on $n \in X$. Prove that

i. If we set $f(n_0) = d$ (which is of no consequence for the definition of T), the value of $T(n)$ is given by a simple summation when $n \in X$:

$$T(n) = \sum_{i=0}^{\log_b(n/n_0)} a^i f(n/b^i) \ .$$

ii. Let q be any strictly positive real constant; then

$$T(n) \in \begin{cases} \Theta(n^p) & \text{if } f(n) \in O(n^p/(\log n)^{1+q}) \\ \Theta(f(n) \log n \log \log n) & \text{if } f(n) \in \Theta(n^p/\log n) \\ \Theta(f(n) \log n) & \text{if } f(n) \in \Theta(n^p (\log n)^{q-1}) \\ \Theta(f(n)) & \text{if } f(n) \in \Theta(n^{p+q}) \ . \end{cases}$$

Note that the third alternative includes $f(n) \in \Theta(n^p)$ by choosing $q=1$.

iii. As a special case of the first alternative, $T(n) \in \Theta(n^p)$ whenever $f(n) \in O(n^r)$ for some real constant $r < p$.

iv. The last alternative can be generalized to include cases such as $f(n) \in \Theta(n^{p+q} \log n)$ or $f(n) \in \Theta(n^{p+q}/\log n)$; we also get $T(n) \in \Theta(f(n))$ if there exist a function $g : X \to \mathbb{R}^*$ and a real constant $\alpha > a$ such that $f(n) \in \Theta(g(n))$ and $g(bn) \geq \alpha g(n)$ for all $n \in X$.

** v. Prove or disprove that the third alternative can be generalized as follows: $T(n) \in \Theta(f(n) \log n)$ whenever there exist two strictly positive real constants $q_1 \leq q_2$ such that $f(n) \in O(n^p (\log n)^{q_2-1})$ and $f(n) \in \Omega(n^p (\log n)^{q_1-1})$. If you disprove it, find the simplest but most general additional constraint on $f(n)$ that suffices to imply $T(n) \in \Theta(f(n) \log n)$. $\qquad \square$

Problem 2.3.14.   Solve the following recurrence exactly:

$$t_n = 1/(4 - t_{n-1}) \qquad n > 1$$

subject to $t_1 = {}^1\!/_4$. $\qquad \square$

Problem 2.3.15.   Solve the following recurrence exactly as a function of the initial conditions a and b:

$$T(n+2) = (1 + T(n+1))/T(n) \qquad n \geq 2$$

subject to $T(0) = a$, $T(1) = b$. $\qquad \square$

Problem 2.3.16.   Solve the following recurrence exactly:

$$T(n) = \tfrac{3}{2} T(n/2) - \tfrac{1}{2} T(n/4) - 1/n \qquad n \geq 3$$

subject to $T(1) = 1$ and $T(2) = 3/2$.   □

2.4 REFERENCES AND FURTHER READING

The asymptotic notation has existed for some while in mathematics: see Bachmann (1894) and de Bruijn (1961). Knuth (1976) gives an account of its history and proposes a standard form for it. Conditional asymptotic notation and its use in Problem 2.1.20 are introduced by Brassard (1985), who also suggests that "one-way inequalities" should be abandoned in favour of a notation based on sets. For information on calculating limits and on de l'Hôpital's rule, consult any book on mathematical analysis, Rudin (1953), for instance.

The book by Purdom and Brown (1985) presents a number of techniques for analysing algorithms. The main mathematical aspects of the analysis of algorithms can also be found in Greene and Knuth (1981).

Example 2.1.1 corresponds to the algorithm of Dixon (1981). Problem 2.2.3 comes from Williams (1964). The analysis of disjoint set structures given in Example 2.2.10 is adapted from Hopcroft and Ullman (1973). The more precise analysis making use of Ackermann's function can be found in Tarjan (1975, 1983). Buneman and Levy (1980) and Dewdney (1984) give a solution to Problem 2.2.15.

Several techniques for solving recurrences, including the characteristic equation and change of variable, are explained in Lueker (1980). For a more rigorous mathematical treatment see Knuth (1968) or Purdom and Brown (1985). The paper by Bentley, Haken, and Saxe (1980) is particularly relevant for recurrences occurring from the analysis of divide-and-conquer algorithms (see Chapter 4).

3

Greedy Algorithms

3.1 INTRODUCTION

Greedy algorithms are usually quite simple. They are typically used to solve optimization problems: find the best order to execute a certain set of jobs on a computer, find the shortest route in a graph, and so on. In the most common situation we have

- a set (or a list) of *candidates*: the jobs to be executed, the nodes of the graph, or whatever;
- the set of candidates that have already been used;
- a function that checks whether a particular set of candidates *provides a solution* to our problem, ignoring questions of optimality for the time being;
- a function that checks whether a set of candidates is *feasible*, that is, whether or not it is possible to complete the set in such a way as to obtain at least one solution (not necessarily optimal) to our problem (we usually expect that the problem has at least one solution making use of candidates from the set initially available);
- a *selection function* that indicates at any time which is the most promising of the candidates not yet used; and
- an *objective function* that gives the value of a solution (the time needed to execute all the jobs in the given order, the length of the path we have found, and so on); this is the function we are trying to optimize.

To solve our optimization problem, we look for a set of candidates constituting a solution that optimizes (minimizes or maximizes, as the case may be) the value of the objective function. A greedy algorithm proceeds step by step. Initially, the set of chosen candidates is empty. Then at each step, we try to add to this set the best remaining candidate, our choice being guided by the selection function. If the enlarged set of chosen candidates is no longer feasible, we remove the candidate we just added; the candidate we tried and removed is never considered again. However, if the enlarged set is still feasible, then the candidate we just added stays in the set of chosen candidates from now on. Each time we enlarge the set of chosen candidates, we check whether the set now constitutes a solution to our problem. When a greedy algorithm works correctly, the first solution found in this way is always optimal.

> **function** *greedy* $(C : set) : set$
>   {C is the set of all the candidates}
> $S \leftarrow \varnothing$ {S is a set in which we construct the solution}
>   **while not** *solution* (S) **and** $C \neq \varnothing$ **do**
> $x \leftarrow$ an element of C maximizing *select* (x)
> $C \leftarrow C \setminus \{x\}$
>   **if** *feasible* $(S \cup \{x\})$ **then** $S \leftarrow S \cup \{x\}$
>   **if** *solution* (S) **then return** S
>   **else return** *"there are no solutions"* .

It is easy to see why such algorithms are called "greedy": at every step, the procedure chooses the best morsel it can swallow, without worrying about the future. It never changes its mind: once a candidate is included in the solution, it is there for good; once a candidate is excluded from the solution, it is never reconsidered.

The selection function is usually based on the objective function; they may even be identical. However, we shall see in the following examples that at times there may be several plausible selection functions, so that we have to choose the right one if we want our algorithm to work properly.

Example 3.1.1. We want to give change to a customer using the smallest possible number of coins. The elements of the problem are

- the candidates: a finite set of coins, representing for instance 1, 5, 10, and 25 units, and containing at least one coin of each type;
- a solution: the total value of the chosen set of coins is *exactly* the amount we have to pay;
- a feasible set: the total value of the chosen set *does not exceed* the amount to be paid;
- the selection function: choose the highest-valued coin remaining in the set of candidates; and
- the objective function: the number of coins used in the solution.   □

*Problem 3.1.1.   Prove that with the values suggested for the coins in the preceding example the greedy algorithm will always find an optimal solution provided one exists.

Prove, on the other hand, by giving specific counterexamples, that the greedy algorithm no longer gives an optimal solution in every case if there also exist 12-unit coins, or if one type of coin is missing from the initial set. Show that it can even happen that the greedy algorithm fails to find a solution at all despite the fact that one exists.   □

It is obviously more efficient to reject all the remaining 25-unit coins (say) at once when the remaining amount to be represented falls below this value. Using integer division is also more efficient than proceeding by successive subtractions.

3.2 GREEDY ALGORITHMS AND GRAPHS

3.2.1 Minimal Spanning Trees

Let $G = <N, A>$ be a connected undirected graph where N is the set of nodes and A is the set of edges. Each edge has a given non-negative *length*. The problem is to find a subset T of the edges of G such that all the nodes remain connected when only the edges in T are used, and the sum of the lengths of the edges in T is as small as possible. (Instead of talking about length, we can associate a *cost* to each edge. In this case the problem is to find a subset T whose total cost is as small as possible. Obviously, this change of terminology does not affect the way we solve the problem.)

Problem 3.2.1.   Prove that the partial graph $<N, T>$ formed by the nodes of G and the edges in T is a tree.   □

The graph $<N, T>$ is called a *minimal spanning tree* for the graph G. This problem has many applications. For instance, if the nodes of G represent towns, and the cost of an edge $\{a, b\}$ is the cost of building a road from a to b, then a minimal spanning tree of G shows us how to construct at the lowest possible cost a road system linking all the towns in question.

We give two greedy algorithms to solve this problem. In the terminology we have used for greedy algorithms, a set of edges is a *solution* if it constitutes a spanning tree, and it is *feasible* if it does not include a cycle. Moreover, a feasible set of edges is *promising* if it can be completed so as to form an optimal solution. In particular, the empty set is always promising since G is connected. Finally, an edge *touches* a given set of nodes if exactly one end of the edge is in the set. The following lemma is crucial for proving the correctness of the forthcoming algorithms.

Lemma 3.2.1.   Let $G = <N, A>$ be a connected undirected graph where the length of each edge is given. Let $B \subset N$ be a strict subset of the nodes of G. Let

$T \subseteq A$ be a promising set of edges such that no edge in T touches B. Let e be the shortest edge that touches B (or any one of the shortest if ties exist). Then $T \cup \{e\}$ is promising.

Proof. Let U be a minimal spanning tree of G such that $T \subseteq U$ (such a U must exist since T is promising by assumption). If $e \in U$, there is nothing to prove. Otherwise when we add the edge e to U, we create exactly one cycle (this is one of the properties of a tree). In this cycle, since e touches B, there necessarily exists at least one other edge, e', say, that also touches B (otherwise the cycle could not close — see Figure 3.2.1). If we now remove e', the cycle disappears and we obtain a new tree U' that spans G. But since the length of e is by definition no greater than the length of e', the total length of the edges in U' does not exceed the total length in U. Therefore U' is also a minimal spanning tree, and it includes e. To complete the proof, we note that $T \subseteq U'$ because the edge e', which touches B, cannot be in T.   □

The initial set of candidates is the set of all the edges. A greedy algorithm selects the edges one by one in some given order. Each edge is either included in the set that will eventually form the solution or eliminated from further consideration. The main difference between the various greedy algorithms to solve this problem lies in the order in which the edges are selected.

Kruskal's algorithm.   The set T of edges is initially empty. As the algorithm progresses, edges are added to T. At every instant the partial graph formed by the nodes of G and the edges in T consists of several connected components. (Initially, when T is empty, each node of G forms a distinct trivial connected component.) The elements of T that are included in a given connected component form a minimal spanning tree for the nodes in this component. At the end of the algorithm only one connected component remains, so that T is then a minimal spanning tree for all the nodes of G.

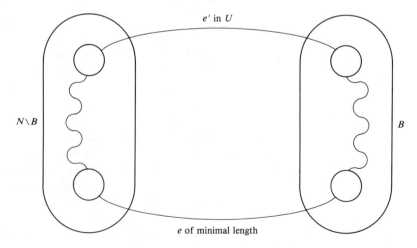

Figure 3.2.1.   A cycle is created if we add edge e to U.

To build bigger and bigger connected components, we examine the edges of G in order of increasing length. If an edge joins two nodes in different connected components, we add it to T, and consequently, the two connected components now form only one component. Otherwise the edge is rejected: it joins two nodes in the same connected component and cannot therefore be added to T without forming a cycle since the edges in T form a minimal spanning tree for each component. The algorithm stops when only one connected component remains.

To illustrate how this algorithm works, consider the graph in figure 3.2.2. In increasing order of length the edges are: {1,2}, {2,3}, {4,5}, {6,7}, {1,4}, {2,5}, {4,7}, {3,5}, {2,4}, {3,6}, {5,7}, {5,6}. The algorithm proceeds as follows.

Step	Edge considered	Connected components
Initialization	—	{1} {2} {3} {4} {5} {6} {7}
1	{1,2}	{1,2} {3} {4} {5} {6} {7}
2	{2,3}	{1,2,3} {4} {5} {6} {7}
3	{4,5}	{1,2,3} {4,5} {6} {7}
4	{6,7}	{1,2,3} {4,5} {6,7}
5	{1,4}	{1,2,3,4,5} {6,7}
6	{2,5}	rejected
7	{4,7}	{1,2,3,4,5,6,7}

T contains the chosen edges {1,2}, {2,3}, {4,5}, {6,7}, {1,4}, and {4,7}. This minimal spanning tree is shown by the heavy lines in Figure 3.2.2; its total length is 17.

Problem 3.2.2.   Prove that Kruskal's algorithm works correctly. The proof, which uses lemma 3.2.1, is by induction on the number of edges selected until now. □

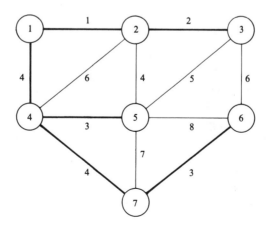

Figure 3.2.2.   A graph and its minimal spanning tree.

Problem 3.2.3.   A graph may have several different minimal spanning trees. Is this the case in our example, and if so, where is this possibility reflected in the algorithm? □

To implement the algorithm, we have to handle a certain number of sets: the nodes in each connected component. We have to carry out rapidly the two operations *find*(x), which tells us in which component the node x is to be found, and *merge*(A,B) to merge two disjoint sets. We therefore use disjoint set structures (Section 1.9.5). For this algorithm it is preferable to represent the graph as a vector of edges with their associated lengths rather than as a matrix of distances. Here is the algorithm.

function *Kruskal*(G = <N,A> : graph ; length : A → $\mathbb{R}^*$) : set of edges
    {initialization}
    Sort A by increasing *length*
    n ← #N
    T ← ∅   {will contain the edges of the minimal spanning tree}
    initialize n sets, each containing one distinct element of N
    {greedy loop}
    repeat
        {u,v} ← shortest edge not yet considered
        ucomp ← find(u)
        vcomp ← find(v)
        if ucomp ≠ vcomp **then**
            merge(ucomp, vcomp)
            T ← T ∪ {{u,v}}
    until #T = n − 1
    return T

Problem 3.2.4.   What happens if, by mistake, we run the algorithm on a graph that is not connected? □

We can estimate the execution time of the algorithm as follows. On a graph with n nodes and a edges the number of operations is in

- $O(a \log a)$ to sort the edges, which is equivalent to $O(a \log n)$ since $n-1 \le a \le n(n-1)/2$;
- $O(n)$ to initialize the n disjoint sets;
- in the worst case $O((2a+n-1)\lg^* n)$ for all the *find* and *merge* operations, by the analysis given in example 2.2.10, since there are at most 2a *find* operations and n − 1 *merge* operations on a universe containing n elements; and
- at worst, $O(a)$ for the remaining operations.

For a connected graph we know that $a \ge n-1$. We conclude that the total time for the algorithm is in $O(a \log n)$ because $O(\lg^* n) \subseteq O(\log n)$. Although this does not change the worst-case analysis, it is preferable to keep the edges in a heap (Section 1.9.4 — here the heap property should be inverted so that the value of each internal node is *less than* or equal to the values of its children). This allows the initialization to be carried out in a time in $O(a)$, although each search for a minimum in the **repeat** loop will now take a time in $O(\log a) = O(\log n)$. This is particularly advantageous in cases when the minimal spanning tree is found at a moment when a considerable number of edges remain to be tried. In such cases, the original algorithm wastes time sorting all these useless edges.

Problem 3.2.5. What can you say about the time required by Kruskal's algorithm if, instead of providing a list of edges, the user supplies a matrix of distances, leaving to the algorithm the job of working out which edges exist?   □

Prim's algorithm. In Kruskal's algorithm we choose promising edges without worrying too much about their connection to previously chosen edges, except that we are careful never to form a cycle. There results a forest of trees that grows somewhat haphazardly. In Prim's algorithm, on the other hand, the minimal spanning tree grows "naturally", starting from an arbitrary root. At each stage we add a new branch to the tree already constructed, and the algorithm stops when all the nodes have been reached.

Initially, the set B of nodes contains a single arbitrary node, and the set T of edges is empty. At each step Prim's algorithm looks for the shortest possible edge $\{u, v\}$ such that $u \in N \setminus B$ and $v \in B$. It then adds u to B and $\{u, v\}$ to T. In this way the edges in T form at any instant a minimal spanning tree for the nodes in B. We continue thus as long as $B \neq N$. Here is an informal statement of the algorithm.

> **function** $Prim(G = <N, A>: graph ; length : A \to \mathbb{R}^*): set\ of\ edges$
> {initialization}
> $T \leftarrow \varnothing$ {will contain the edges of the minimal spanning tree}
> $B \leftarrow$ {an arbitrary member of N }
> **while** $B \neq N$ **do**
>   find $\{u, v\}$ of minimum *length* such that $u \in N \setminus B$ and $v \in B$
> $T \leftarrow T \cup \{\{u, v\}\}$
> $B \leftarrow B \cup \{u\}$
> **return** T

Problem 3.2.6. Prove that Prim's algorithm works correctly. The proof, which again uses Lemma 3.2.1, is by induction on the number of nodes in B.   □

To illustrate how the algorithm works, consider once again the graph in Figure 3.2.2. We arbitrarily choose node 1 as the starting node.

Step	$\{u, v\}$	B
Initialization	—	$\{1\}$
1	$\{2, 1\}$	$\{1, 2\}$
2	$\{3, 2\}$	$\{1, 2, 3\}$
3	$\{4, 1\}$	$\{1, 2, 3, 4\}$
4	$\{5, 4\}$	$\{1, 2, 3, 4, 5\}$
5	$\{7, 4\}$	$\{1, 2, 3, 4, 5, 7\}$
6	$\{6, 7\}$	$\{1, 2, 3, 4, 5, 6, 7\}$

T contains the chosen edges $\{2, 1\}$, $\{3, 2\}$, $\{4, 1\}$, $\{5, 4\}$, $\{7, 4\}$, and $\{6, 7\}$.

Problem 3.2.7.   A graph may have several different minimal spanning trees. Where is this possibility reflected in the algorithm?   □

To obtain a simple implementation on a computer, suppose that the nodes of G are numbered from 1 to n, $N = \{1, 2, \ldots, n\}$, and that a symmetric matrix L gives the (nonnegative) length of each edge, with $L[i, j] = \infty$ if the corresponding edge does not exist. We use two arrays. For each node $i \in N \setminus B$, *nearest*$[i]$ gives the node in B that is nearest to i, and *mindist*$[i]$ gives the distance from i to *nearest*$[i]$; for a node $i \in B$, we set *mindist*$[i] = -1$. The set B, arbitrarily initialized to $\{1\}$, is not represented explicitly; *nearest*$[1]$ and *mindist*$[1]$ are not used. Here is the algorithm.

```
function Prim (L [1 .. n, 1 .. n]): set of edges
    {initialization — only node 1 is in B }
    T ← ∅  {will contain the edges of the minimal spanning tree}
    for i ← 2 to n do
        nearest [i] ← 1
        mindist [i] ← L [i, 1]
    {greedy loop}
    repeat n − 1 times
        min ← ∞
        for j ← 2 to n do
            if 0 ≤ mindist [ j ] < min then min ← mindist [ j ]
                                             k ← j
        T ← T ∪ {{k, nearest [k]}}
        mindist [k] ← −1  {add k to B }
        for j ← 2 to n do
            if L [k, j] < mindist [ j ] then mindist [ j ] ← L [k, j ]
                                            nearest [ j ] ← k
    return T
```

Problem 3.2.8.   What happens if, by mistake, we run the algorithm on a graph that is not connected?   □

The main loop of the algorithm is executed $n-1$ times; at each iteration the enclosed **for** loops take a time in $O(n)$. It is thus clear that Prim's algorithm takes a time in $O(n^2)$.

We saw that Kruskal's algorithm takes a time in $O(a \log n)$, where a is the number of edges in the graph. For a very dense graph, a tends toward $n(n-1)/2$. In this case Kruskal's algorithm takes a time in $O(n^2 \log n)$, and Prim's algorithm is probably better. For a very sparse graph, a tends towards n. In this case Kruskal's algorithm takes a time in $O(n \log n)$, and Prim's algorithm is probably less efficient. For sparse graphs there exist other algorithms that are more efficient than the one invented by Kruskal.

Problem 3.2.9.   What happens

  i. in the case of Prim's algorithm
  ii. in the case of Kruskal's algorithm

if we allow edges with negative lengths? Is it still sensible to talk about minimal spanning trees if edges with negative lengths are allowed?   □

3.2.2 Shortest Paths

Consider now the case of a directed graph $G = <N, A>$ where N is the set of nodes of G and A is the set of directed edges. Each edge has a nonnegative *length*. One of the nodes is designated as the *source* node. The problem is to determine the length of the shortest path from the source to each of the other nodes of the graph. (Once again we could talk about the *cost* of an edge instead of its length, and pose the problem of finding the cheapest path from the source to each other node.)

This problem can be solved by a greedy algorithm often called *Dijkstra's algorithm*. The sets C and S are respectively the set of available candidate nodes and the set of nodes already chosen. At every moment S contains those nodes whose minimal distance from the source is already known, whereas C contains all the others. At the outset, S contains only the source itself; when the algorithm ends, S contains all the nodes of the graph and our problem is solved. At each step we choose the node in C whose distance to the source is least, and we add it to S.

We say that a path from the source to some other node is *special* if all the intermediate nodes along the path belong to S. At each step of the algorithm, an array D contains the length of the shortest special path to each node of the graph. At the moment we add a new node v to S, the shortest special path to v is also the shortest of all the paths to v (this is proved later). When the algorithm ends, all the nodes of the graph are in S, and hence all the paths from the source to some other node are special. Consequently, the values in D give the solution to our problem.

To make life simple, we assume once again that the nodes of the graph are numbered from 1 to n, $N = \{1, 2, \ldots, n\}$, that node 1 is the source, and that a matrix L

gives the length of each directed edge: $L[i,j] \ge 0$ if the edge (i,j) exists and $L[i,j] = \infty$ otherwise. Here is the algorithm.

function $Dijkstra(L[1..n, 1..n])$: **array**$[2..n]$
   {initialization}
   $C \leftarrow \{2, 3, \ldots, n\}$ {$S = N \setminus C$ exists only by implication}
   for $i \leftarrow 2$ **to** n **do** $D[i] \leftarrow L[1, i]$
   {greedy loop}
   repeat $n-2$ **times**
      $v \leftarrow$ some element of C minimizing $D[v]$
      $C \leftarrow C \setminus \{v\}$ {and implicitly $S \leftarrow S \cup \{v\}$}
      for each $w \in C$ **do**
         $D[w] \leftarrow \min(D[w], D[v] + L[v, w])$
   return D

The algorithm proceeds as follows on the graph in Figure 3.2.3.

Step	v	C	D
Initialization	—	$\{2, 3, 4, 5\}$	$[50, 30, 100, 10]$
1	5	$\{2, 3, 4\}$	$[50, 30, 20, 10]$
2	4	$\{2, 3\}$	$[40, 30, 20, 10]$
3	3	$\{2\}$	$[35, 30, 20, 10]$

Clearly, D would not change if we did one more iteration to remove the last element of C, which is why the main loop is only repeated $n-2$ times.

If we want not only to know the length of the shortest paths but also where they pass, it suffices to add a second array $P[2..n]$, where $P[v]$ contains the number of the node that precedes v in the shortest path. To find the complete path, simply follow the pointers P backwards from a destination to the source. The modifications to the algorithm are simple:

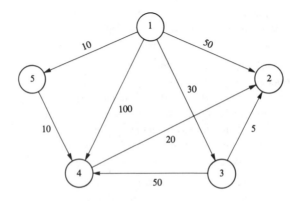

Figure 3.2.3 A directed graph.

- initialize $P[i]$ to 1 for $i = 2, 3, \ldots, n$;
- replace the contents of the inner **for** loop by

$$\textbf{if } D[w] > D[v] + L[v,w] \quad \textbf{then } D[w] \leftarrow D[v] + L[v,w]$$
$$P[w] \leftarrow v .$$

Problem 3.2.10.   Show how the modified algorithm works on the graph of Figure 3.2.3.   □

Proof of correctness.   We prove by mathematical induction that

i. if a node i is in S, then $D[i]$ gives the length of the shortest path from the source to i ;

ii. if a node i is not in S, then $D[i]$ gives the length of the shortest *special* path from the source to i.

Look at the initialization of D and S to convince yourself that these two conditions hold at the outset; the base for our induction is thus obtained. Next, consider the inductive step, and suppose by the induction hypothesis that these two conditions hold just before we add a new node v to S.

i. This follows immediately from the induction hypothesis for each node i that was already in S before the addition of v. As for node v, it will now belong to S. We must therefore check that $D[v]$ gives the length of the shortest path from the source to v. By the induction hypothesis $D[v]$ certainly gives the length of the shortest special path. We therefore have to verify that the shortest path from the source to v does not pass through a node that does not belong to S. Suppose the contrary: when we follow the shortest path from the source to v, the first node encountered that does not belong to S is some node x distinct from v (see Figure 3.2.4).

   The initial section of the path, as far as x, is a special path. Consequently, the total distance to v via x is

   $\geq$ distance to x (since edge lengths are non-negative)
   $\geq D[x]$ (by part (ii) of the induction)
   $\geq D[v]$ (because the algorithm chose v before x)

   and the path via x cannot be shorter than the special path leading to v.

   We have thus verified that when v is added to S, part (i) of the induction remains true.

ii. Consider now a node $w \notin S$ different from v. When v is added to S, there are two possibilities for the shortest special path from the source to w : either it does not change, or else it now passes through v. In the latter case it seems at first

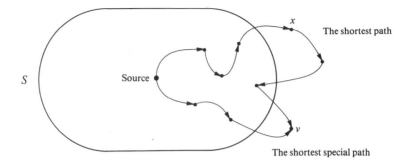

Figure 3.2.4. The shortest path from the source to v cannot go through node x.

glance that there are again two possibilities: either v is the last node in S visited before arriving at w or it is not. We have to compare explicitly the length of the old special path leading to w and the length of the special path that visits v just before arriving at w; the algorithm does this. However, we can ignore the possibility (see Figure 3.2.5) that v is visited, but not just before arriving at w: a path of this type cannot be shorter than the path of length $D[x] + L[x, w]$ that we examined at a previous step when x was added to S, because $D[x] \le D[v]$.

Thus the algorithm ensures that part (ii) of the induction also remains true when a new node v is added to S.

To complete the proof that the algorithm works, we need only note that when its execution stops all the nodes but one are in S (even though the set S is not constructed explicitly). At this point it is clear that the shortest path from the source to the remaining node is a special path.

Problem 3.2.11. Show by giving an explicit example that if the edge lengths can be negative, then Dijkstra's algorithm does not always work correctly. Is it still sensible to talk about shortest paths if negative distances are allowed?   □

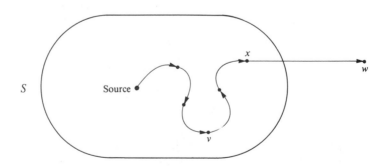

Figure 3.2.5 The shortest path from the source to w cannot visit x between v and w.

Analysis of the algorithm.   Suppose Dijkstra's algorithm is applied to a graph having n nodes and a edges. Using the representation suggested up to now, the instance is given in the form of a matrix $L[1..n, 1..n]$. Initialization takes a time in $O(n)$. In a straightforward implementation, choosing v in the **repeat** loop requires all the elements of C to be examined, so that we look at $n-1, n-2, \ldots, 2$ values of D on successive iterations, giving a total time in $O(n^2)$. The inner **for** loop does $n-2$, $n-1, \ldots, 1$ iterations, for a total also in $O(n^2)$. The time required by this version of the algorithm is therefore in $O(n^2)$.

If $a \ll n^2$, it seems we might be able to avoid looking at the many entries containing ∞ in the matrix L. With this in mind, it could be preferable to represent the graph by an array of n lists, giving for each node its direct distance to adjacent nodes (like the type *lisgraph* of Section 1.9.2). This allows us to save time in the inner **for** loop, since we only have to consider those nodes w adjacent to v, but how are we to avoid taking a time in $\Omega(n^2)$ to determine in succession the $n-2$ values taken by v?

The answer is to use a heap containing one node for each element v of C, ordered by the value of $D[v]$. If we remember to invert the heap, the element v of C that minimizes $D[v]$ will always be found at the root. Initialization of the heap takes a time in $O(n)$. The instruction "$C \leftarrow C \setminus \{v\}$" consists of eliminating the root from the heap, which takes a time in $O(\log n)$. As for the inner **for** loop, it now consists of looking, for each element w of C adjacent to v, to see whether $D[v] + L[v, w] < D[w]$. If so, we must modify $D[w]$ and percolate w up the heap, which again takes a time in $O(\log n)$. This does not happen more than once for each edge of the graph.

To sum up, we have to remove the root of the heap exactly $n-2$ times and to percolate at most a nodes, giving a total time in $O((a+n)\log n)$. If the graph is connected, $a \geq n-1$ and this time is in $O(a \log n)$. The straightforward implementation is therefore preferable if the graph is dense, whereas it is preferable to use a heap if the graph is sparse. If $a \in \Theta(n^2/\log n)$, the choice of algorithm may depend on the specific implementation.

*Problem 3.2.12.   In the preceding analysis, we saw that up to a nodes can be percolated, whereas less than n roots are eliminated. This is interesting when we remember that eliminating the root has for effect to sift down the node that takes its place, and that percolating up is somewhat quicker than sifting down (at each level, we compare the value of a node to the value of its parent rather than making comparisons with both children). We might therefore consider modifying the definition of a heap slightly to allow percolation to run faster still, at the cost of slowing down sifting.

   i. What modification to the definition of a heap do you suggest?
   ii. Let $k = \max(2, \lfloor a/n \rfloor)$. Show how your modification allows you to calculate the shortest paths from a source to all the other nodes of a graph in a time in

$O(a \log_k n)$. Problem 2.1.17(i) does not apply here since k is not a constant. Note that this gives $O(n^2)$ if $a \approx n^2$ and $O(a \log n)$ if $a \approx n$; it therefore gives the best of both worlds. (Still faster algorithms exist.)   □

Problem 3.2.13.   Show that Prim's algorithm to find minimal spanning trees can also be implemented through the use of heaps. Show that it then takes a time in $O(a \log n)$, just as Kruskal's algorithm would. Finally, show that the modification suggested in the previous problem applies just as well to Prim's algorithm.   □

3.3 GREEDY ALGORITHMS FOR SCHEDULING

3.3.1 Minimizing Time in the System

A single server (a processor, a petrol pump, a cashier in a bank, and so on) has n customers to serve. The service time required by each customer is known in advance: customer i will take time t_i, $1 \le i \le n$. We want to minimize

$$T = \sum_{i=1}^{n} (\text{time in system for customer } i).$$

Since the number of customers is fixed, minimizing the total time in the system is equivalent to minimizing the average time. For example, if we have three customers with

$$t_1 = 5, \quad t_2 = 10, \quad t_3 = 3,$$

then six orders of service are possible.

Order	T	
1 2 3 :	$5 + (5+10) + (5+10+3) = 38$	
1 3 2 :	$5 + (5+3) + (5+3+10) = 31$	
2 1 3 :	$10 + (10+5) + (10+5+3) = 43$	
2 3 1 :	$10 + (10+3) + (10+3+5) = 41$	
3 1 2 :	$3 + (3+5) + (3+5+10) = 29$	← optimal
3 2 1 :	$3 + (3+10) + (3+10+5) = 34$	

In the first case, customer 1 is served immediately, customer 2 waits while customer 1 is served and then gets his turn, and customer 3 waits while both 1 and 2 are served and then is served himself: the total time passed in the system by the three customers is 38.

Imagine an algorithm that builds the optimal schedule step by step. Suppose that after scheduling customers $i_1, i_2, \ldots, i_m$ we add customer j. The increase in T at this stage is

$$t_{i_1} + t_{i_2} + \cdots + t_{i_m} + t_j \ .$$

To minimize this increase, we need only minimize t_j. This suggests a simple greedy algorithm: at each step, add to the end of the schedule the customer requiring the least service among those who remain. In the preceding example this algorithm gives the correct answer 3, 1, 2.

We now prove that this algorithm is always optimal. Let $I = (i_1 i_2 \cdots i_n)$ be any permutation of the integers $\{1, 2, \ldots, n\}$. If customers are served in the order I, the total time passed in the system by all the customers is

$$T(I) = t_{i_1} + (t_{i_1} + t_{i_2}) + (t_{i_1} + t_{i_2} + t_{i_3}) + \cdots$$

$$= nt_{i_1} + (n-1)t_{i_2} + (n-2)t_{i_3} + \cdots$$

$$= \sum_{k=1}^{n} (n-k+1)t_{i_k} \ .$$

Suppose now that I is such that we can find two integers a and b with $a < b$ and $t_{i_a} > t_{i_b}$: in other words, the ath customer is served before the bth customer even though the former needs more service time than the latter (see Figure 3.3.1). If we exchange the positions of these two customers, we obtain a new order of service I' obtained from I by interchanging the items i_a and i_b. This new order is preferable because

$$T(I') = (n-a+1)t_{i_b} + (n-b+1)t_{i_a} + \sum_{\substack{k=1 \\ k \neq a,b}}^{n} (n-k+1)t_{i_k}$$

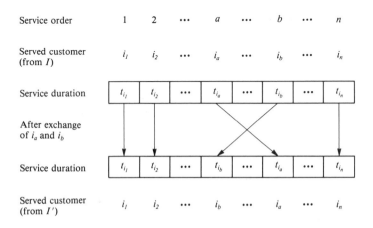

Figure 3.3.1 Exchanging the positions of two customers.

and thus

$$T(I) - T(I') = (n - a + 1)(t_{i_a} - t_{i_b}) + (n - b + 1)(t_{i_b} - t_{i_a})$$

$$= (b - a)(t_{i_a} - t_{i_b})$$

$$> 0.$$

We can therefore improve any schedule in which a customer is served before someone else who requires less service. The only schedules that remain are those obtained by putting the customers in nondecreasing order of service time. All such schedules are clearly equivalent, and therefore they are all optimal.

Problem 3.3.1.   How much time (use the O notation) is required by a greedy algorithm that accepts n and $t[1..n]$ as data and produces an optimal schedule?   □

The problem can be generalized to a system with s servers, as can the algorithm. Without loss of generality, suppose the customers are numbered so that $t_1 \leq t_2 \leq \cdots \leq t_n$. In this context, server i, $1 \leq i \leq s$, must serve customers number $i, i+s, i+2s, \ldots$ in that order.

Problem 3.3.2.   Prove that this algorithm always yields an optimal schedule.   □

Problem 3.3.3.   A magnetic tape contains n programs of length $l_1, l_2, \ldots, l_n$. We know how often each program is used: a fraction p_i of requests to load a program concern program i, $1 \leq i \leq n$. (This of course implies that $\sum_{i=1}^n p_i = 1$.) Information is recorded along the tape at constant density, and the speed of the tape drive is also constant. Each time a program has been loaded, the tape is rewound to the beginning.

If the programs are held in the order $i_1, i_2, \ldots, i_n$, the average time required to load a program is

$$\bar{T} = c \sum_{j=1}^n \left[p_{i_j} \sum_{k=1}^j l_{i_k} \right],$$

where the constant c depends on the recording density and the speed of the drive. We want to minimize $\bar{T}$.

i. Prove by giving an explicit example that it is not necessarily optimal to hold the programs in order of increasing values of l_i.

ii. Prove by giving an explicit example that it is not necessarily optimal to hold the programs in order of decreasing values of p_i.

iii. Prove that $\bar{T}$ is minimized if the programs are held in order of decreasing p_i/l_i.   □

3.3.2 Scheduling with Deadlines

We have a set of n jobs to execute, each of which takes unit time. At any instant $t = 1, 2, \ldots,$ we can execute exactly one job. Job i, $1 \leq i \leq n$, earns us a profit g_i if and only if it is executed no later than time d_i.

For example, with $n = 4$ and the following values:

i	1	2	3	4
g_i	50	10	15	30
d_i	2	1	2	1

the schedules to consider and the corresponding profits are

Sequence:	Profit:	
1	50	
2	10	
3	15	
4	30	
1, 3	65	
2, 1	60	
2, 3	25	
3, 1	65	
4, 1	80	← optimum
4, 3	45 .	

The sequence 3, 2, for instance, is not considered because job 2 would be executed at time $t = 2$, after its deadline $d_2 = 1$. To maximize our profit in this example, we should execute the schedule 4, 1.

A set of jobs is *feasible* if there exists at least one sequence (also called feasible) that allows all the jobs in the set to be executed in time for their respective deadlines. An obvious greedy algorithm consists of constructing the schedule step by step, adding at each step the job with the highest value of g_i among those not yet considered, provided that the chosen set of jobs remains feasible.

In the preceding example we first choose job 1. Next, we choose job 4: the set $\{1, 4\}$ is feasible because it can be executed in the order 4, 1. Next, we try the set $\{1, 3, 4\}$, which turns out not to be feasible; job 3 is therefore rejected. Finally we try $\{1, 2, 4\}$, which is also infeasible; so job 2 is also rejected. Our solution — optimal in this case — is therefore to execute the set of jobs $\{1, 4\}$, which in fact can only be done in the order 4, 1. It remains to prove that this algorithm always finds an optimal schedule and to find an efficient way of implementing it.

Let J be a set of k jobs. At first glance it seems we might have to try all the $k!$ possible permutations of these jobs to see whether J is feasible. Happily this is not the case.

Lemma 3.3.1. Let J be a set of k jobs, and let $\sigma = (s_1 s_2 \cdots s_k)$ be a permutation of these jobs such that $d_{s_1} \leq d_{s_2} \leq \cdots \leq d_{s_k}$. Then the set J is feasible if and only if the sequence σ is feasible.

Proof. The "if" is obvious. For the "only if":

If J is feasible, there exists at least one sequence of jobs $\rho = (r_1 r_2 \cdots r_k)$ such that $d_{r_i} \geq i$, $1 \leq i \leq k$. Suppose $\sigma \neq \rho$. Let a be the smallest index such that $s_a \neq r_a$, and let b be defined by $r_b = s_a$; it is clear that $b > a$. Also

$$d_{r_a} \geq d_{s_a} \quad \text{(by the construction of } \sigma \text{ and the minimality of } a)$$
$$= d_{r_b} \quad \text{(by the definition of } b).$$

The job r_a could therefore be executed later, at the time when r_b is at present scheduled. Since r_b can certainly be executed earlier than planned, we can interchange the items r_a and r_b in ρ. The result is a new feasible sequence, which is the same as σ at least in positions $1, 2, \ldots, a$. Continuing thus, we obtain a series of feasible sequences, each having at least one more position in agreement with σ. Finally, after a maximum of $k-1$ steps of this type, we obtain σ itself, which is therefore feasible. $\square$

This shows that it suffices to check a single sequence, in order of increasing deadlines, to know whether a set of jobs is or is not feasible. We now prove that the greedy algorithm outlined earlier always finds an optimal schedule.

Proof of optimality. Suppose that the greedy algorithm chooses to execute a set of jobs I whereas in fact the set $J \neq I$ is optimal. Consider two feasible sequences S_I and S_J for the two sets of jobs in question. By making appropriate interchanges of jobs in S_I and S_J, we can obtain two feasible sequences S_I' and S_J' such that every job common to both I and J is scheduled for execution at the same time in the two sequences. (We may have to leave gaps in the schedule. The necessary interchanges are easily found if we scan the two sequences S_I and S_J from right to left. See Figure 3.3.2 for an example.) S_I' and S_J' are distinct sequences since $I \neq J$. Let us consider an arbitrary time when the task scheduled in S_I' is different from that scheduled in S_J'.

- If some task a is scheduled in S_I' whereas there is a gap in S_J' (and therefore task a does not belong to J), the set $J \cup \{a\}$ is feasible and would be more profitable than J. This is not possible since J is optimal by assumption.
- If some task b is scheduled in S_J' whereas there is a gap in S_I', the set $I \cup \{b\}$ is feasible, hence the greedy algorithm should have included b in I. This is also impossible since it did not do so.
- The only remaining possibility is that some task a is scheduled in S_I' whereas a different task b is scheduled in S_J'. Again, this implies that a does not appear in J and that b does not appear in I.
  - If $g_a > g_b$, one could substitute a for b in J and improve it. This goes against the optimality of J.

— If $g_a < g_b$, the greedy algorithm should have chosen b before even considering a since $(I \setminus \{a\}) \cup \{b\}$ would be feasible. This is not possible either since it did not include b in I.

— The only remaining possibility is therefore that $g_a = g_b$.

In conclusion, for each time slot, sequences S'_I and S'_J either schedule no tasks, the same task, or two distinct tasks yielding the same profit. This implies that the total worth of I is identical with that of the optimal set J, and thus I is optimal as well. □

For our first implementation of the algorithm suppose without loss of generality that the jobs are numbered so that $g_1 \geq g_2 \geq \cdots \geq g_n$. To allow us to use sentinels, suppose further that $n > 0$ and that $d_i > 0$, $1 \leq i \leq n$.

```
function sequence (d [0 .. n ]) : k , array[1 .. k ]
    array j [0 .. n ]
    d [0], j [0] ← 0  {sentinels}
    k , j [1] ← 1  {task 1 is always chosen}
    {greedy loop}
    for i ← 2 to n do {in decreasing order of g }
        r ← k
        while d [ j [r ]] > max(d [i ], r ) do r ← r − 1
        if d [ j [r ]] ≤ d [i ] and d [i ] > r then
            for l ← k step −1 to r + 1 do j [l + 1] ← j [l ]
            j [r + 1] ← i
            k ← k + 1
    return k , j [1 .. k ]
```

after reorganization,

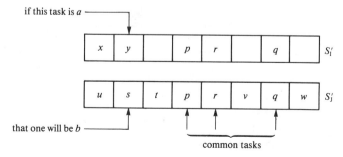

Figure 3.3.2.   Rearranging schedules to bring identical tasks together.

The exact values of the g_i are unnecessary provided the jobs are correctly numbered in order of decreasing profit.

Problem 3.3.4.   Verify that the algorithm works, and show that it requires quadratic time in the worst case.   □

A second, more efficient algorithm is obtained if we use a different technique to verify whether a given set of jobs is feasible.

Lemma 3.3.2.   A set of jobs J is feasible if and only if we can construct a feasible sequence including all the jobs in J as follows: for each job $i \in J$, execute i at time t, where t is the largest integer such that $0 < t \leq \min(n, d_i)$ and the job to be executed at time t has not yet been decided.   □

In other words, consider each job $i \in J$ in turn, and add it to the schedule being built as late as possible, but no later than its deadline. If a job cannot be executed in time for its deadline, then the set J is infeasible. It may happen that there are gaps in the schedule thus constructed when no job is to be executed. This obviously does not affect the feasibility of a set of jobs. If we so wish, the schedule can be compressed once all the jobs are included.

Problem 3.3.5.   Prove Lemma 3.3.2.   □

The lemma suggests that we should consider an algorithm that tries to fill one by one the positions in a sequence of length $l = \min(n, \max(\{d_i \mid 1 \leq i \leq n\}))$. For any position t, define $n_t = \max\{k \leq t \mid \text{position } k \text{ is free}\}$. Also define certain sets of positions: two positions i and j are in the same set if $n_i = n_j$ (see Figure 3.3.3). For a given set K of positions, let $F(K)$ be the smallest member of K. Finally, define a fictitious position 0, which is always free.

Clearly, as we assign new jobs to vacant positions, these sets will merge to form larger sets; disjoint set structures are intended for just this purpose. We obtain an algorithm whose essential steps are the following:

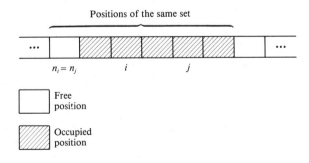

Figure 3.3.3.   Sets of positions.

i. initialization: each position $0, 1, 2, \ldots, l$ is in a different set and $F(\{i\}) = i, 0 \le i \le l$;

ii. addition of a job with deadline d:
- find the set that contains $\min(n, d)$; let this be set K;
- if $F(K) = 0$, reject the job;
- if $F(K) \ne 0$,
  - assign the job to position $F(K)$;
  - find the set that contains $F(K) - 1$; let this be set L (it is necessarily different from K);
  - merge K and L; the value of F for this new set is the old value of $F(L)$.

Example 3.3.1.   Consider a problem involving six jobs:

i	1	2	3	4	5	6
g_i	20	15	10	7	5	3
d_i	3	1	1	3	1	3 .

Figures 3.3.4 and 3.3.5 illustrate the workings of the slow and fast algorithms, respectively. $\square$

Here is a more precise statement of the fast algorithm. To simplify the description, we have assumed that the label of the set produced by a *merge* operation is necessarily the label of one of the sets that were merged.

```
function sequence 2(d[1..n]): k, array[1..k]
    array j, F[0..l]
    l ← min(n, max{d[i] | 1 ≤ i ≤ n})
    {initialization}
    for i ← 0 to l do j[i] ← 0
                     F[i] ← i
                     initialize set {i}
    {greedy loop}
    for i ← 1 to n do  {in decreasing order of g}
        k ← find(min(n, d[i]))
        m ← F[k]
        if m ≠ 0 then
            j[m] ← i
            l ← find(m − 1)
            F[k] ← F[l]
            merge(k, l)
    {it remains to compress the solution}
    k ← 0
    for i ← 1 to l do
        if j[i] > 0 then k ← k + 1
                        j[k] ← j[i]
    return k, j[1..k]
```

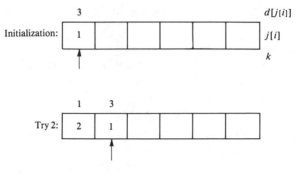

Try 3: unchanged

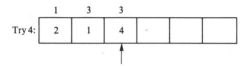

Try 5: unchanged

Try 6: unchanged

Optimal sequence: 2, 1, 4; value = 42

Figure 3.3.4.   Illustration of the slow algorithm.

If the instance is given to us with the jobs already ordered by decreasing profit, so that an optimal sequence can be obtained merely by calling the preceding algorithm, most of the time will be spent manipulating disjoint sets. Since there are at most $n+l$ *find* operations and l *merge* operations to execute, and since $n \geq l$, the required time is in $O(n \lg^* l)$, which is essentially linear. If, on the other hand, the jobs are given to us in arbitrary order, so that we have to begin by sorting them, we need a time in $O(n \log n)$ to obtain the initial sequence.

3.4 GREEDY HEURISTICS

Because they are so simple, greedy algorithms are often used as heuristics in situations where we can (or must) accept an approximate solution instead of an exact optimal solution. We content ourselves with giving two examples of this technique. These examples also serve to illustrate that the greedy approach does not always yield an optimal solution.

Initialization: $l = \min(6, \max(d_i)) = 3$

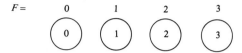

Try 1: $d_1 = 3$, assign task 1 to position 3

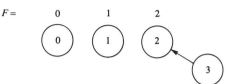

Try 2: $d_2 = 1$, assign task 2 to position 1

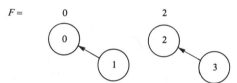

Try 3: $d_3 = 1$, no free position available since the F value is 0

Try 4: $d_4 = 3$, assign task 4 to position 2

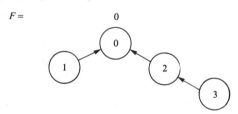

Try 5: $d_5 = 1$, no free position available

Try 6: $d_6 = 3$, no free position available

Optimal sequence: 2, 4, 1; value = 42

Figure 3.3.5.   Illustration of the fast algorithm.

3.4.1 Colouring a Graph

Let $G = <N, A>$ be an undirected graph whose nodes are to be coloured. If two nodes are joined by an edge, then they must be of different colours. Our aim is to use as few different colours as possible. For instance, the graph in Figure 3.4.1 can be coloured using only two colours: red for nodes 1, 3 and 4, and blue for nodes 2 and 5.

An obvious greedy algorithm consists of choosing a colour and an arbitrary starting node, and then considering each other node in turn, painting it with this colour if possible. When no further nodes can be painted, we choose a new colour and a new

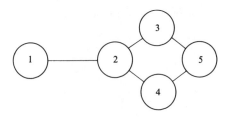

Figure 3.4.1. A graph to be coloured.

starting node that has not yet been painted, we paint as many nodes as we can with this second colour, and so on.

In our example if node 1 is painted red, we are not allowed to paint node 2 with the same colour, nodes 3 and 4 can be red, and lastly node 5 may not be painted. If we start again at node 2 using blue paint, we can colour nodes 2 and 5 and finish the job using only two colours; this is an optimal solution. However, if we systematically consider the nodes in the order 1, 5, 2, 3, 4, we get a different answer: nodes 1 and 5 are painted red, then node 2 is painted blue, but now nodes 3 and 4 require us to use a third colour; in this case the result is not optimal.

The algorithm is therefore no more than a heuristic that may possibly, but not certainly, find a "good" solution. Why should we be interested by such algorithms? For the colouring problem and many others the answer is that all the exact algorithms known require exponential computation time. This is an example of the **NP**-complete problems that we shall study in Chapter 10. For a large-scale instance these algorithms cannot be used in practice, and we are forced to make do with an approximate method (of which this greedy heuristic is among the least effective).

Problem 3.4.1.   For a graph G and an ordering σ of the nodes of G, let $c_\sigma(G)$ be the number of colours used by the greedy algorithm. Let $\hat{c}(G)$ be the optimal (smallest) number of colours. Prove the following assertions:

    i. $(\forall G)(\exists \sigma)[c_\sigma(G) = \hat{c}(G)]$,
    ii. $(\forall \alpha \in \mathbb{R}^+)(\exists G)(\exists \sigma)[c_\sigma(G)/\hat{c}(G) > \alpha]$.

In other words, the greedy heuristic may find the optimal solution, but it may also give an arbitrarily bad answer. □

Problem 3.4.2.   Find two or three practical problems that can be expressed in terms of the graph colouring problem. □

3.4.2 The Travelling Salesperson Problem

We know the distances between a certain number of towns. The travelling salesperson wants to leave one of these towns, to visit each other town exactly once, and to arrive back at the starting point, having travelled the shortest total distance possible. We assume that the distance between two towns is never negative. As for the previous problem, all the known exact algorithms for this problem require exponential time (it is also **NP**-complete). Hence they are impractical for large instances.

The problem can be represented using a complete undirected graph with n nodes. (The graph can also be directed if the distance matrix is not symmetric: see Section 5.6.) One obvious greedy algorithm consists of choosing at each step the shortest remaining edge provided that

    i. it does not form a cycle with the edges already chosen (except for the very last edge chosen, which completes the salesperson's tour);

    ii. if chosen, it will not be the third chosen edge incident on some node.

For example, if our problem concerns six towns with the following distance matrix:

From	To:	2	3	4	5	6
1		3	10	11	7	25
2			6	12	8	26
3				9	4	20
4					5	15
5						18

edges are chosen in the order $\{1,2\}$, $\{3,5\}$, $\{4,5\}$, $\{2,3\}$, $\{4,6\}$, $\{1,6\}$ to make the circuit (1, 2, 3, 5, 4, 6, 1) whose total length is 58. Edge $\{1,5\}$, for example, was not kept when we looked at it because it would have completed a circuit (1, 2, 3, 5, 1), and also because it would have been the third edge incident on node 5. In this instance the greedy algorithm does not find an optimal tour since the tour (1, 2, 3, 6, 4, 5, 1) has a total length of only 56.

Problem 3.4.3.   What happens to this greedy algorithm if the graph is not complete, that is, if it is not possible to travel directly between certain pairs of towns? □

Problem 3.4.4.   In some instances it is possible to find a shorter optimal tour if the salesperson is allowed to pass through the same town several times. Give an explicit example illustrating this. On the other hand, a distance matrix is said to be *Euclidean* if the triangle inequality holds: for any towns i, j, and k, it is true that $distance(i,j) \leq distance(i,k) + distance(k,j)$. Show that in this case it is never advantageous to pass through the same town several times.   □

Problem 3.4.5.   Give a heuristic greedy algorithm to solve the travelling salesperson problem in the case when the distance matrix is Euclidean. Your algorithm must find a solution whose length is not more than double the length of an optimal tour. (*Hint*: start by constructing a minimal spanning tree and then use the answer to the previous problem.)   □

Problem 3.4.6.   Invent a heuristic greedy algorithm for the case when the distance matrix is not symmetric.   □

Problem 3.4.7.   There exists a greedy algorithm (which perhaps would better be called "abstinent") for solving the problem of the knight's tour on a chessboard: at each step move the knight to the square that threatens the least possible number of squares not yet visited. Try it !   □

Problem 3.4.8.   In a directed graph a path is said to be *Hamiltonian* if it passes exactly once through each node of the graph, but without coming back to the starting node. Prove that if a directed graph is complete (that is, if each pair of nodes is joined in at least one direction) then it has a Hamiltonian path, and give an algorithm for finding such a path in this case.   □

3.5 REFERENCES AND FURTHER READING

A discussion of topics connected with Problem 3.1.1 can be found in Wright (1975) and Chang and Korsh (1976); see also Problem 5.8.5 of this book.

The problem of minimal spanning trees has a long history, which is discussed in Graham and Hell (1985). The first algorithm proposed (which we have not described) is due to Borůvka (1926). The algorithm to which Prim's name is attached was invented by Jarník (1930) and rediscovered by Prim (1957) and Dijkstra (1959). Kruskal's algorithm comes from Kruskal (1956). Other more sophisticated algorithms are described in Yao (1975), Cheriton and Tarjan (1976), and Tarjan (1983).

The implementation of Dijkstra's algorithm that takes a time in $O(n^2)$ is from Dijkstra (1959). The details of the improvement suggested in Problem 3.2.12 can be found in Johnson (1977). Similar improvement for the minimal spanning tree problem (Problem 3.2.13) is from Johnson (1975). Faster algorithms for both these problems are given in Fredman and Tarjan (1984); in particular, use of the Fibonacci heap allows them to implement Dijkstra's algorithm in a time in $O(a + n \log n)$. Other ideas concerning shortest paths can be found in Tarjan (1983).

The solution to Problem 3.4.5 is given in Christofides (1976); the same reference gives an efficient heuristic for finding a solution to the travelling salesperson problem with a Euclidean distance matrix that is not more than 50% longer than the optimal tour.

An important greedy algorithm that we have not discussed is used to derive optimal Huffman codes; see Schwartz (1964). Other greedy algorithms for a variety of problems are described in Horowitz and Sahni (1978).

4

Divide-and-Conquer

4.1 INTRODUCTION

Divide-and-conquer is a technique for designing algorithms that consists of decomposing the instance to be solved into a number of smaller subinstances of the same problem, solving successively and independently each of these subinstances, and then combining the subsolutions thus obtained in such a way as to obtain the solution of the original instance. The first question that springs to mind is naturally "How should we solve the subinstances?". The efficiency of the divide-and-conquer technique lies in the answer to this question.

Suppose you already have some algorithm A that requires quadratic time. Let c be a constant such that your particular implementation requires a time $t_A(n) \leq cn^2$ to solve an instance of size n. You discover that it would be possible to solve such an instance by decomposing it into three subinstances each of size $\lceil n/2 \rceil$, solving these subinstances, and combining the results. Let d be a constant such that the time needed to carry out the decomposition and the recombination is $t(n) \leq dn$. By using both your old algorithm and your new idea, you obtain a new algorithm B whose implementation takes time

$$t_B(n) = 3t_A(\lceil n/2 \rceil) + t(n) \leq 3c((n+1)/2)^2 + dn = \tfrac{3}{4}cn^2 + (\tfrac{3}{2}c + d)n + \tfrac{3}{4}c \ .$$

The term $\tfrac{3}{4}cn^2$ dominates the others when n is sufficiently large, which means that algorithm B is essentially 25% faster than algorithm A. Although this improvement is not to be sneezed at, nevertheless, you have not managed to change the order of the time required: algorithm B still takes quadratic time.

To do better than this, we come back to the question posed in the opening paragraph: how should the subinstances be solved? If they are small, it is possible that algorithm A may still be the best way to proceed. However, when the subinstances are sufficiently large, might it not be better to use our new algorithm *recursively*? The idea is analogous to profiting from a bank account that compounds interest payments! We thus obtain a third algorithm C whose implementation runs in time

$$t_C(n) = \begin{cases} t_A(n) & \text{if } n \leq n_0 \\ 3t_C(\lceil n/2 \rceil) + t(n) & \text{otherwise} \end{cases}$$

where n_0 is the *threshold* above which the algorithm is called recursively. This equation, which is similar to the one in Example 2.3.10, gives us a time in the order of $n^{\lg 3}$, which is approximately $n^{1.59}$. The improvement compared to the order of n^2 is therefore quite substantial, and the bigger n is, the more this improvement is worth having. We shall see in the following section how to choose n_0 in practice. Although this choice does not affect the order of the execution time of our algorithm, we are also concerned to make the hidden constant that multiplies $n^{\lg 3}$ as small as possible.

Here then is the general outline of the divide-and-conquer method:

function $DQ(x)$
    { returns a solution to instance x }
    if x is sufficiently small or simple **then return** $ADHOC(x)$
    decompose x into smaller subinstances $x_1, x_2, \cdots, x_k$
    for $i \leftarrow 1$ **to** k **do** $y_i \leftarrow DQ(x_i)$
    recombine the y_i's to obtain a solution y for x
    return y ,

where $ADHOC$, the *basic subalgorithm*, is used to solve small instances of the problem in question.

The number of subinstances, k, is usually both small and also independent of the particular instance to be solved. When $k = 1$, it is hard to justify calling the technique *divide*-and-conquer, and in this case it goes by the name of *simplification* (see sections 4.3, 4.8, and 4.10). We should also mention that some divide-and-conquer algorithms do not follow the preceding outline exactly, but instead, they require that the first subinstance be solved even before the second subinstance is formulated (Section 4.6).

For this approach to be worthwhile a number of conditions are usually required: it must be possible to decompose an instance into subinstances and to recombine the subsolutions fairly efficiently, the decision when to use the basic subalgorithm rather than to make recursive calls must be taken judiciously, and the subinstances should be as far as possible of about the same size.

After looking at the question of how to choose the optimal threshold, this chapter shows how divide-and-conquer is used to solve a variety of important problems and how the resulting algorithms can be analysed. We shall see that it is sometimes possible to replace the recursivity inherent in divide-and-conquer by an iterative loop. When implemented in a conventional language such as Pascal on a conventional machine, an iterative algorithm is likely to be somewhat faster than the recursive version, although only by a constant multiplicative factor. On the other hand, it may be possible to save a substantial amount of memory space in this way: for an instance of size n, the recursive algorithm uses a stack whose depth is often in $\Omega(\log n)$ and in bad cases even in $\Omega(n)$.

4.2 DETERMINING THE THRESHOLD

An algorithm derived by divide-and-conquer must avoid proceeding recursively when the size of the subinstances no longer justifies this. In this case, it is better to apply the basic subalgorithm. To illustrate this, consider once again algorithm C from the previous section, whose execution time is given by

$$t_C(n) = \begin{cases} t_A(n) & \text{if } n \leq n_0 \\ 3t_C(\lceil n/2 \rceil) + t(n) & \text{otherwise}, \end{cases}$$

where $t_A(n)$ is the time required by the basic subalgorithm, and $t(n)$ is the time taken to do the decomposition and recombination. To determine the value of the threshold n_0 that minimizes $t_C(n)$, it is not sufficient to know that $t_A(n) \in \Theta(n^2)$ and that $t(n) \in \Theta(n)$.

For instance, consider an implementation for which the values of $t_A(n)$ and $t(n)$ are given respectively by n^2 and $16n$ milliseconds. Suppose we have an instance of size 1024 to solve. If the algorithm proceeds recursively until it obtains subinstances of size 1, that is, if $n_0 = 1$, it takes more than half an hour to solve this instance. This is ridiculous, since the instance can be solved in little more than a quarter of an hour by using the basic subalgorithm directly, that is, by setting $n_0 = \infty$. Must we conclude that divide-and-conquer allows us to go from a quadratic algorithm to an algorithm whose execution time is in $O(n^{\lg 3})$, but only at the cost of an increase in the hidden constant so enormous that the new algorithm is never economic on instances that can be solved in a reasonable time? Fortunately, the answer is no: in our example, the instance of size 1024 can be solved in less than 8 minutes, provided we choose the threshold n_0 intelligently.

Problem 4.2.1.   Prove that if we set $n_0 = 2^k$ for some given integer $k \geq 0$, then for all $l \geq k$ the implementation considered previously takes

$$2^k 3^{l-k}(32 + 2^k) - 2^{l+5} \text{ milliseconds}$$

to solve an instance of size 2^l.   □

Problem 4.2.2.   Find all the values of the threshold that allow an instance of size 1024 to be solved in less than 8 minutes.   □

This example shows that the choice of threshold can have a considerable influence on the efficiency of a divide-and-conquer algorithm. Choosing the threshold is complicated by the fact that the best value does not generally depend only on the algorithm concerned, but also on the particular implementation. Moreover, the preceding problem shows that, over a certain range, changes in the value of the threshold may have no effect on the efficiency of the algorithm when only instances of some specific size are considered. Finally, there is in general no uniformly best value of the threshold: in our example, a threshold larger than 66 is optimal for instances of size 67, whereas it is best to use a threshold between 33 and 65 for instances of size 66. We shall in future abuse the term "optimal threshold" to mean *nearly* optimal.

So how shall we choose n_0? One easy condition is that we must have $n_0 \geq 1$ to avoid the infinite recursion that results if the solution of an instance of size 1 requires us first to solve a few other instances of the same size. This remark may appear trivial, but Section 4.6 describes an algorithm for which the *ultimate threshold* is less obvious, as Problem 4.6.8 makes clear.

Given a particular implementation, the optimal threshold can be determined empirically. We vary the value of the threshold and the size of the instances used for our tests and time the implementation on a number of cases. Obviously, we must avoid thresholds below the ultimate threshold. It is often possible to estimate an optimal threshold simply by tabulating the results of these tests or by drawing a few diagrams. Problem 4.2.2 makes it clear, however, that it is not usually enough simply to vary the threshold for an instance whose size remains fixed. This approach may require considerable amounts of computer time. We once asked the students in an algorithmics course to implement the algorithm for multiplying large integers given in Section 4.7, in order to compare it with the classic algorithm from Section 1.1. Several groups of students tried to estimate the optimal threshold empirically, each group using in the attempt more than 5,000 (1982) Canadian dollars worth of machine time! On the other hand, a purely theoretical calculation of the optimal threshold is rarely possible, given that it varies from one implementation to another.

The hybrid approach, which we recommend, consists of determining theoretically the form of the recurrence equations, and then finding empirically the values of the constants used in these equations for the implementation at hand. The optimal threshold can then be estimated by finding the value of n at which it makes no difference, for an instance of size n, whether we apply the basic subalgorithm directly or whether we go on for one more level of recursion.

Coming back to our example, the optimal threshold can be found by solving $t_A(n) = 3t_A(\lceil n/2 \rceil) + t(n)$, because $t_C(\lceil n/2 \rceil) = t_A(\lceil n/2 \rceil)$ if $\lceil n/2 \rceil \leq n_0$. The presence of a ceiling in this equation complicates things. If we neglect this difficulty, we obtain $n = 64$. On the other hand, if we systematically replace $\lceil n/2 \rceil$ by $(n+1)/2$, we find $n \approx 70$. There is nothing surprising in this, since we saw in Problem 4.2.2 that in fact no uniformly optimal threshold exists. A reasonable compromise, corresponding to the fact that the average value of $\lceil n/2 \rceil$ is $(2n+1)/4$, is to choose $n_0 = 67$ for our threshold.

* Problem 4.2.3.   Show that this choice of $n_0 = 67$ has the merit of being suboptimal for only two values of n in the neighbourhood of the threshold. Furthermore, prove that there are no instances that take more than 1% longer with threshold 67 than they would with any other threshold.   □

The following problem shows that a threshold of 64 would be optimal were it always possible to decompose an instance of size n into three subinstances exactly of size $n/2$, that is, if instances of fractional size were allowed. (Notice that this would *not* cause an infinite recursion because the threshold is strictly larger than zero.)

Problem 4.2.4.   Let a and b be real positive constants. For each positive real number s, consider the function $f_s : \mathbb{R}^* \to \mathbb{R}^*$ defined by the recurrence

$$f_s(x) = \begin{cases} ax^2 & \text{if } x \le s \\ 3f_s(x/2) + bx & \text{otherwise .} \end{cases}$$

Prove by mathematical induction that if $u = 4b/a$ and if v is an arbitrary positive real number, then $f_u(x) \le f_v(x)$ for every real number x. Notice that this u is chosen so that $au^2 = 3a(u/2)^2 + bu$. (For purists: even if the domain of f_u and f_v is not countable, the problem can be solved without recourse to transfinite induction, precisely because infinite recursion is not a worry.)   □

In practice, one more complication arises. Supposing, for instance, that $t_A(n)$ is quadratic, it may happen that $t_A(n) = an^2 + bn + c$ for some constants a, b, and c depending on the implementation. Although $bn + c$ becomes negligible compared to an^2 when n is large, the basic subalgorithm is used in fact *precisely* on instances of moderate size. It is therefore usually insufficient merely to estimate the constant a. Instead, measure $t_A(n)$ a number of times for several different values of n, and then estimate all the necessary constants, probably using a regression technique.

4.3 BINARY SEARCHING

Binary searching predates computers. In essence, it is the algorithm used to look up a word in a dictionary or a name in a telephone directory. It is probably the simplest application of divide-and-conquer.

Let $T[1 .. n]$ be an array sorted into increasing order; that is, $1 \le i < j \le n \Rightarrow T[i] \le T[j]$, and let x be some item. The problem consists of finding x in the array T if indeed it is there. If the item we are looking for is not in the array, then instead we want to find the position where it might be inserted. Formally, we wish to find the index i such that $0 \le i \le n$ and $T[i] \le x < T[i+1]$, with the logical convention that $T[0] = -\infty$ and $T[n+1] = +\infty$. (By *logical* convention, we mean that these values are not in fact present as sentinels in the array.) The obvious approach to this problem is to look sequentially at each element of T until we either come to the end of the array or find an item bigger than x.

> **function** *sequential* $(T [1 .. n], x)$
>   { sequential search for x in array T }
>   **for** $i \leftarrow 1$ **to** n **do**
>   **if** $T[i] > x$ **then return** $i - 1$
>   **return** n

This algorithm clearly takes a time in $\Theta(1+r)$, where r is the index returned: this is $\Omega(n)$ in the worst case and $O(1)$ in the best case. If we assume that all the elements of T are distinct, that x is indeed somewhere in the array, and that it is to be found with equal probability at each possible position, then the average number of trips round the loop is $(n^2+3n-2)/2n$. On the average, therefore, as well as in the worst case, sequential search takes a time in $\Theta(n)$.

To speed up the search, divide-and-conquer suggests that we should look for x either in the first half of the array or in the second half. To find out which of these searches is appropriate, we compare x to an element in the middle of the array: if $x < T[1+\lfloor n/2\rfloor]$, then the search for x can be confined to $T[1 .. \lfloor n/2\rfloor]$; otherwise it is sufficient to search $T[1+\lfloor n/2\rfloor .. n]$. We obtain the following algorithm.

> **function** *binsearch* $(T [1 .. n], x)$
>   { binary search for x in array T }
>   **if** $n = 0$ **or** $x < T[1]$ **then return** 0
>   **return** *binrec* (T, x)

> **function** *binrec* $(T [i .. j], x)$
>   { binary search for x in subarray $T[i .. j]$;
>   this procedure is only called if $T[i] \le x < T[j+1]$ and $i \le j$ }
>   **if** $i = j$ **then return** i
> $k \leftarrow (i+j+1)$ **div** 2
>   **if** $x < T[k]$ **then return** *binrec* $(T[i .. k-1], x)$
>   **else return** *binrec* $(T[k .. j], x)$

Problem 4.3.1. Prove that the function *binrec* is never called on $T[i .. j]$ with $j < i$. Prove too that when *binrec* $(T[i .. j], x)$ makes a recursive call *binrec* $(T[u .. v], x)$, it is always true that $v - u < j - i$. Conclude from these two results that a call on *binsearch* always terminates. Show finally that the values $T[0]$ and $T[n+1]$ are never used (except in the comments!). □

Problem 4.3.2. Show that the algorithm takes a time in $\Theta(\log n)$ to find x in $T[1 .. n]$ whatever the position of x in T. □

The algorithm in fact executes only one of the two recursive calls, so that technically it is an example of simplification rather than of divide-and-conquer. Because the recursive call is situated dynamically at the very end of the algorithm, it is easy to produce an iterative version.

function *iterbin* $(T\,[1\,..\,n\,],x)$
   { iterative binary search for x in array T }
   if $n = 0$ **or** $x < T\,[1]$ **then return** 0
   $i \leftarrow 1\,;\, j \leftarrow n$
   while $i < j$ **do**
     $\{\,T[i] \le x < T\,[\,j+1]\,\}$
     $k \leftarrow (i+j+1)$ **div** 2
     if $x < T\,[k]$ **then** $j \leftarrow k - 1$
              else $i \leftarrow k$
   return i

Problem 4.3.3.   It is easy to go wrong when programming the concept of binary searching, simple though this is. Show by examples that the preceding algorithm would be incorrect if we replaced

  i. " $k \leftarrow (i+j+1)$ **div** 2 " by " $k \leftarrow (i+j)$ **div** 2 ",
  ii. " $i \leftarrow k$ " by " $i \leftarrow k+1$ ", or
  iii. " $j \leftarrow k-1$ " by " $j \leftarrow k$ ".   □

A first inspection of this algorithm shows what is apparently an inefficiency. Suppose T contains 17 distinct elements and that $x = T\,[13]$. On the first trip round the loop, $i = 1$, $j = 17$, and $k = 9$. The comparison between x and $T\,[9]$ causes the assignment $i \leftarrow 9$ to be executed. On the second trip round the loop $i = 9$, $j = 17$, and $k = 13$. A comparison is made between x and $T\,[13]$. This comparison could allow us to end the search immediately, but no test is made for equality, and so the assignment $i \leftarrow 13$ is carried out. Two more trips round the loop are necessary before we leave with $i = j = 13$. The following algorithm leaves the loop immediately after we find the element we are looking for.

function *iterbin* $2(T\,[1\,..\,n\,],x)$
   { variant on iterative binary search }
   if $n = 0$ **or** $x < T\,[1]$ **then return** 0
   $i \leftarrow 1\,;\, j \leftarrow n$
   while $i < j$ **do**
     $\{T[i] \le x < T\,[\,j+1]\}$
     $k \leftarrow (i+j)$ **div** 2
     case $x < T\,[k]:\ j \leftarrow k - 1$
           $x \ge T\,[k+1]:\ i \leftarrow k + 1$
           otherwise $:\ i\,,j \leftarrow k$
   return i

Which of these algorithms is better? The first systematically makes a number of trips round the loop in $\Theta(\log n)$, regardless of the position of x in T, while it is possible that the variant will only make one or two trips round the loop if x is favourably situated. On the other hand, a trip round the loop in the variant will take a little longer to execute on the average than a trip round the loop in the first algorithm. To compare them, we shall analyse exactly the average number of trips round the loop that each version makes. Suppose to make life simpler that T contains n distinct elements and that x is indeed somewhere in T, occupying each possible position with equal probability. Let $A(n)$ and $B(n)$ be the average number of trips round the loop made by the first and the second iterative versions, respectively.

Analysis of the First Version. Let $k = 1 + \lfloor n/2 \rfloor$. With probability $(k-1)/n$, $x < T[k]$, which causes the assignment $j \leftarrow k-1$, after which the algorithm starts over on an instance reduced to $k-1$ elements. With probability $1 - (k-1)/n$, $x \geq T[k]$, which causes the assignment $i \leftarrow k$, after which the algorithm starts over on an instance reduced to $n-k+1$ elements. One trip round the loop is carried out before the algorithm starts over, so the average number of trips round the loop is given by the recurrence

$$A(n) = 1 + \frac{\lfloor n/2 \rfloor}{n} A(\lfloor n/2 \rfloor) + \frac{\lceil n/2 \rceil}{n} A(\lceil n/2 \rceil), \quad n \geq 2$$

$$A(1) = 0 .$$

Analysis of the Second Version. In a similar way, taking $k = \lceil n/2 \rceil$, we obtain the recurrence

$$B(n) = 1 + \frac{\lceil n/2 \rceil - 1}{n} B(\lceil n/2 \rceil - 1) + \frac{\lfloor n/2 \rfloor}{n} B(\lfloor n/2 \rfloor), \quad n \geq 3$$

$$B(1) = 0, \ B(2) = 1 .$$

Define $a(n)$ and $b(n)$ as $nA(n)$ and $nB(n)$, respectively. The equations then become

$$\left. \begin{array}{l} a(n) = n + a(\lfloor n/2 \rfloor) + a(\lceil n/2 \rceil), \quad n \geq 2 \\ b(n) = n + b(\lceil n/2 \rceil - 1) + b(\lfloor n/2 \rfloor), \quad n \geq 3 \\ a(1) = b(1) = 0, \ b(2) = 2. \end{array} \right\} \qquad (*)$$

The first equation is easy in the case when n is a power of 2, since it then reduces to

$$a(n) = 2a(n/2) + n, \quad n \geq 2$$

$$a(1) = 0 ,$$

which yields $a(n) = n \lg n$ using the techniques of Section 2.3. Exact analysis for arbitrary n is harder. We proceed by constructive induction, guessing the likely form of the answer and determining the missing parameters in the course of a tentative proof by mathematical induction. A likely hypothesis, already shown to hold when n is a power of 2, is that $n \lfloor \lg n \rfloor \leq a(n) \leq n \lceil \lg n \rceil$. What might we add to $n \lfloor \lg n \rfloor$ to arrive at $a(n)$? Let n^* denote the largest power of 2 that is less than or equal to n. In particular, $\lfloor \lg n \rfloor = \lg n^*$. It seems reasonable to hope that

$a(n) = n \lg n^* + cn + dn^* + e \lg n^* + f$ for appropriate constants c, d, e, and f. Denote this hypothesis by $HI(n)$.

When $n > 1$ is not of the form $2^m - 1$, then $(\lfloor n/2 \rfloor)^* = (\lceil n/2 \rceil)^* = n^*/2$. To prove $HI(n)$ in this case, using the recurrence (*), $HI(\lfloor n/2 \rfloor)$ and $HI(\lceil n/2 \rceil)$, it is thus necessary and sufficient that

$$n \lg n^* + cn + dn^* + e \lg n^* + f \;=\; n \lg n^* + cn + dn^* + 2e \lg n^* + (2f - 2e)$$

that is, we need $e = 0$ and $f = 0$, c and d being still unconstrained. If our hypothesis is correct, we therefore know

$$a(n) = n \lg n^* + cn + dn^* .$$

When $n > 1$ is of the form $2^m - 1$, then $\lfloor n/2 \rfloor = (n-1)/2$, $(\lfloor n/2 \rfloor)^* = (n+1)/4$ and $\lceil n/2 \rceil = (\lceil n/2 \rceil)^* = n^* = (n+1)/2$. To prove $HI(n)$ in this case, it is necessary and sufficient that

$$n \lg \tfrac{n+1}{2} + (c + \tfrac{d}{2})n + \tfrac{d}{2} \;=\; n \lg \tfrac{n+1}{2} + (c + \tfrac{3d}{4} + \tfrac{1}{2})n + (\tfrac{3d}{4} + \tfrac{1}{2}) ;$$

that is

$$4c + 2d \;=\; 4c + 3d + 2 \quad \text{and} \quad 2d \;=\; 3d + 2 .$$

These two equations are not linearly independent. They allow us to conclude that $d = -2$, there still being no constraints on c.

At this point we know that if only we can make the hypothesis

$$a(n) = n \lg n^* + cn - 2n^*$$

true for the base $n = 1$, then we shall have proved by mathematical induction that it is true for every positive integer n. Our final constraint is therefore

$$0 = a(1) = c - 2 ,$$

which gives $c = 2$ and implies that the general solution of the recurrence (*) for $a(n)$ is

$$a(n) = n \lg n^* + 2(n - n^*) .$$

The average number of trips round the loop executed by the first iterative algorithm for binary searching, when looking for an element that is in fact present with uniform probability distribution among the n different elements of an array sorted into increasing order, is given by $A(n) = a(n)/n$, that is

$$A(n) = \lfloor \lg n \rfloor + 2(1 - n^*/n) .$$

In particular, our initial guess holds:

$$\lfloor \lg n \rfloor \leq A(n) \leq \lceil \lg n \rceil .$$

Problem 4.3.4.   A seemingly simpler approach to determine the constants would be to substitute the values 1, 2, 3, and 4 for n in the hypothesis $HI(n)$, thus obtaining four linear equations in four unknowns, which can easily be solved to give the same c, d, e, and f. Explain why this is insufficient.   □

Problem 4.3.5.   Using the techniques presented in Section 2.3, solve equation (*) exactly for $b(n)$ when n is of the form $2^m - 1$.   □

The general solution of the recurrence for $b(n)$ is more difficult than the one we have just obtained. It seems reasonable to formulate the same incompletely specified induction hypothesis: $b(n) = n \lg n^* + cn + dn^* + e \lg n^* + f$ for some new constants c, d, e, and f. Constructive induction yields $e = 1$ and $f = 1 + c$ to take account of the case when $n \geq 3$ is not a power of 2. The case when $n \geq 4$ is a power of 2 obliges us to choose $d = -2$. The hypothesis therefore becomes $b(n) = n \lg n^* + cn - 2n^* + \lg n^* + (1 + c)$. Unfortunately, $b(1) = 0 \Rightarrow c = \frac{1}{2}$, whereas $b(2) = 2 \Rightarrow c = \frac{2}{3}$, which is inconsistent and shows that the original hypothesis was wrong. The problem arises because the two basic cases are incompatible in the sense that it is impossible to obtain them both from the recurrence starting from some artificial definition of $b(0)$. Nonetheless, our efforts were not entirely in vain. A simple modification of the argument allows us to conclude that

$$n \lg n^* + n/2 - 2n^* + \lg n^* + 3/2 \leq b(n) \leq n \lg n^* + 2n/3 - 2n^* + \lg n^* + 5/3.$$

Stated more elegantly:

*Problem 4.3.6.   Show that there exists a function $\pi: \mathbb{N}^+ \to \mathbb{N}^+$ such that $(n+1)/3 \leq \pi(n) \leq (n+1)/2$ for every positive integer n, and such that the exact solution of the recurrence is $b(n) = n \lg n^* + n - 2n^* + \lg n^* + 2 - \pi(n)$.   □

*Problem 4.3.7.   Show that the function $\pi(n)$ of the previous exercise is given by

$$\pi(n-1) = \begin{cases} n^*/2 & \text{if } 2n < 3n^* \\ n - n^* & \text{otherwise} \end{cases}$$

for all $n \geq 2$. Equivalently

$$\pi(n-1) = [n^* + (\lfloor 2n / n^* \rfloor - 2)(2n - 3n^*)]/2.$$ □

We are finally in a position to answer the initial question: which of the two algorithms for binary searching is preferable? By combining the preceding analysis of the function $a(n)$ with the solution to Problem 4.3.6, we obtain

$$A(n) - B(n) = 1 + \frac{\pi(n) - \lfloor \lg n \rfloor - 2}{n} < \frac{3}{2}.$$

Thus we see that the first algorithm makes on the average less than one and a half trips round the loop more than the second. Given that the first algorithm takes less time on the average than the variant to execute one trip round the loop, we conclude that the first algorithm is more efficient than the second on the average whenever n is sufficiently large. The situation is similar if the element we are looking for is not in fact in the array. However, the threshold beyond which the first algorithm is preferable to the variant can be very high for some implementations.

4.4 SORTING BY MERGING

Let $T[1 .. n]$ be an array of n elements for which there exists a total ordering. We are interested in the problem of sorting these elements into ascending order. We have already seen that the problem can be solved by selection sorting and insertion sorting (Section 1.4), or by heapsort (Example 2.2.4 and Problem 2.2.3). Recall that an analysis both in the worst case and on the average shows that the latter method takes a time in $\Theta(n \log n)$, whereas both the former methods take quadratic time.

The obvious divide-and-conquer approach to this problem consists of separating the array T into two parts whose sizes are as nearly equal as possible, sorting these parts by recursive calls, and then merging the solutions for each part, being careful to preserve the order. We obtain the following algorithm:

> **procedure** *mergesort* $(T[1 .. n])$
>   { sorts array T into increasing order }
>   **if** n is small **then** *insert* (T)
>   **else arrays** $U[1 .. n \textbf{ div } 2], V[1 .. (n+1) \textbf{ div } 2]$
> $U \leftarrow T[1 .. n \textbf{ div } 2]$
> $V \leftarrow T[1 + (n \textbf{ div } 2) .. n]$
>   *mergesort* (U); *mergesort* (V)
>   *merge* (T, U, V),

where *insert* (T) is the algorithm for sorting by insertion from Section 1.4, and *merge* (T, U, V) merges into a single sorted array T two arrays U and V that are already sorted.

Problem 4.4.1.   Give an algorithm capable of merging two sorted arrays U and V in linear time, that is, in a time in the exact order of the sum of the lengths of U and V.   □

** Problem 4.4.2.   Repeat the previous problem, but without using an auxiliary array: the sections $T[1 .. k]$ and $T[k+1 .. n]$ of an array are sorted independently, and you wish to sort the whole array $T[1 .. n]$. You may only use a fixed number of working variables to solve the problem, and your algorithm must work in linear time. □

This sorting algorithm is a good illustration of all the facets of divide-and-conquer. When the number of elements to be sorted is small, a relatively simple algorithm is used. On the other hand, when this is justified by the number of elements, *mergesort* separates the instance into two subinstances half the size, solves each of these recursively, and then combines the two sorted half-arrays to obtain the solution to the original instance.

Let $t(n)$ be the time taken by this algorithm to sort an array of n elements. Separating T into U and V takes linear time. By the result of Problem 4.4.1, the final merge also takes linear time. Consequently, $t(n) \in t(\lfloor n/2 \rfloor) + t(\lceil n/2 \rceil) + \Theta(n)$. This equation, which we analysed in Section 2.1.6, allows us to conclude that the time required by the algorithm for sorting by merging is in $\Theta(n \log n)$.

Problem 4.4.3.   Rather than separate T into two half-size arrays, we might choose to separate it into three arrays of size $\lfloor n/3 \rfloor$, $\lfloor (n+1)/3 \rfloor$, and $\lfloor (n+2)/3 \rfloor$, respectively, to sort each of these recursively, and then to merge the three sorted arrays. Give a more formal description of this algorithm, and analyse its execution time.   □

** Problem 4.4.4.   Following up the previous problem, we might choose to separate T into about $\lfloor \sqrt{n} \rfloor$ arrays, each containing approximately $\lfloor \sqrt{n} \rfloor$ elements. Develop this idea, and analyse its performance.   □

The merge sorting algorithm we gave, and those suggested by the two previous problems, have two points in common. The fact that the sum of the sizes of the subinstances is equal to the size of the original instance is not typical of algorithms derived using divide-and-conquer, as we shall see in several subsequent examples. On the other hand, the fact that the original instance is divided into subinstances whose sizes are as nearly as possible equal is crucial if we are to arrive at an efficient algorithm. To see why, look at what happens if instead we decide to separate T into an array U with $n-1$ elements and an array V containing only 1 element. Let $t'(n)$ be the time required by this variant to sort n items. We obtain $t'(n) \in t'(n-1) + t'(1) + \Theta(n)$.

Problem 4.4.5.   Show that $t'(n) \in \Theta(n^2)$.   □

Simply forgetting to balance the sizes of the subinstances can therefore be disastrous for the efficiency of an algorithm obtained using divide-and-conquer.

Problem 4.4.6.   This poor sorting algorithm is very like one we have already seen in this book. Which one, and why?   □

4.5 QUICKSORT

The sorting algorithm invented by Hoare, usually known as "quicksort", is also based on the idea of divide-and-conquer. Unlike sorting by merging, the nonrecursive part of the work to be done is spent constructing the subinstances rather than combining their solutions. As a first step, this algorithm chooses one of the items in the array to be sorted as the *pivot*. The array is then partitioned on either side of the pivot: elements are moved in such a way that those greater than the pivot are placed on its right, whereas all the others are moved to its left. If now the two sections of the array on either side of the pivot are sorted independently by recursive calls of the algorithm, the final result is a completely sorted array, no subsequent merge step being necessary. To balance the sizes of the two subinstances to be sorted, we would like to use the median element as the pivot. (For a definition of the median, see Section 4.6.) Unfortunately, finding the median takes more time than it is worth. For this reason we simply use the first element of the array as the pivot. Here is the algorithm.

> **procedure** *quicksort* $(T [i .. j])$
> { sorts array $T [i .. j]$ into increasing order }
> **if** $j - i$ is small **then** *insert* $(T [i .. j])$ { Section 1.4 }
> **else** *pivot* $(T [i .. j], l)$
> > { after pivoting, $i \le k < l \;\Rightarrow\; T [k] \le T [l]$
> > > and $l < k \le j \;\Rightarrow\; T [k] > T [l]$ }
> > *quicksort* $(T [i .. l - 1])$
> > *quicksort* $(T [l + 1 .. j])$

Designing a linear time pivoting algorithm is no challenge. It is, however, crucial in practice that the hidden constant be small. Let $p = T [i]$ be the pivot. One good way of pivoting consists of scanning the array $T [i .. j]$ just once, but starting at both ends. Pointers k and l are initialized to i and $j + 1$, respectively. Pointer k is then incremented until $T [k] > p$, and pointer l is decremented until $T [l] \le p$. Now $T [k]$ and $T [l]$ are interchanged. This process continues as long as $k < l$. Finally, $T [i]$ and $T [l]$ are interchanged to put the pivot in its correct position.

> **procedure** *pivot* $(T [i .. j];$ **var** $l)$
> > { permutes the elements in array $T [i .. j]$ in such a way that, at the end,
> > $i \le l \le j$, the elements of $T [i .. l-1]$ are not greater than p,
> > $T [l] = p$, and the elements of $T [l+1 .. j]$ are greater than p,
> > where p is the initial value of $T [i]$ }
> > $p \leftarrow T [i]$
> > $k \leftarrow i ; l \leftarrow j + 1$
> > **repeat** $k \leftarrow k + 1$ **until** $T [k] > p$ **or** $k \ge j$
> > **repeat** $l \leftarrow l - 1$ **until** $T [l] \le p$
> > **while** $k < l$ **do**
> > > interchange $T [k]$ and $T [l]$
> > > **repeat** $k \leftarrow k + 1$ **until** $T [k] > p$
> > > **repeat** $l \leftarrow l - 1$ **until** $T [l] \le p$
> > interchange $T [i]$ and $T [l]$

Problem 4.5.1.   Invent several examples representing different situations that might arise, and simulate the pivoting algorithm on these examples.   □

Quicksort is inefficient if it happens systematically on most recursive calls that the subinstances $T [i .. l - 1]$ and $T [l + 1 .. j]$ to be sorted are severely unbalanced.

Problem 4.5.2.   Show that in the worst case quicksort requires quadratic time. Give an explicit example of an array to be sorted that causes such behaviour.   □

On the other hand, if the array to be sorted is initially in random order, it is likely that most of the time the subinstances to be sorted will be sufficiently well balanced. To determine the average time required by quicksort to sort an array of n items, we assume that all the elements of T are distinct and that each of the $n!$ possible initial permutations of the elements has the same probability of occurring. Let $t(m)$ be the average time taken by a call on *quicksort* $(T[a+1 .. a+m])$ for $0 \le m \le n$ and $0 \le a \le n-m$. The pivot chosen by the algorithm is situated with equal probability in any position with respect to the other elements of T. The value of l returned by the pivoting algorithm after the initial call *pivot* $(T[1 .. n], l)$ can therefore be any integer between 1 and n, each value having probability $1/n$. This pivoting operation takes a time in $\Theta(n)$. It remains to sort recursively two subarrays of size $l-1$ and $n-l$, respectively. The average time required to execute these recursive calls is $t(l-1) + t(n-l)$. Consequently,

$$t(n) \in \Theta(n) + \frac{1}{n} \sum_{l=1}^{n} (t(l-1) + t(n-l)) .$$

A little manipulation yields

$$t(n) \in \Theta(n) + \frac{2}{n} \sum_{k=0}^{n-1} t(k) .$$

To make this more explicit, let d and n_0 be two constants such that

$$t(n) \le dn + \frac{2}{n} \sum_{k=0}^{n-1} t(k) \quad \text{for } n > n_0 .$$

An equation of this type is more difficult to analyse than the linear recurrences we saw in Section 2.3. By analogy with sorting by merging, it is, nevertheless, reasonable to hope that $t(n)$ will be in $O(n \log n)$ and to apply constructive induction to look for a constant c such that $t(n) \le cn \lg n$.

To use this approach we need an upper bound on $\sum_{i=n_0+1}^{n-1} i \lg i$. This is obtained with the help of a simple lemma. (We suggest you find a graphical interpretation of the lemma.) Let a and b be real numbers, $a < b$, and let $f : [a, b] \to \mathbb{R}$ be a nondecreasing function. Let j and k be two integers such that $a \le j < k \le b$. Then

$$\sum_{i=j}^{k-1} f(i) \le \int_{x=j}^{k} f(x) \, dx .$$

In particular, taking $f(x) = x \lg x$, $j = n_0$ and $k = n$, we obtain

$$\sum_{i=n_0+1}^{n-1} i \lg i \le \int_{x=n_0+1}^{n} x \lg x \, dx$$

$$= \left[\frac{x^2}{2} \lg x - \frac{\lg e}{4} x^2 \right]_{x = n_0 + 1}^{n}$$

$$\le \frac{n^2}{2} \lg n - \frac{\lg e}{4} n^2$$

provided $n_0 \ge 1$.

Problem 4.5.3.   Complete the proof by mathematical induction that $t(n) \le cn \lg n$ for all $n > n_0 \ge 1$, where

$$c = \frac{2d}{\lg e} + \frac{4}{(n_0 + 1)^2 \lg e} \sum_{k=0}^{n_0} t(k) .$$ $\square$

Quicksort can therefore sort an array of n distinct elements in an average time in $O(n \log n)$. The hidden constant is in practice smaller than those involved in heapsort or in merge sort. If an occasional long execution time can be tolerated, this is the sorting algorithm to be preferred among all those presented in this book. The probability of suffering an execution time in $\Omega(n^2)$ can be greatly diminished, at the price of a small increase in the hidden constant, by choosing as pivot the median of $T[i]$, $T[(i+j) \textbf{ div } 2]$ and $T[j]$.

Problem 4.5.4.   Show by a simple argument that, whatever the choice of pivot, quicksort as described here always takes a time in $\Omega(n^2)$ in the worst case. Outline a modification to the algorithm to avoid this. $\square$

By combining the modification hinted at in the previous problem with the linear algorithm from the following section, we can obtain a version of quicksort that takes a time in $O(n \log n)$ even in the worst case. We mention this possibility only to point out that it should be shunned: the hidden constant associated with the "improved" version of quicksort is so large that it results in an algorithm worse than heapsort in every case.

4.6 SELECTION AND THE MEDIAN

Let $T[1 .. n]$ be an array of integers. What could be easier than to find the smallest element or to calculate the mean of all the elements? However, it is not obvious that the median can be found so easily. Intuitively, the median of T is that element m in T such that there are as many items in T smaller than m as there are items larger than m. The formal definition takes care of the possibility that n may be even, or that the elements of T may not all be distinct. Thus we define m to be the *median* of T if and only if m is in T and

$$\#\{ i \in [1 .. n] \mid T[i] < m \} < n/2 \text{ and } \#\{ i \in [1 .. n] \mid T[i] \le m \} \ge n/2 .$$

The naive algorithm for determining the median of T consists of sorting the array into ascending order and then extracting the $\lceil n/2 \rceil$th entry. If we use heapsort or merge sort, this algorithm takes a time in $O(n \log n)$ to determine the median of n elements. Can we do better? To answer this question, we consider a more general problem: *selection*. Let T be an array of n elements, and let k be an integer between 1 and n. The kth smallest element of T is that element m such that $\#\{i \in [1..n] \mid T[i] < m\} < k$, whereas $\#\{i \in [1..n] \mid T[i] \le m\} \ge k$. In other words, it is the kth item in T if the array is sorted into ascending order. For instance, the median of T is its $\lceil n/2 \rceil$th smallest element.

The following algorithm, which is not yet completely specified, solves the selection problem in a way suggested by quicksort.

> **function** *selection*$(T[1..n], k)$
>   { finds the kth smallest element of T;
>   this algorithm assumes that $1 \le k \le n$ }
>   **if** n is small **then** sort T
>   **return** $T[k]$
> $p \leftarrow$ some element of $T[1..n]$ { to be specified later }
> $u \leftarrow \#\{i \in [1..n] \mid T[i] < p\}$
> $v \leftarrow \#\{i \in [1..n] \mid T[i] \le p\}$
>   **if** $k \le u$ **then**
>   **array** $U[1..u]$
> $U \leftarrow$ the elements of T smaller than p
>   { the kth smallest element of T is
>   also the kth smallest element of U }
>   **return** *selection*(U, k)
>   **if** $k \le v$ **then** { got it! } **return** p
>   **otherwise** { $k > v$ }
>   **array** $V[1..n-v]$
> $V \leftarrow$ the elements of T larger than p
>   { the kth smallest element of T is
>   also the $(k-v)$th smallest of V }
>   **return** *selection*$(V, k-v)$

Problem 4.6.1. Generalize the notion of pivoting from Section 4.5 to partition the array T into *three* sections, $T[1..i-1]$, $T[i..j]$, and $T[j+1..n]$, containing the elements of T that are smaller than p, equal to p, and greater than p, respectively. The values i and j should be returned by the pivoting procedure, not calculated beforehand. Your algorithm should scan T once only, and no auxiliary arrays should be used. □

Problem 4.6.2. Using ideas from the iterative version of binary searching seen in Section 4.3 and the pivoting procedure of the previous problem, give a nonrecursive version of the selection algorithm. Do not use any auxiliary arrays. Your algorithm is allowed to alter the initial order of the elements of T. □

Which element of T should we use as the pivot p? The natural choice is surely the median of T, so that the sizes of the arrays U and V will be as similar as possible (even if at most one of these two arrays will actually be used in a recursive call).

Problem 4.6.3.   What happens to the selection algorithm if the choice of p is made using "$p \leftarrow selection(T, (n+1) \text{ div } 2)$"?   ☐

Suppose first that the median can be obtained by magic at unit cost. For the time being, therefore, the algorithm works by simplification, not by divide-and-conquer. To analyse the efficiency of the selection algorithm, notice first that, by definition of the median, $u < \lceil n/2 \rceil$ and $v \geq \lceil n/2 \rceil$. Consequently, $n - v \leq \lfloor n/2 \rfloor$. If there is a recursive call, the arrays U and V therefore contain a maximum of $\lfloor n/2 \rfloor$ elements. The remaining operations, still supposing that the median can be obtained magically, take a time in $O(n)$. Let $t_m(n)$ be the time required by this method in the worst case to find the kth smallest element of an array of at most n elements, independently of the value of k. We have $t_m(n) \in O(n) + \max\{ t_m(i) \mid i \leq \lfloor n/2 \rfloor \}$.

Problem 4.6.4.   Show that $t_m(n)$ is in $O(n)$.   ☐

Thus we have an algorithm capable of finding the kth smallest element of an array in a time linear in the size of the array. But what shall we do if there is no magic way of getting the median? If we are willing to sacrifice speed in the worst case in order to obtain an algorithm reasonably fast on the average, we can once again borrow an idea from quicksort and choose simply

$p \leftarrow T[1]$.

When we do this, we hope that on the average the sizes of the arrays U and V will not be too unbalanced, even if occasionally we meet an instance where $u = 0$, $v = 1$, and $k > 1$, which causes a recursive call on $n - 1$ elements.

*Problem 4.6.5.   In this problem, consider the selection algorithm obtained by choosing "$p \leftarrow T[1]$". Assume the n elements of the array T are distinct and that each of the $n!$ permutations of the elements has the same probability of occurring. Let $E(n, k)$ stand for the expected size of the subarray involved in the first recursive call produced by a call on $selection(T[1 .. n], k)$, taking the size to be zero if there is no recursive call. Prove that

$$E(n, k) < \frac{n}{2} + \frac{k(n-k)}{n} \leq \frac{3n}{4}.$$

Assuming that the pivoting algorithm preserves random ordering in the subarrays it produces for the recursive calls, prove that this selection algorithm takes linear time on the average, whatever the value of k. (The hidden constant must not depend on k.) Note that the technique hinted at in this exercise *only* applies because the average time turns out to be linear: the average time taken on several instances is not otherwise equal to the time taken on an average instance.   ☐

Problem 4.6.6. Show, however, that in the worst case this algorithm requires quadratic time. ☐

This quadratic worst case can be avoided without sacrificing linear behaviour on the average: the idea is to find quickly a good approximation to the median. This can be done with a little cunning. Assuming $n \geq 5$, consider the following algorithm:

function *pseudomed* $(T[1 .. n])$
    { finds an approximation to the median of array T }
    $s \leftarrow n$ **div** 5
    array $S[1 .. s]$
    for $i \leftarrow 1$ **to** s **do** $S[i] \leftarrow adhocmed\,5(T[5i-4 .. 5i])$
    return $selection(S, (s+1)$ **div** $2)$,

where *adhocmed* 5 is an algorithm specially designed to find the median of exactly five elements. Note that the time taken by *adhocmed* 5 is bounded above by a constant.

We look first at the value of the approximation to the median found by the algorithm *pseudomed*. Let m be this approximation. Since m is the exact median of the array S, we have

$$\#\{ i \in [1 .. s] \mid S[i] \leq m \} \geq \lceil s/2 \rceil .$$

But each element of S is the median of five elements of T. Consequently, for every i such that $S[i] \leq m$, there are three i_1, i_2, i_3 between $5i-4$ and $5i$ such that $T[i_1] \leq T[i_2] \leq T[i_3] = S[i] \leq m$. Therefore

$$\#\{ i \in [1 .. n] \mid T[i] \leq m \} \geq 3\lceil s/2 \rceil = 3\lceil \lfloor n/5 \rfloor /2 \rceil \geq (3n-12)/10 .$$

Problem 4.6.7. Show similarly that

$$\#\{ i \in [1 .. n] \mid T[i] < m \} \leq (7n-3)/10.$$ ☐

The conclusion is that although m is perhaps not the exact median of T, yet its rank is approximately between $3n/10$ and $7n/10$. To visualize how these factors arise, although nothing in the execution of the algorithm *pseudomed* really corresponds to this illustration, imagine that all the elements of T are arranged in five rows, with the possible exception of one to four elements left aside (Figure 4.6.1). Now suppose that the middle row is sorted by magic, as is each of the $\lfloor n/5 \rfloor$ columns, the smallest elements going to the left and to the top, respectively. The middle row corresponds to the array S in the algorithm. Similarly, the element in the circle corresponds to the median of this array, that is, to the value of m returned by the algorithm. By the transitivity of "$\leq$", each of the elements in the box is less than or equal to m. Notice that the box contains approximately three-fifths of one-half of the elements of T, that is, about $3n/10$ elements.

We now look at the efficiency of the selection algorithm given at the beginning of this section when we use

$$p \leftarrow pseudomed(T) .$$

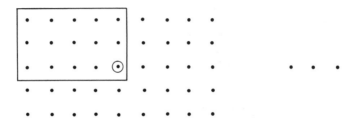

Figure 4.6.1.   Visualization of the pseudomedian.

Let n be the number of elements in T, and let $t(n)$ be the time required in the worst case by this algorithm to find the kth smallest element of T, still independently of the value of k. At the first step, calculating *pseudomed* (T) takes a time in $O(n) + t(\lfloor n/5 \rfloor)$, because the array S can be constructed in linear time. Calculating u and v also takes linear time. Problem 4.6.7 and the preceding discussion show that $u \leq (7n - 3)/10$ and $v \geq (3n - 12)/10$, so $n - v \leq (7n + 12)/10$. The recursive call that may follow therefore takes a time bounded above by

$$\max\{ t(i) \mid i \leq (7n + 12)/10 \} .$$

The initial preparation of the arrays U and V takes linear time. Hence, there exists a constant c such that

$$t(n) \leq t(\lfloor n/5 \rfloor) + \max\{ t(i) \mid i \leq (7n + 12)/10 \} + cn$$

for every sufficiently large n.

Problem 4.6.8.   We have to ensure that $n > 4$ in the preceding equation. What is it in the algorithm that makes this restriction necessary?   □

This equation looks quite complicated. First let us solve a more general, yet simpler problem of the same type.

*Problem 4.6.9.   Let p and q be two positive real constants such that $p + q < 1$, let n_0 be a positive integer, and let b be some positive real constant. Let $f : \mathbb{N} \rightarrow \mathbb{R}^*$ be any function such that

$$f(n) = f(\lfloor pn \rfloor) + f(\lfloor qn \rfloor) + bn$$

for every $n > n_0$. Use constructive induction to prove that $f(n) \in \Theta(n)$.   □

Problem 4.6.10.   Let $t(n)$ be the time required in the worst case to find the kth smallest element in an array of n elements using the selection algorithm discussed earlier. Give explicitly a nondecreasing function $f(n)$ defined as in Problem 4.6.9 (with $p = 1/5$ and $q = 3/4$) such that $t(n) \leq f(n)$ for every integer n. Conclude that $t(n) \in O(n)$. Argue that $t(n) \in \Omega(n)$, and thus $t(n) \in \Theta(n)$.   □

In particular, it is possible to find the median of an array of n elements in linear time. The version of the algorithm suggested by Problem 4.6.2 is preferable in practice, even though this does not constitute an iterative algorithm: it avoids calculating u and v beforehand and using two auxiliary arrays U and V. To use still less auxiliary space, we can also construct the array S (needed to calculate the pseudomedian) by exchanging elements inside the array T itself.

4.7 ARITHMETIC WITH LARGE INTEGERS

In most of the preceding analyses we have taken it for granted that addition and multiplication are elementary operations, that is, the time required to execute these operations is bounded above by a constant that depends only on the speed of the circuits in the computer being used. This is only reasonable if the size of the operands is such that they can be handled directly by the hardware. For some applications we have to consider very large integers. Representing these numbers in floating-point is not useful unless we are concerned solely with the order of magnitude and a few of the most significant figures of our results. If results have to be calculated exactly and all the figures count, we are obliged to implement the arithmetic operations in software.

This was necessary, for instance, when the Japanese calculated the first 134 million digits of π in early 1987. (At the very least, this feat constitutes an excellent aerobic exercice for the computer!) The algorithm developed in this section is not, alas, sufficiently efficient to be used with such operands (see Chapter 9 for more on this). From a more practical point of view, large integers are of crucial importance in cryptology (Section 4.8).

Problem 4.7.1.   Design a good data structure for representing large integers on a computer. Your representation should use a number of bits in $O(n)$ for an integer that can be expressed in n decimal digits. It must also allow negative numbers to be represented, and it must be possible to carry out in linear time multiplications and integer divisions by positive powers of 10 (or another base if you prefer), as well as additions and subtractions.   □

Problem 4.7.2.   Give an algorithm able to add an integer with m digits and an integer with n digits in a time in $\Theta(m+n)$.   □

Problem 4.7.3.   Implement your solution to Problem 4.7.2 on a computer using the representation you invented for Problem 4.7.1. Also implement the classic algorithm for multiplying large integers (see Section 1.1).   □

Although an elementary multiplication takes scarcely more time than an addition on most computers, this is no longer true when the operands involved are very large. Your solution to Problem 4.7.2 shows how to add two integers in linear time. On the other hand, the classic algorithm and multiplication *à la russe* both take quadratic time to multiply these same operands. Can we do better? Let u and v be two integers of n decimal digits to be multiplied. Divide-and-conquer suggests that we should separate each of these operands into two parts of as near the same size as possible: $u = 10^s w + x$ and $v = 10^s y + z$, where $0 \le x < 10^s$, $0 \le z < 10^s$, and $s = \lfloor n/2 \rfloor$. The integers w and y therefore both have $\lceil n/2 \rceil$ digits. See Figure 4.7.1. (For convenience, we say that an integer has j digits if it is smaller than 10^j, even if it is not greater than or equal to 10^{j-1}.)

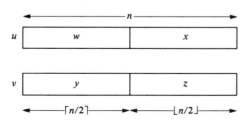

Figure 4.7.1. Splitting the operands for large integer multiplication.

The product that interests us is

$$uv = 10^{2s} wy + 10^s (wz + xy) + xz \ .$$

We obtain the following algorithm.

function $mult (u, v : large\text{-}integers) : large\text{-}integer$
    $n \leftarrow$ smallest integer so that u and v are of size n
    if n is small **then** multiply u by v using the classic algorithm
                    return the product thus computed
    $s \leftarrow n$ **div** 2
    $w \leftarrow u$ **div** 10^s ; $x \leftarrow u$ **mod** 10^s
    $y \leftarrow v$ **div** 10^s ; $z \leftarrow v$ **mod** 10^s
    return $mult (w, y) \times 10^{2s}$
            $+ (mult (w, z) + mult (x, y)) \times 10^s$
            $+ mult (x, z)$

Let $t_b(n)$ be the time required by this algorithm in the worst case to multiply two n digit integers. If we use the representation suggested in Problem 4.7.1 and the algorithms of Problem 4.7.2, the integer divisions and multiplications by 10^{2s} and 10^s, as well as the additions, are executed in linear time. The same is true of the modulo operations, since these are equivalent to an integer division, a multiplication, and a subtraction. The last statement of the algorithm consists of four recursive calls, each of which serves to multiply two integers whose size is about $n/2$. Thus $t_b(n) \in 3t_b(\lceil n/2 \rceil) + t_b(\lfloor n/2 \rfloor) + \Theta(n)$.   This equation becomes $t_b(n) \in 4t_b(n/2) + \Theta(n)$ when n is a power of 2. By Example 2.3.7 the time taken by the preceding algorithm is therefore quadratic, so we have not made any improvement compared to the classic algorithm. In fact, we have only managed to increase the hidden constant!

The trick that allows us to speed things up consists of calculating wy, $wz + xy$, and xz by executing less than four half-length multiplications, even if this means that we have to do more additions. This is sensible because addition is much faster than multiplication when the operands are large. Consider the product

$$r = (w + x)(y + z) = wy + (wz + xy) + xz \ .$$

After only one multiplication this includes the three terms we need in order to calculate uv. Two other multiplications are needed to isolate these terms. This suggests we should replace the last statement of the algorithm by

$$r \leftarrow mult(w + x, y + z)$$
$$p \leftarrow mult(w, y); q \leftarrow mult(x, z)$$
$$\textbf{return } 10^{2s} p + 10^{s}(r - p - q) + q \ .$$

Let $t(n)$ be the time required by the modified algorithm to multiply two integers of size at most n. Taking account of the fact that $w + x$ and $y + z$ may have up to $1 + \lceil n/2 \rceil$ digits, we find that there exist constants $c \in \mathbb{R}^{+}$ and $n_0 \in \mathbb{N}$ such that

$$t(n) \le t(\lfloor n/2 \rfloor) + t(\lceil n/2 \rceil) + t(1 + \lceil n/2 \rceil) + cn$$

for every $n \ge n_0$. This resembles the equation $t_C(n) \in 3t_C(\lceil n/2 \rceil) + O(n)$, which we met at the beginning of the chapter. Using Problem 2.3.11 (notice that $t(n)$ is nondecreasing by definition), we conclude that $t(n) \in O(n^{\lg 3})$.

It is thus possible to multiply two n digit integers in a time in $O(n^{\lg 3})$, which is in $O(n^{1.59})$. However, the hidden constants are such that this algorithm only becomes interesting in practice when n is quite large. A good implementation will probably not use base 10, but rather, the largest base for which the hardware allows two "digits" to be multiplied directly. Recall that the performance of this algorithm and of the classic algorithm are compared empirically at the end of Section 1.7.2.

Example 4.7.1. We require to multiply $u = 2,345$ and $v = 6,789$. The initial decomposition of the operands gives $n = 4$, $s = 2$, $w = 23$, $x = 45$, $y = 67$, and $z = 89$. We obtain successively $p = 23 \times 67 = 1541$, $q = 45 \times 89 = 4,005$, and $r = (23 + 45)(67 + 89) = 68 \times 156 = 10,608$. Finally, the required product uv is obtained by calculating

$$1,541 \times 10^4 + (10,608 - 1,541 - 4,005) \times 10^2 + 4,005$$

$$= 15,410,000 + 506,200 + 405 = 15,920,205 \ .$$

Of course, our example is so small that it would be quicker to use the classic multiplication algorithm in this case.   □

Problem 4.7.4. Let u and v be integers of exactly m and n digits, respectively. Suppose without loss of generality that $m \le n$. The classic algorithm obtains the product of u by v in a time in $O(mn)$. Even when $m \ne n$, the algorithm we have just seen can multiply u and v, since it will simply treat u as though it too were of size n. This gives an execution time in $O(n^{\lg 3})$, which is unacceptable if m is very much smaller than n. Show that it is possible in this case to multiply u and v in a time in $O(nm^{\lg(3/2)})$. □

** Problem 4.7.5.   Rework Problem 2.2.11 (analysis of the algorithm *fib*3) in the context of our new multiplication algorithm, and compare your answer once again to Example 2.2.8. □

* Problem 4.7.6.   Following up Problem 4.7.3, implement on your machine the algorithm we have just discussed. Your algorithm must allow for the possibility that the operands may differ in size. Compare empirically this implementation and the one you made for Problem 4.7.3. □

* Problem 4.7.7.   Show that it is possible to separate each of the operands to be multiplied into three parts rather than two, so as to obtain the required product using five multiplications of integers about one-third as long (not nine as would seem necessary at first sight). Analyse the efficiency of the algorithm suggested by this idea. (*Remark*: Integer division of an n digit integer by an arbitrary constant k can be carried out in a time in $\Theta(n)$, although the value of the hidden constant may depend on k.)   □

** Problem 4.7.8.   Generalize the algorithm suggested by Problem 4.7.7. by showing that there exists, for every real number $\alpha > 1$, an algorithm A_α that can multiply two n digit integers in a time in the order of n^α. □

Problem 4.7.9.   Show by a simple argument that the preceding problem is impossible if we insist that the algorithm A_α must take a time in the *exact* order of n^α. □

Problem 4.7.10.   Following up Problem 4.7.8, consider the following algorithm for multiplying large integers.

> **function** *supermul* $(u, v : large\text{-}integers)$: *large-integer*
> $n \leftarrow$ smallest integer so that u and v are of size n
>   **if** n is small **then** multiply u by v using the classic algorithm
>   **else** $\alpha \leftarrow 1 + (\lg\lg n) / \lg n$
>   multiply u and v using algorithm A_α
>   **return** the product thus computed

At first glance this algorithm seems to multiply two n digit numbers in a time in the order of n^α, where $\alpha = 1 + (\lg\lg n)/\lg n$, that is, in a time in $O(n \log n)$. Find two fundamental errors in this analysis of *supermul*. □

Although the idea tried in Problem 4.7.10 does not work, it is nevertheless possible to multiply two n digit integers in a time in $O(n \log n \log\log n)$ by separating each operand to be multiplied into about $\sqrt{n}$ parts of about the same size and using Fast Fourier Transforms (Section 9.5).

Multiplication is not the only interesting operation involving large integers. Integer division, modulo operations, and the calculation of the integer part of a square root can all be carried out in a time whose order is the same as that required for multiplication (Section 10.2.3). Some other important operations, such as calculating the greatest common divisor, may well be inherently harder to compute; they are not treated here.

4.8 EXPONENTIATION: AN INTRODUCTION TO CRYPTOLOGY

Alice and Bob do not initially share any common secret information. For some reason they wish to establish such a secret. Their problem is complicated by the fact that the only way they can communicate is by using a telephone, which they suspect is being tapped by Eve, a malevolent third party. They do not want Eve to be privy to their newly exchanged secret. To simplify the problem, we assume that, although Eve can overhear conversations, she can neither add nor modify messages on the communications line.

** Problem 4.8.1. Find a protocol by which Alice and Bob can attain their ends. (If you wish to think about the problem, delay reading the rest of this section!)　　□

A first solution to this problem was given in 1976 by Diffie and Hellman. Several other protocols have been proposed since. As a first step, Alice and Bob agree openly on some integer p with a few hundred decimal digits, and on some other integer g between 2 and $p-1$. The security of the secret they intend to establish is not compromised should Eve learn these two numbers.

At the second step Alice and Bob choose randomly and independently of each other two positive integers A and B less than p. Next Alice computes $a = g^A \bmod p$ and transmits this result to Bob; similarly, Bob sends Alice the value $b = g^B \bmod p$. Finally, Alice computes $x = b^A \bmod p$ and Bob calculates $y = a^B \bmod p$. Now $x = y$ since both are equal to $g^{AB} \bmod p$. This value is therefore a piece of information shared by Alice and Bob. Clearly, neither of them can control directly what this value will be. They cannot therefore use this protocol to exchange directly a message chosen beforehand by one or the other of them. Nevertheless, the secret value exchanged can now be used as the key in a conventional cryptographic system.

* Problem 4.8.2. Let p be an odd prime number. The *cyclic multiplicative group* $\mathbf{Z}_p^*$ is defined as $\{ x \in \mathbb{N} \mid 1 \le x < p \}$ under multiplication modulo p. An integer g in this group is called a *generator* if each member of the group can be obtained as some integral power of g. Such a generator always exists. Clearly, the condition that g be a generator is necessary if Alice and Bob require that the secret exchanged by the protocol could take on any value between 1 and $p - 1$. Prove, however, that regardless of the choice of p and g some secrets are more likely to be chosen than others, even if A and B are chosen randomly with uniform probability between 1 and $p - 1$.   □

At the end of the exchange Eve has been able to obtain directly the values of p, g, a, and b only. One way for her to deduce x would be to find an integer A' such that $a = g^{A'} \bmod p$, and then to proceed like Alice to calculate $x' = b^{A'} \bmod p$. If p is an odd prime, g a generator of $\mathbf{Z}_p^*$, and $1 \leq A' < p$, then A' is necessarily equal to A, and so $x' = x$ and the secret is correctly computed by Eve in this case.

Problem 4.8.3.   Show that even if $A \neq A'$, still $b^A \bmod p = b^{A'} \bmod p$ provided that $g^A \bmod p = g^{A'} \bmod p$ and that there exists a B such that $b = g^B \bmod p$. The value x' calculated by Eve in this way is therefore always equal to the value x shared by Alice and Bob.   □

Calculating A' from p, g and a is called the problem of the *discrete logarithm*. There exists an obvious algorithm to solve it. (If the logarithm does not exist, the algorithm returns the value p. For instance, there is no integer A such that $3 = 2^A \bmod 7$.)

```
function dlog (g,a,p)
    A ← 0; x ← 1
    repeat
        A ← A + 1
        x ← xg
    until (a = x mod p) or (A = p)
    return A
```

This algorithm takes an unacceptable amount of time, since it makes $p/2$ trips round the loop on the average when the conditions of Problem 4.8.2 hold. If each trip round the loop takes 1 microsecond, this average time is more than the age of Earth even if p only has two dozen decimal digits. Although there exist other more efficient algorithms for calculating discrete logarithms, none of them is able to solve a randomly chosen instance in a reasonable amount of time when p is a prime with several hundred decimal digits. Furthermore, there is no known way of recovering x from p, g, a, and b that does not involve calculating a discrete logarithm. For the time being, it seems therefore that this method of providing Alice and Bob with a shared secret is sound, although no one has yet been able to prove this.

An attentive reader may wonder whether we are pulling his (or her) leg. If Eve needs to be able to calculate discrete logarithms efficiently to discover the secret shared by Alice and Bob, it is equally true that Alice and Bob must be able to calculate efficiently exponentiations of the form $a = g^A \bmod p$. The obvious algorithm for this is no more subtle or efficient than the one for discrete logarithms.

function $dexpo1(g,A,p)$
  $a \leftarrow 1$
  for $i \leftarrow 1$ **to** A **do** $a \leftarrow ag$
  return a **mod** p

The fact that $x y z \bmod p = ((x y \bmod p) \times z) \bmod p$ for every x, y, z, and p allows us to avoid accumulation of extremely large integers in the loop. (The same improvement can be made in *dlog*, which is necessary if we hope to execute each trip round the loop in 1 microsecond.)

function $dexpo2(g,A,p)$
  $a \leftarrow 1$
  for $i \leftarrow 1$ **to** A **do** $a \leftarrow ag$ **mod** p
  return a

*Problem 4.8.4.   Analyse and compare the execution times of *dexpo1* and *dexpo2* as a function of the value of A and of the size of p. For simplicity, suppose that g is approximately equal to $p/2$. Use the classic algorithm for multiplying large integers. Repeat the problem using the divide-and-conquer algorithm from Section 4.7 for the multiplications. In both cases, assume that calculating a modulo takes a time in the exact order of that required for multiplication. □

Happily for Alice and Bob, there exists a more efficient algorithm for computing the exponentiation. An example will make the basic idea clear.

$$x^{25} = (((x^2 x)^2)^2)^2 x$$

Thus x^{25} can be obtained with just two multiplications and four squarings. We leave the reader to work out the connection between 25 and the sequence of bits 11001 obtained from the expression $(((x^2 x)^2 \times 1)^2 \times 1)^2 x$ by replacing every x by a 1 and every 1 by a 0.

The preceding formula for x^{25} arises because $x^{25} = x^{24} x$, $x^{24} = (x^{12})^2$, and so on. This idea can be generalized to obtain a divide-and-conquer algorithm.

function $dexpo(g,A,p)$
  if $A = 0$ **then return** 1
  if A is odd **then** $a \leftarrow dexpo(g,A-1,p)$
              return $(ag \bmod p)$
        else $a \leftarrow dexpo(g,A/2,p)$
              return $(a^2 \bmod p)$

Let $h(A)$ be the number of multiplications modulo p carried out when we calculate $dexpo(g,A,p)$, including the squarings. These operations dominate the execution time of the algorithm, which consequently takes a time in $O(h(A) \times M(p))$, where $M(p)$ is an upper bound on the time required to multiply two positive integers less than p and to reduce the result modulo p. By inspection of the algorithm we find

$$h(A) = \begin{cases} 0 & \text{if } A = 0 \\ 1 + h(A-1) & \text{if } A \text{ is odd} \\ 1 + h(A/2) & \text{otherwise .} \end{cases}$$

Problem 4.8.5.   Find an explicit formula for $h(A)$ and prove your answer by mathematical induction. (Do not try to use characteristic equations.)   □

Without answering Problem 4.8.5, let us just say that $h(A)$ is situated between once and twice the length of the binary representation of A, provided $A \geq 1$. This means that Alice and Bob can use numbers p, A and B of 200 decimal digits each and still finish the protocol after less than 3,000 multiplications of 200-digit numbers and 3,000 computations of a 400-digit number modulo a 200-digit number, which is entirely reasonable. More generally, the computation of $dexpo(g, A, P)$ takes a time in $O(M(p) \times \log A)$. As was the case for binary searching, the algorithm $dexpo$ only requires one recursive call on a smaller instance. It is therefore an example of simplification rather than of divide-and-conquer. This recursive call is not at the dynamic end of the algorithm, which makes it harder to find an iterative version. Nonetheless, there exists a similar iterative algorithm, which corresponds intuitively to calculating x^{25} as $x^{16} x^8 x^1$.

```
function dexpoiter (g, A, p)
  n ← A ; y ← g ; a ← 1
  while n > 0 do
    if n is odd then a ← ay mod p
    y ← y² mod p
    n ← n div 2
  return a
```

Problem 4.8.6.   The algorithms $dexpo$ and $dexpoiter$ do not minimize the number of multiplications (including squarings) required. For example, $dexpo$ calculates x^{15} as $(((1x)^2 x)^2 x)^2 x$, that is, with seven multiplications. On the other hand, $dexpoiter$ calculates x^{15} as $1 x x^2 x^4 x^8$, which involves eight multiplications (the last being a useless computation of x^{16}). In both cases the number of multiplications can easily be reduced to six by avoiding pointless multiplications by the constant 1 and the last squaring carried out by $dexpoiter$. Show that in fact x^{15} can be calculated with only five multiplications.   □

*Problem 4.8.7.   By suppressing all the reductions modulo p in the preceding algorithms, we obtain algorithms for handling large integers capable of calculating efficiently all the digits of g^A, for an arbitrary base g and an arbitrary exponent A. The efficiency of these algorithms depends in turn on the efficiency of the algorithm used to multiply large integers. As a function of the size of the base and of the value of the exponent, how much time do the algorithms corresponding to $dexpo2$ and $dexpo$

take when the classic multiplication algorithm is used? Rework the problem using the divide-and-conquer multiplication algorithm from Section 4.7.   □

The preceding problem shows that it is sometimes not sufficient to be only half-clever!

4.9 MATRIX MULTIPLICATION

Let A and B be two $n \times n$ matrices to be multiplied, and let C be their product. The classic algorithm comes directly from the definition:

$$C_{ij} = \sum_{k=1}^{n} A_{ik} B_{kj} .$$

Each entry in C is calculated in a time in $\Theta(n)$, assuming that scalar addition and multiplication are elementary operations. Since there are n^2 entries to compute in order to obtain C, the product of A and B can be calculated in a time in $\Theta(n^3)$.

Towards the end of the 1960s, Strassen caused a considerable stir by improving this algorithm. The basic idea is similar to that used in the divide-and-conquer algorithm of Section 4.7 for multiplying large integers. First we show that two 2×2 matrices can be multiplied using less than the eight scalar multiplications apparently required by the definition. Let

$$A = \begin{bmatrix} a_{11} & a_{12} \\ a_{21} & a_{22} \end{bmatrix} \quad \text{and} \quad B = \begin{bmatrix} b_{11} & b_{12} \\ b_{21} & b_{22} \end{bmatrix}$$

be two matrices to be multiplied. Consider the following operations:

$$m_1 = (a_{21} + a_{22} - a_{11})(b_{22} - b_{12} + b_{11})$$
$$m_2 = a_{11}b_{11}$$
$$m_3 = a_{12}b_{21}$$
$$m_4 = (a_{11} - a_{21})(b_{22} - b_{12})$$
$$m_5 = (a_{21} + a_{22})(b_{12} - b_{11})$$
$$m_6 = (a_{12} - a_{21} + a_{11} - a_{22})b_{22}$$
$$m_7 = a_{22}(b_{11} + b_{22} - b_{12} - b_{21}).$$

We leave the reader to verify that the required product AB is given by the following matrix:

$$\begin{bmatrix} m_2 + m_3 & m_1 + m_2 + m_5 + m_6 \\ m_1 + m_2 + m_4 - m_7 & m_1 + m_2 + m_4 + m_5 \end{bmatrix} .$$

It is therefore possible to multiply two 2×2 matrices using only seven scalar multiplications. At first glance, this algorithm does not look very interesting: it uses a large number of additions and subtractions compared to the four additions that are sufficient for the classic algorithm.

If we now replace each entry of A and B by an $n \times n$ matrix, we obtain an algorithm that can multiply two $2n \times 2n$ matrices by carrying out seven multiplications of $n \times n$ matrices, as well as a number of additions and subtractions of $n \times n$ matrices. This is possible because the basic algorithm does not rely on the commutativity of scalar multiplication. Given that matrix additions can be executed much faster than matrix multiplications, the few additional additions compared to the classic algorithm are more than compensated by saving one multiplication, provided n is sufficiently large.

Problem 4.9.1.   Building on the preceding discussion, show that it is possible to multiply two $n \times n$ matrices in a time in $O(n^{2.81})$. What do you do about matrices whose size is not a power of 2?   □

Problem 4.9.2.   The number of additions and subtractions needed to calculate the product of two 2×2 matrices using this method seems to be 24. Show that this can be reduced to 15 by using auxiliary variables to avoid recalculating terms such as $m_1 + m_2 + m_4$. Strassen's original algorithm takes 18 additions and subtractions as well as 7 multiplications. The algorithm discussed in this section is a variant discovered subsequently by Shmuel Winograd.   □

*Problem 4.9.3.   Assuming that n is a power of 2, find the exact number of scalar additions and multiplications executed by your algorithm for Problem 4.9.1, taking account of the idea from Problem 4.9.2. Your answer will depend on the threshold used to stop making recursive calls. Bearing in mind what you learnt from Section 4.2, propose a threshold that will minimize the number of scalar operations.   □

Following publication of Strassen's algorithm, a number of researchers tried to improve the constant ω such that it is possible to multiply two $n \times n$ matrices in a time in $O(n^\omega)$. Almost a decade passed before Pan discovered a more efficient algorithm, once again based on divide-and-conquer: he found a way to multiply two 70×70 matrices that involves only 143,640 scalar multiplications. (Compare this to $70^3 = 343,000$ and to $70^{2.81}$, which exceeds 150,000.) Numerous algorithms, asymptotically more and more efficient, have been discovered subsequently. The asymptotically fastest matrix multiplication algorithm known at the time of this writing can multiply two $n \times n$ matrices in a time in $O(n^{2.376})$; it was discovered by Coppersmith and Winograd in September 1986. Because of the hidden constants, however, none of the algorithms found after Strassen's is of much practical use.

4.10 EXCHANGING TWO SECTIONS OF AN ARRAY

For this additional example of an algorithm based on simplification, we ignore the recursive formulation and give only an iterative version. Let T be an array of n elements. We wish to interchange the first k elements and the last $n-k$, without making use of an auxiliary array. For instance, if T is the array

| a | b | c | d | e | f | g | h | i | j | k |

and $k = 3$, the required result in T is

| d | e | f | g | h | i | j | k | a | b | c | .

It is easy to invent an algorithm $exchange(i,j,m)$ to interchange the elements $T[i \, .. \, i+m-1]$ and $T[j \, .. \, j+m-1]$ in a time in $\Theta(m)$, provided that $m < i+m \le j \le n-m+1$ (see Fig. 4.10.1). With its help, we can solve our problem as illustrated in Figure 4.10.2. Here the arrows indicate the part of the array where there are still some changes to make. After each exchange this part is smaller than before: thus we can affirm that each exchange simplifies the solution of the instance.

The general algorithm is as follows.

procedure $exchange(i,j,m)$
   for $p \leftarrow 0$ **to** $m-1$ **do**
      interchange $T[i+p]$ and $T[j+p]$

procedure $transpose(T[1 \, .. \, n], k)$
   $i \leftarrow k \,; j \leftarrow n-k \,; k \leftarrow k+1$
   while $i \ne j$ **do**
      if $i > j$ **then**
         $exchange(k-i, k, j)$
         $i \leftarrow i-j$
      else
         $j \leftarrow j-i$
         $exchange(k-i, k+j, i)$
   $exchange(k-i, k, i)$

The analysis of this algorithm is interesting. Let $T(i,j)$ be the number of elementary exchanges that have to be made to transpose a block of i elements and a block of j elements. Then

$$T(i,j) = \begin{cases} i & \text{if } i = j \\ j + T(i-j, j) & \text{if } i > j \\ i + T(i, j-i) & \text{if } i < j \end{cases} .$$

For instance, Figure 4.10.2 shows how a block of three elements and a block of eight elements are transposed. We have

$$T(3,8) = 3 + T(3,5) = 6 + T(3,2) = 8 + T(1,2) = 9 + T(1,1) = 10 .$$

The progression of the parameters of the function T recalls an application of Euclid's algorithm and leads to the following result:

Problem 4.10.1.   Prove that

$$T(i,j) = i + j - \gcd(i,j) ,$$

where $\gcd(i,j)$ denotes the greatest common divisor of i and j (Section 1.7.4).   □

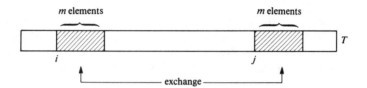

Figure 4.10.1.   Effect of the *exchange* algorithm.

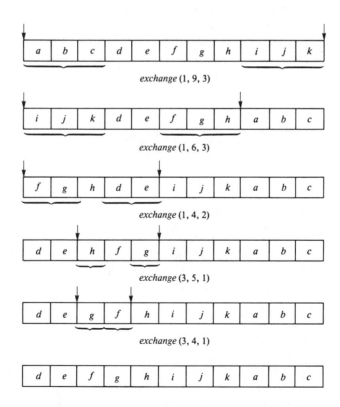

Figure 4.10.2.   Progress of *transpose* $(T,3)$.

Problem 4.10.2.   Can we do better if we are allowed unlimited auxiliary space?   □

4.11 SUPPLEMENTARY PROBLEMS

Problem 4.11.1. The Fibonacci sequence.   Consider the matrix

$$F = \begin{bmatrix} 0 & 1 \\ 1 & 1 \end{bmatrix}.$$

Let i and j be any two integers. What is the product of the vector (i, j) and the matrix F? What happens if i and j are two consecutive numbers from the Fibonacci sequence? Use this idea to invent a divide-and-conquer algorithm to calculate this sequence. Does this help you to understand the algorithm *fib*3 of Section 1.7.5?   □

Problem 4.11.2. Polynomial interpolation.   Represent the polynomial $p(n) = a_0 + a_1 n + a_2 n^2 + \cdots + a_d n^d$ of degree d by an array $P[0 .. d]$ containing its coefficients. Suppose you already have an algorithm capable of multiplying a polynomial of degree i by a polynomial of degree 1 in a time in $O(i)$, as well as another algorithm capable of multiplying two polynomials of degree i in a time in $O(i \log i)$ — see Chapter 9. Let $n_1, n_2, \ldots, n_d$ be any integers. Give an efficient algorithm based on divide-and-conquer to find the unique monic polynomial $p(n)$ of degree d such that $p(n_1) = p(n_2) = \cdots = p(n_d) = 0$. (A polynomial is *monic* if its coefficient of highest degree $a_d = 1$.) Analyse the efficiency of your algorithm.   □

Problem 4.11.3. Smallest and largest elements.   Let $T[1 .. n]$ be an array of n elements. It is easy to find the largest element of T by making exactly $n - 1$ comparisons between elements.

> $max \leftarrow T[1]$; $ind \leftarrow 1$
> **for** $i \leftarrow 2$ **to** n **do**
>   **if** $max < T[i]$ **then** $max \leftarrow T[i]$, $ind \leftarrow i$

We only count comparisons between elements; we exclude implicit comparisons in the control of the **for** loop. We could subsequently find the smallest element of T by making $n - 2$ more comparisons.

> $T[ind] \leftarrow T[1]$
> $min \leftarrow T[2]$
> **for** $i \leftarrow 3$ **to** n **do**
>   **if** $min > T[i]$ **then** $min \leftarrow T[i]$

Find an algorithm that can find both the largest and the smallest elements of an array of n elements by making less than $2n - 3$ comparisons between elements. You may assume that n is a power of 2. Exactly how many comparisons does your algorithm require? How would you handle the situation when n is not a power of 2?   □

Problem 4.11.4.   Let $T[1 .. n]$ be a sorted array of integers, some of which may be negative but all of which are different. Give an algorithm that is able to find an index i such that $1 \le i \le n$ and $T[i] = i$, provided such an index exists. Your algorithm should take a time in $O(\log n)$ in the worst case.   □

Problem 4.11.5. Majority element.   Let $T[1 .. n]$ be an array of n elements. An element x is said to be a *majority element* in T if $\#\{i \mid T[i] = x\} > n/2$. Give an algorithm that can decide whether an array $T[1 .. n]$ includes a majority element (it cannot have more than one), and if so find it. Your algorithm must run in linear time.   □

*Problem 4.11.6.   Rework Problem 4.11.5 with the supplementary constraint that the only comparisons allowed between elements are tests of equality. You may therefore not assume that an order relation exists between the elements.   □

Problem 4.11.7.   If you could not manage the previous problem, try again, but allow your algorithm to take a time in $O(n \log n)$.   □

Problem 4.11.8. Tally circuit.   An *n-tally* is a circuit that takes n bits as inputs and produces $1 + \lfloor \lg n \rfloor$ bits as output. It counts (in binary) the number of bits equal to 1 among the inputs. For example, if $n = 9$ and the inputs are 011001011, the output is 0101. An (i, j)-*adder* is a circuit that has one i bit input, one j bit input, and one $[1 + \max(i, j)]$-bit output. It adds its two inputs in binary. For example, if $i = 3$, $j = 5$, and the inputs are 101 and 10111 respectively, the output is 011100. It is always possible to construct an (i, j)-adder using exactly $\max(i, j)$ 3-tallies. For this reason the 3-tally is often called a *full adder*.

   i. Using full adders and (i, j)-adders as primitive elements, show how to build an efficient n-tally. You may not suppose that n has any special form.

   ii. Give the recurrence, including the initial conditions, for the number of 3-tallies needed to build your n-tally. Do not forget to count the 3-tallies that are part of any (i, j)-adders you might have used.

   iii. Using the Θ notation, give the simplest expression you can for the number of 3-tallies needed in the construction of your n-tally. Justify your answer.   □

*Problem 4.11.9. Telephone switching.   A *switch* is a circuit with two inputs, a control, and two outputs. It connects input A with output A and input B with output B, or input A with output B and input B with output A, depending on the position of the control — see Figure 4.11.1. Use these switches to construct a network with n inputs and n outputs able to implement any of the $n!$ possible permutations of the inputs. The number of switches used must be in $O(n \log n)$.   □

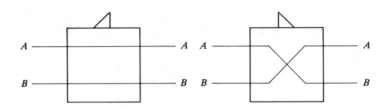

Figure 4.11.1.   Telephone switches.

*Problem 4.11.10. Merge circuit.   A *comparator* is a circuit with two inputs and two outputs. The smaller input appears on the upper output, and the larger input appears on the lower output. For a given integer n, a *merge circuit* F_n has two groups of n inputs and a single group of $2n$ outputs. Provided each of the two groups of inputs is already sorted, then each input appears on one of the outputs, and the outputs are also sorted. For instance, Figure 4.11.2 shows an F_4 circuit, illustrating how the inputs are transmitted to the outputs. Each rectangle represents a comparator. By convention, inputs are on the left and outputs are on the right.

There are two ways to measure the complexity of such a circuit: the *size* of the circuit is the number of comparators it includes, and the *depth* is the largest number of comparators an input may have to pass before it reaches the corresponding output. The depth is interesting because it determines the reaction time of the circuit. For example, the F_4 merge circuit shown in Figure 4.11.2 has size 9 and depth 3.

For n a power of 2, show how to construct a merge circuit F_n whose size and depth are exactly $1 + n \lg n$ and $1 + \lg n$, respectively.   □

Problem 4.11.11. Batcher's sorting circuit.   Following up the previous problem, a *sorting circuit* S_n has n inputs and n outputs; it sorts the inputs presented to it. Figure 4.11.3 gives an example of S_4, which is of size 5 and depth 3.

For n a power of 2, show how to construct an efficient sorting circuit S_n. By "efficient" we mean that the depth of your circuit must be significantly less than n

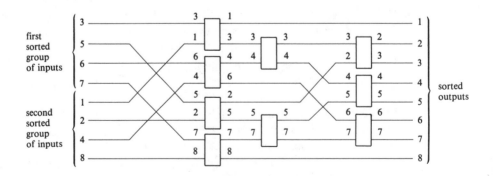

Figure 4.11.2.   A merge circuit.

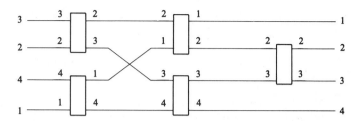

Figure 4.11.3. A sorting circuit.

whenever n is sufficiently large. You may use merge circuits to your heart's content, but their depth and size must then be taken into account. Give recurrences, including the initial conditions, for the size T_n and the depth P_n of your circuit S_n. Solve these equations exactly, and express T_n and P_n in Θ notation as simply as possible.   □

∗∗ Problem 4.11.12.   Continuing the two previous problems, show that it is possible to construct a sorting circuit for n elements whose size and depth are in $\Theta(n \log n)$ and $\Theta(\log n)$, respectively.   □

Problem 4.11.13. Tournaments.   You are to organize a tournament involving n competitors. Each competitor must play exactly once against each of his opponents.

		Player				
$n = 5$	Day	1	2	3	4	5
	1	2	1	-	5	4
	2	3	5	1	-	2
	3	4	3	2	1	-
	4	5	-	4	3	1
	5	-	4	5	2	3

		Player					
$n = 6$	Day	1	2	3	4	5	6
	1	2	1	6	5	4	3
	2	3	5	1	6	2	4
	3	4	3	2	1	6	5
	4	5	6	4	3	1	2
	5	6	4	5	2	3	1

Figure 4.11.4. Timetables for five and six players.

Moreover, each competitor must play exactly one match every day, with the possible exception of a single day when he does not play at all.

1. If n is a power of 2, give an algorithm to construct a timetable allowing the tournament to be finished in $n - 1$ days.
2. For any integer $n > 1$ give an algorithm to construct a timetable allowing the tournament to be finished in $n - 1$ days if n is even, or in n days if n is odd. For example, Figure 4.11.4 gives possible timetables for tournaments involving five and six players.

** Problem 4.11.14. Closest pair of points.   You are given the coordinates of n points in the plane. Give an algorithm capable of finding the closest pair of points in a time in $O(n \log n)$.   □

4.12 REFERENCES AND FURTHER READING

Quicksort is from Hoare (1962). Mergesort and quicksort are discussed in detail in Knuth (1973). Problem 4.4.2 was solved by Kronrod; see the solution to Exercise 18 of Section 5.2.4 of Knuth (1973). The algorithm linear in the worst case for selection and for finding the median is due to Blum, Floyd, Pratt, Rivest, and Tarjan (1972). The algorithm for multiplying large integers in a time in $O(n^{1.59})$ is attributed to Karatsuba and Ofman (1962). The answer to Problems 4.7.7 and 4.7.8 is discussed in Knuth (1969). The survey article by Brassard, Monet, and Zuffellato (1986) covers computation with very large integers.

The algorithm that multiplies two $n \times n$ matrices in a time in $O(n^{2.81})$ comes from Strassen (1969). Subsequent efforts to do better than Strassen's algorithm began with the proof by Hopcroft and Kerr (1971) that seven multiplications are necessary to multiply two 2×2 matrices in a non-commutative structure; the first positive success was obtained by Pan (1978), and the algorithm that is asymptotically the most efficient known at present is by Coppersmith and Winograd (1987). The algorithm of Section 4.10 comes from Gries (1981).

The original solution to Problem 4.8.1 is due to Diffie and Hellman (1976). The importance for cryptology of the arithmetic of large integers and of the theory of numbers was pointed out by Rivest, Shamir, and Adleman (1978). For more information about cryptology, consult the introductory papers by Gardner (1977) and Hellman (1980) and the books by Kahn (1967), Denning (1983), Kranakis (1986), and Brassard (1988). Bear in mind, however, that the cryptosystem based on the knapsack problem, as described in Hellman (1980), has since been broken. Notice too that the integers involved in these applications are not sufficiently large for the algorithms of Section 4.7 to be worthwhile. On the other hand, efficient exponentiation as described in Section 4.8 is crucial. The natural generalization of Problem 4.8.6 is examined in Knuth (1969).

The solution to Problem 4.11.1 can be found in Gries and Levin (1980) and Urbanek (1980). Problem 4.11.3 is discussed in Pohl (1972) and Stinson (1985). Problems 4.11.10 and 4.11.11 are solved in Batcher (1968). Problem 4.11.12 is solved, at least in principle, in Ajtai, Komlós, and Szemerédi (1983). Problem 4.11.14 is solved in Bentley and Shamos (1976), but consult Section 8.7 for more on this problem.

5

Dynamic Programming

5.1 INTRODUCTION

In the last chapter we saw that it is often possible to divide an instance into subinstances, to solve the subinstances (perhaps by further dividing them), and then to combine the solutions of the subinstances so as to solve the original instance. It sometimes happens that the natural way of dividing an instance suggested by the structure of the problem leads us to consider several overlapping subinstances. If we solve each of these independently, they will in turn create a large number of identical subinstances. If we pay no attention to this duplication, it is likely that we will end up with an inefficient algorithm. If, on the other hand, we take advantage of the duplication and solve each subinstance only once, saving the solution for later use, then a more efficient algorithm will result. The underlying idea of dynamic programming is thus quite simple: avoid calculating the same thing twice, usually by keeping a table of known results, which we fill up as subinstances are solved.

Dynamic programming is a *bottom-up* technique. We usually start with the smallest, and hence the simplest, subinstances. By combining their solutions, we obtain the answers to subinstances of increasing size, until finally we arrive at the solution of the original instance. Divide-and-conquer, on the other hand, is a *top-down* method. When a problem is solved by divide-and-conquer, we immediately attack the complete instance, which we then divide into smaller and smaller subinstances as the algorithm progresses.

142

Example 5.1.1.   Consider calculating the binomial coefficient

$$
\binom{n}{k} = \begin{cases} \binom{n-1}{k-1} + \binom{n-1}{k}, & 0 < k < n \\ 1 & \text{otherwise}. \end{cases}
$$

If we calculate $\binom{n}{k}$ directly by

> **function** $C(n,k)$
>   **if** $k = 0$ **or** $k = n$ **then return** 1
>   **else return** $C(n-1,k-1) + C(n-1,k)$,

many of the values $C(i,j), i < n, j < k$ are calculated over and over. Since the final result is obtained by adding up a certain number of 1s, the execution time of this algorithm is certainly in $\Omega\left(\binom{n}{k}\right)$. We have already met a similar phenomenon in algorithm *fib*1 for calculating the Fibonacci sequence (see Section 1.7.5 and Example 2.2.7).

If, on the other hand, we use a table of intermediate results (this is of course Pascal's triangle; see Figure 5.1.1), we obtain a more efficient algorithm. The table should be filled line by line. In fact, it is not even necessary to store a matrix: it is sufficient to keep a vector of length k, representing the current line, which we update from left to right. Thus the algorithm takes a time in $O(nk)$ and space in $O(k)$, if we assume that addition is an elementary operation.   □

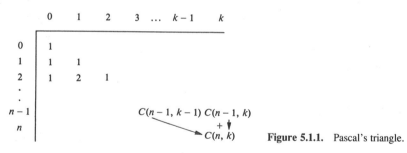

Figure 5.1.1. Pascal's triangle.

Problem 5.1.1.   Prove that the total number of recursive calls made during the computation of $C(n,k)$ is exactly $2\binom{n}{k} - 2$.   □

Problem 5.1.2.   Calculating the Fibonacci sequence affords another example of this kind of technique. Which of the algorithms presented in Section 1.7.5 uses dynamic programming?   □

Dynamic programming is often used to solve optimization problems that satisfy the *principle of optimality*: in an optimal sequence of decisions or choices, each subsequence must also be optimal. Although this principle may appear obvious, it does not always apply.

Example 5.1.2. If the shortest route from Montréal to Toronto goes via Kingston, then that part of the journey from Montréal to Kingston must also follow the shortest route between these two cities: the principle of optimality applies. However, if the fastest way to drive from Montréal to Toronto takes us first to Kingston, it does not follow that we should drive from Montréal to Kingston as quickly as possible: if we use too much petrol on the first half of the trip, maybe we have to stop to fill up somewhere on the second half, losing more time than we gained by driving hard. The subtrips Montréal-Kingston and Kingston-Toronto are not independent, and the principle of optimality does not apply.   □

Problem 5.1.3. Show that the principle of optimality does not apply to the problem of finding the longest simple path between two cities. Argue that this is due to the fact that one cannot in general splice two simple paths together and expect to obtain a simple path. (A path is *simple* if it never passes through the same place twice. Without this restriction the longest path might be an infinite loop.)   □

The principle of optimality can be restated as follows for those problems for which it applies: the optimal solution to any nontrivial instance is a combination of optimal solutions to *some* of its subinstances. The difficulty in turning this principle into an algorithm is that it is not usually obvious which subinstances are relevant to the instance under consideration. Coming back to Example 5.1.2, it is not immediately obvious that the subinstance consisting of finding the shortest route from Montréal to Ottawa is irrelevant to the shortest route from Montréal to Toronto. This difficulty prevents us from using a divide-and-conquer approach that would start from the original instance and recursively find optimal solutions precisely to those relevant subinstances. Instead, dynamic programming efficiently solves every possible subinstance in order to figure out which are in fact relevant, and only then are these combined into an optimal solution to the original instance.

5.2 THE WORLD SERIES

As our first example of dynamic programming, let us not worry about the principle of optimality, but rather concentrate on the control structure and the order of resolution of the subinstances. For this reason the problem considered in this section is not one of optimization.

Imagine a competition in which two teams A and B play not more than $2n-1$ games, the winner being the first team to achieve n victories. We assume that there are no tied games, that the results of each match are independent, and that for any given match there is a constant probability p that team A will be the winner and hence a constant probability $q = 1-p$ that team B will win.

Let $P(i,j)$ be the probability that team A will win the series given that they still need i more victories to achieve this, whereas team B still needs j more victories if they are to win. For example, before the first game of the series the probability that team A will be the overall winner is $P(n,n)$: both teams still need n victories. If team A has already won all the matches it needs, then it is of course certain that they will win the series: $P(0,i) = 1$, $1 \leq i \leq n$. Similarly $P(i,0) = 0$, $1 \leq i \leq n$. $P(0,0)$ is undefined. Finally, since team A wins any given match with probability p and loses it with probability q,

$$P(i,j) = pP(i-1,j) + qP(i,j-1) \quad i \geq 1, j \geq 1 .$$

Thus we can compute $P(i,j)$ using

> **function** $P(i,j)$
> **if** $i = 0$ **then return** 1
> **else if** $j = 0$ **then return** 0
> **else return** $pP(i-1,j) + qP(i,j-1)$.

Let $T(k)$ be the time needed in the worst case to calculate $P(i,j)$, where $k = i+j$. With this method, we see that

$$T(1) = c$$
$$T(k) \leq 2T(k-1) + d , \quad k > 1$$

where c and d are constants. $T(k)$ is therefore in $O(2^k) = O(4^n)$ if $i = j = n$. In fact, if we look at the way the recursive calls are generated, we find the pattern shown in Figure 5.2.1, which is identical to that followed by the naive calculation of the binomial coefficient $C(i+j,j) = C((i-1)+j,j) + C(i+(j-1),j-1)$. The total number of recursive calls is therefore exactly $2\left[{i+j \atop j}\right] - 2$ (Problem 5.1.1). To calculate the probability $P(n,n)$ that team A will win given that the series has not yet started, the required time is thus in $\Omega\left(\left[{2n \atop n}\right]\right)$.

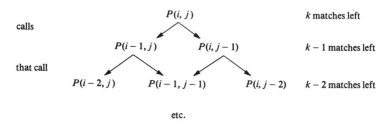

Figure 5.2.1.   Recursive calls made by a call on **function** $P(i,j)$.

Problem 5.2.1.   Prove that $\left[{2n \atop n}\right] \geq 4^n/(2n+1)$.   □

Combining these results, we see that the time required to calculate $P(n,n)$ is in $O(4^n)$ and $\Omega(4^n/n)$. This method is therefore not practical for large values of n. (Although sporting competitions with $n > 4$ are the exception, this problem does have other applications!)

*Problem 5.2.2.   Prove that in fact the time needed to calculate $P(n,n)$ using the preceding algorithm is in $\Theta(4^n/\sqrt{n})$.   □

To speed up the algorithm, we proceed more or less as with Pascal's triangle: we declare an array of the appropriate size and then fill in the entries. This time, however, instead of filling the array line by line, we work diagonal by diagonal. Here is the algorithm to calculate $P(n,n)$.

```
function series (n,p)
  array P [0..n, 0..n]
  q ← 1 − p
  for s ← 1 to n do
    P[0,s] ← 1 ; P[s,0] ← 0
    for k ← 1 to s−1 do
      P[k,s−k] ← pP[k−1,s−k] + qP[k,s−k−1]
  for s ← 1 to n do
    for k ← 0 to n−s do
      P[s+k,n−k] ← pP[s+k−1,n−k] + qP[s+k,n−k−1]
  return P[n,n]
```

Problem 5.2.3.   Using this algorithm, calculate the probability that team A will win the series if $p = 0.45$ and if four victories are needed to win.   □

Since in essence the algorithm has to fill up an $n \times n$ array, and since a constant time is required to calculate each entry, its execution time is in $\Theta(n^2)$.

Problem 5.2.4.   Show that a memory space in $\Theta(n)$ is sufficient to implement this algorithm.   □

Problem 5.2.5.   Show how to compute $P(n,n)$ in a time in $\Theta(n)$. (*Hint*: use a completely different approach — see Section 8.6.)   □

5.3 CHAINED MATRIX MULTIPLICATION

We wish to compute the matrix product

$$M = M_1 M_2 \cdots M_n \ .$$

Matrix multiplication is associative, so we can compute the product in a number of ways:

$$M = (\cdots ((M_1 M_2) M_3) \cdots M_n)$$

$$= (M_1 (M_2 (M_3 \cdots (M_{n-1} M_n) \cdots)))$$

$$= ((M_1 M_2)(M_3 M_4) \cdots), \text{ and so on .}$$

The choice of a method of computation can have a considerable influence on the time required.

Problem 5.3.1.   Show that calculating the product AB of a $p \times q$ matrix A and a $q \times r$ matrix B by the direct method requires pqr scalar multiplications.   □

Example 5.3.1.   We wish to calculate the product $ABCD$ of four matrices: A is 13×5, B is 5×89, C is 89×3, and D is 3×34. To measure the efficiency of the different methods, we count the number of scalar multiplications that are involved. For example, if we calculate $M = ((AB)C)D$, we obtain successively

(AB)	5,785 multiplications
$(AB)C$	3,471 multiplications
$((AB)C)D$	1,326 multiplications

for a total of 10,582 multiplications. There are five essentially different ways of calculating this product. (In the second case that follows, we do not differentiate between the method that first calculates AB and the one that starts with CD.) In each case, here is the corresponding number of scalar multiplications.

$((AB)C)D$	10,582
$(AB)(CD)$	54,201
$(A(BC))D$	2,856
$A((BC)D)$	4,055
$A(B(CD))$	26,418

The most efficient method is almost 19 times faster than the slowest.   □

To find directly the best way to calculate the product, we could simply parenthesize the expression in every possible fashion and count each time how many scalar multiplications will be required. Let $T(n)$ be the number of essentially different ways to parenthesize a product of n matrices. Suppose we decide to make the first cut between the ith and the $(i+1)$st matrices of the product:

$$M = (M_1 M_2 \cdots M_i)(M_{i+1} M_{i+2} \cdots M_n) .$$

There are now $T(i)$ ways to parenthesize the left-hand term and $T(n-i)$ ways to parenthesize the right-hand term. Since i can take any value from 1 to $n-1$, we obtain the following recurrence for $T(n)$:

$$T(n) = \sum_{i=1}^{n-1} T(i)T(n-i) .$$

Adding the initial condition $T(1) = 1$, we can thus calculate all the values of T. Among other values, we find

n	1	2	3	4	5	10	15
$T(n)$	1	1	2	5	14	4,862	2,674,440.

The values of $T(n)$ are called *Catalan numbers*.

*Problem 5.3.2.   Prove that

$$T(n) = \frac{1}{n}\binom{2n-2}{n-1}.$$ □

For each way that parentheses can be inserted it takes a time in $\Omega(n)$ to count the number of scalar multiplications required (at least if we do not try to be subtle). Since $T(n)$ is in $\Omega(4^n/n^2)$ (from Problems 5.2.1 and 5.3.2), finding the best way to calculate M using the direct approach requires a time in $\Omega(4^n/n)$. This method is therefore impracticable for large values of n: there are too many ways in which parentheses can be inserted for us to look at them all.

Fortunately, the principle of optimality applies to this problem. For instance, if the best way of multiplying all the matrices requires us to make the first cut between the ith and the $(i+1)$st matrices of the product, then both the subproducts $M_1 M_2 \cdots M_i$ and $M_{i+1}M_{i+2} \cdots M_n$ must also be calculated in an optimal way. This suggests that we should consider using dynamic programming. We construct a table m_{ij}, $1 \le i \le j \le n$, where m_{ij} gives the optimal solution — that is, the required number of scalar multiplications — for the part $M_i M_{i+1} \cdots M_j$ of the required product. The solution to the original problem is thus given by m_{1n}.

Suppose the dimensions of the matrices M_i are given by a vector d_i, $0 \le i \le n$, such that the matrix M_i is of dimension d_{i-1} by d_i. We build the table m_{ij} diagonal by diagonal: diagonal s contains the elements m_{ij} such that $j - i = s$. We thus obtain in succession:

$$s = 0: m_{ii} = 0, \quad i = 1, 2, \ldots, n$$

$$s = 1: m_{i,i+1} = d_{i-1}d_i d_{i+1}, \quad i = 1, 2, \ldots, n-1 \quad \text{(see Problem 5.3.1)}$$

$$1 < s < n: m_{i,i+s} = \min_{i \le k < i+s} (m_{ik} + m_{k+1,i+s} + d_{i-1}d_k d_{i+s}), \quad i = 1, 2, \ldots, n-s.$$

The third case represents the fact that to calculate $M_i M_{i+1} \cdots M_{i+s}$ we try all the possibilities $(M_i M_{i+1} \cdots M_k)(M_{k+1} \cdots M_{i+s})$ and choose the best for $i \le k < i+s$. It is only for clarity that the second case is written out explicitly, as it falls under the general case with $s = 1$.

Example 5.3.2.   Continuation of Example 5.3.1.   We have $d = (13,5,89,3,34)$. For $s = 1$, we find $m_{12} = 5,785$, $m_{23} = 1,335$, $m_{34} = 9,078$. Next, for $s = 2$ we obtain

$$m_{13} = \min(m_{11} + m_{23} + 13 \times 5 \times 3, \, m_{12} + m_{33} + 13 \times 89 \times 3)$$
$$= \min(1530, 9256) = 1{,}530$$

$$m_{24} = \min(m_{22} + m_{34} + 5 \times 89 \times 34, \, m_{23} + m_{44} + 5 \times 3 \times 34)$$
$$= \min(24208, 1845) = 1{,}845 \ .$$

Finally, for $s = 3$

$$m_{14} = \min(\{k = 1\} \ m_{11} + m_{24} + 13 \times 5 \times 34 \ ,$$
$$\{k = 2\} \ m_{12} + m_{34} + 13 \times 89 \times 34 \ ,$$
$$\{k = 3\} \ m_{13} + m_{44} + 13 \times 3 \times 34)$$

$$= \min(4055, 54201, 2856) = 2{,}856 \ .$$

The array m is thus given in Figure 5.3.1.   □

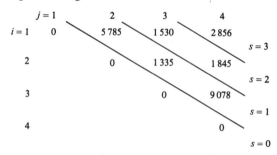

Figure 5.3.1.   Example of the chained matrix multiplication algorithm.

Problem 5.3.3.   Write the algorithm to calculate m_{1n} .   □

Problem 5.3.4.   How must the algorithm be modified if we want not only to calculate the value of m_{1n} , but also to know how to calculate the product M in the most efficient way ?   □

For $s > 0$, there are $n - s$ elements to be computed in the diagonal s ; for each, we must choose between s possibilities (the different possible values of k). The execution time of the algorithm is therefore in the exact order of

$$\sum_{s=1}^{n-1} (n - s)s = n \sum_{s=1}^{n-1} s - \sum_{s=1}^{n-1} s^2$$

$$= n^2(n-1)/2 - n(n-1)(2n-1)/6$$

$$= (n^3 - n)/6 \ .$$

The execution time is thus in $\Theta(n^3)$.

*Problem 5.3.5. Prove that with the following recursive algorithm

function *minmat* (i, j)
    if $i = j$ **then return** 0
    $ans \leftarrow \infty$
    for $k \leftarrow i$ **to** $j-1$ **do**
        $ans \leftarrow \min(ans, d[i-1]d[k]d[j] + minmat(i,k) + minmat(k+1, j))$
    return ans ,

where the array $d[0..n]$ is global, a call on *minmat* $(1,n)$ takes a time in $\Theta(3^n)$. (*Hint:* for the "*O*" part, use constructive induction to find constants a and b such that the time taken by a call on *minmat* $(1,n)$ is no greater than $a\,3^n - b$.)   □

Although a call on the recursive *minmat* $(1,n)$ of Problem 5.3.5 is faster than naively trying all possible ways to parenthesize the desired product, it is still much slower than the dynamic programming algorithm described previously. This behaviour illustrates a point made in the first paragraph of this chapter. In order to decide on the best way to parenthesize the product *ABCDEFG*, *minmat* recursively solves 12 subinstances, including the overlapping *ABCDEF* and *BCDEFG*, both of which recursively solve *BCDEF* from scratch. It is this duplication of effort that causes the inefficiency of *minmat*.

5.4 SHORTEST PATHS

Let $G = <N,A>$ be a directed graph; N is the set of nodes and A is the set of edges. Each edge has an associated nonnegative length. We want to calculate the length of the shortest path between each pair of nodes. (Compare this to Section 3.2.2, where we were looking for the length of the shortest paths from one particular node, the source, to all the others.)

As before, suppose that the nodes of G are numbered from 1 to n, $N = \{1, 2, \ldots, n\}$, and that a matrix L gives the length of each edge, with $L[i,i] = 0$, $L[i,j] \geq 0$ if $i \neq j$, and $L[i,j] = \infty$ if the edge (i,j) does not exist.

The principle of optimality applies: if k is a node on the shortest path from i to j, then that part of the path from i to k, and that from k to j, must also be optimal.

We construct a matrix D that gives the length of the shortest path between each pair of nodes. The algorithm initializes D to L. It then does n iterations. After iteration k, D gives the length of the shortest paths that only use nodes in $\{1, 2, \ldots, k\}$ as intermediate nodes. After n iterations we therefore obtain the result we want. At iteration k, the algorithm has to check for each pair of nodes (i, j) whether or not there exists a path passing through node k that is better than the present optimal path passing only through nodes in $\{1, 2, \ldots, k-1\}$. Let D_k be the matrix D after the kth iteration. The necessary check can be written as

$$D_k[i,j] = \min(D_{k-1}[i,j], D_{k-1}[i,k] + D_{k-1}[k,j]),$$

where we make use of the principle of optimality to compute the length of the shortest path passing through k. We have also implicitly made use of the fact that an optimal path through k does not visit k twice.

At the kth iteration the values in the kth row and the kth column of D do not change, since $D[k,k]$ is always zero. It is therefore not necessary to protect these values when updating D. This allows us to get away with using only a two-dimensional matrix D, whereas at first sight a matrix $n \times n \times 2$ (or even $n \times n \times n$) seems necessary.

The algorithm, known as *Floyd's algorithm*, follows.

```
procedure Floyd (L [1..n, 1..n]): array[1..n, 1..n]
    array D[1..n, 1..n]
    D ← L
    for k ← 1 to n do
        for i ← 1 to n do
            for j ← 1 to n do
                D[i, j] ← min(D[i, j], D[i, k] + D[k, j])
    return D
```

Figure 5.4.1 gives an example of the way the algorithm works.

It is obvious that this algorithm takes a time in $\Theta(n^3)$. We can also use Dijkstra's algorithm (Section 3.2.2) to solve the same problem. In this case we have to apply the algorithm n times, each time choosing a different node as the source. If we use the version of Dijkstra's algorithm that works with a matrix of distances, the total computation time is in $n \times \Theta(n^2)$, that is, in $\Theta(n^3)$. The order is the same as for Floyd's algorithm, but the simplicity of Floyd's algorithm means that it will probably be faster in practice. On the other hand, if we use the version of Dijkstra's algorithm that works with a heap, and hence with lists of the distances to adjacent nodes, the total time is in $n \times O((a+n)\log n)$, that is, in $O((an+n^2)\log n)$, where a is the number of edges in the graph. If the graph is not very dense ($a \ll n^2$), it may be preferable to use Dijkstra's algorithm n times; if the graph is dense ($a \approx n^2$), it is better to use Floyd's algorithm.

We usually want to know where the shortest path goes, not just its length. In this case we use a second matrix P initialized to 0. The innermost loop of the algorithm becomes

$$\textbf{if } D[i,k] + D[k,j] < D[i,j] \textbf{ then } D[i,j] \leftarrow D[i,k] + D[k,j]$$
$$P[i,j] \leftarrow k \quad .$$

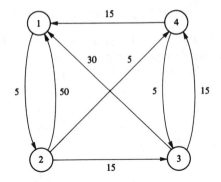

$$D_0 = L = \begin{bmatrix} 0 & 5 & \infty & \infty \\ 50 & 0 & 15 & 5 \\ 30 & \infty & 0 & 15 \\ 15 & \infty & 5 & 0 \end{bmatrix}$$

$$D_1 = \begin{bmatrix} 0 & 5 & \infty & \infty \\ 50 & 0 & 15 & 5 \\ 30 & 35 & 0 & 15 \\ 15 & 20 & 5 & 0 \end{bmatrix} \quad D_2 = \begin{bmatrix} 0 & 5 & 20 & 10 \\ 50 & 0 & 15 & 5 \\ 30 & 35 & 0 & 15 \\ 15 & 20 & 5 & 0 \end{bmatrix}$$

$$D_3 = \begin{bmatrix} 0 & 5 & 20 & 10 \\ 45 & 0 & 15 & 5 \\ 30 & 35 & 0 & 15 \\ 15 & 20 & 5 & 0 \end{bmatrix} \quad D_4 = \begin{bmatrix} 0 & 5 & 15 & 10 \\ 20 & 0 & 10 & 5 \\ 30 & 35 & 0 & 15 \\ 15 & 20 & 5 & 0 \end{bmatrix}$$

Figure 5.4.1. Floyd's algorithm at work.

When the algorithm stops, $P[i, j]$ contains the number of the last iteration that caused a change in $D[i, j]$. To recover the shortest path from i to j, look at $P[i, j]$. If $P[i, j] = 0$, the shortest path is directly along the edge (i, j); otherwise, if $P[i, j] = k$, the shortest path from i to j passes through k. Look recursively at $P[i, k]$ and $P[k, j]$ to find any other intermediate nodes along the shortest path.

Example 5.4.1.   For the graph of Figure 5.4.1, P becomes

$$P = \begin{bmatrix} 0 & 0 & 4 & 2 \\ 4 & 0 & 4 & 0 \\ 0 & 1 & 0 & 0 \\ 0 & 1 & 0 & 0 \end{bmatrix}.$$

Since $P[1,3] = 4$, the shortest path from 1 to 3 passes through 4. Looking now at $P[1,4]$ and $P[4,3]$, we discover that between 1 and 4 we have to go via 2, but that from 4 to 3 we proceed directly. Finally we see that the trips from 1 to 2 and from 2 to 4 are also direct. The shortest path from 1 to 3 is thus $1, 2, 4, 3$.   □

Problem 5.4.1.   Suppose we allow edges to have negative lengths. If G includes cycles whose total length is negative, the notion of "shortest path" loses much of its meaning: the more often we go round the negative cycle, the shorter our path will be! Does Floyd's algorithm work

i. On a graph that includes a negative cycle?

ii. On a graph that has some edges whose lengths are negative, but that does not include a negative cycle?

Prove your answer.   □

Even if a graph has edges with negative length, the notion of a shortest *simple* path still makes sense. No efficient algorithm is known for finding shortest simple paths in graphs that may have edges of negative length. This is the situation we encountered in Problem 5.1.3. These two problems are **NP**-complete (see Section 10.3).

Problem 5.4.2. Warshall's algorithm.   In this case, the length of the edges is of no interest; only their existence is important. Initially, $L[i, j] = true$ if the edge (i, j) exists, and $L[i, j] = false$ otherwise. We want to find a matrix D such that $D[i, j] = true$ if there exists at least one path from i to j, and $D[i, j] = false$ otherwise. (We are looking for the *reflexive transitive closure* of the graph G.) Adapt Floyd's algorithm for this slightly different case. (We shall see an asymptotically more efficient algorithm for this problem in Section 10.2.2.)   □

∗Problem 5.4.3.   Find a significantly better algorithm for Problem 5.4.2 in the case when the matrix L is symmetric ($L[i, j] = L[j, i]$).   □

5.5 OPTIMAL SEARCH TREES

We begin by recalling the definition of a binary search tree. A binary tree each of whose nodes contains a key is a *search tree* if the value contained in every internal node is greater than or equal to (numerically or lexicographically) the values contained in its left-hand descendants, and less than or equal to the values contained in its right-hand descendants.

Problem 5.5.1. Show by an example that the following definition will not do: "A binary tree is a search tree if the key contained in each internal node is greater than or equal to the key of its left-hand child, and less than or equal to the key of its right-hand child." □

Figure 5.5.1 shows an example of a binary search tree containing the keys A, B, C , . . . , H. (For the rest of this section, search trees will be understood to be binary.) To determine whether a key X is present in the tree, we first examine the key held in the root. Suppose this key is R. If $X=R$, we have found the key we want, and the search stops; if $X < R$, we only need look at the left-hand subtree; and if $X > R$, we only need look at the right-hand subtree. A recursive implementation of this technique is obvious. (It provides an example of simplification: see chapter 4.)

Problem 5.5.2. Write a procedure that looks for a given key in a search tree and returns *true* if the key is present and *false* otherwise. □

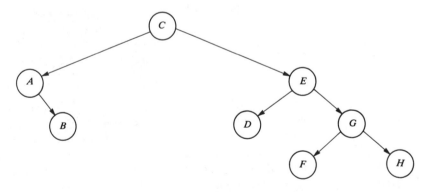

Figure 5.5.1. A binary search tree.

The nodes may also contain further information related to the keys: in this case a search procedure does not simply return *true* or *false*, but rather, the information attached to the key we are looking for.

For a given set of keys, several search trees may be possible: for instance, the tree in Figure 5.5.2 contains the same keys as those in Figure 5.5.1.

Problem 5.5.3. How many different search trees can be made with eight distinct keys? □

*Problem 5.5.4.   If $T(n)$ is the number of different search trees we can make with n distinct keys, find either an explicit formula for $T(n)$ or else an algorithm to calculate this value. (*Hint* : reread Section 5.3.)   □

In Figure 5.5.1 two comparisons are needed to find the key E; in Figure 5.5.2, on the other hand, a single comparison suffices. If all the keys are sought with the same probability, it takes $(2+3+1+3+2+4+3+4)/8 = 22/8$ comparisons on the average to find a key in Figure 5.5.1, and $(4+3+2+3+1+3+2+3)/8 = 21/8$ comparisons on the average in Figure 5.5.2.

Problem 5.5.5.   For the case when the keys are equiprobable, give a tree that minimizes the average number of comparisons needed. Repeat the problem for the general case of n equiprobable keys.   □

In fact, we shall solve a more general problem still. Suppose we have an ordered set $c_1 < c_2 < \cdots < c_n$ of n distinct keys. Let the probability that a request refers to key c_i be p_i, $i = 1, 2, \ldots, n$. For the time being, suppose that $\sum_{i=1}^{n} p_i = 1$, that is, all the requests refer to keys that are indeed present in the search tree.

Recall that the depth of the root of a tree is 0, the depth of its children is 1, and so on. If some key c_i is held in a node at depth d_i, then $d_i + 1$ comparisons are necessary to find it. For a given tree the average number of comparisons needed is

$$C = \sum_{i=1}^{n} p_i (d_i + 1).$$

This is the function we seek to minimize.

Consider the sequence of successive keys c_i, $c_{i+1}, \ldots, c_j$, $j \geq i$. Suppose that in an optimal tree containing all the n keys this sequence of $j - i + 1$ keys occupies the nodes of a subtree. If the key c_k, $i \leq k \leq j$, is held in a node of depth d_k^* in the subtree, the average number of comparisons carried out in this subtree when we look

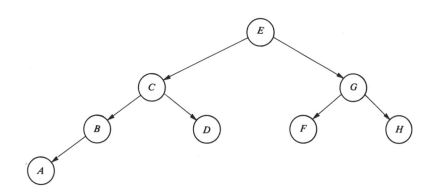

Figure 5.5.2   Another binary search tree.

for a key in the main tree (the key in question is not necessarily one of those held in the subtree) is

$$C^* = \sum_{k=i}^{j} p_k \, (d_k^* + 1) \, .$$

We observe that

- this expression has the same form as that for C, and

- a change in the subtree does not affect the contribution made to C by other subtrees of the main tree disjoint from the one under consideration.

We thus arrive at the principle of optimality: in an optimal tree all the subtrees must also be optimal with respect to the keys they contain.

Let $m_{ij} = \sum_{k=i}^{j} p_k$, and let C_{ij} be the average number of comparisons carried out in an optimal subtree containing the keys c_i, $c_{i+1}, \ldots,$ c_j when a key is sought in the main tree. (It is convenient to define $C_{ij} = 0$ if $j = i - 1$.) One of these keys, k, say, must occupy the root of the subtree. In Figure 5.5.3, L is an optimal subtree containing the keys c_i, $c_{i+1}, \ldots,$ c_{k-1} and R is an optimal subtree containing $c_{k+1}, \ldots,$ c_j. When we look for a key in the main tree, the probability that it is in the sequence c_i, $c_{i+1}, \ldots,$ c_j is m_{ij}. In this case one comparison is made with c_k, and others may then be made in L or R. The average number of comparisons carried out is therefore

$$C_{ij}^k = m_{ij} + C_{i,k-1} + C_{k+1,j}$$

where the three terms are the contributions of the root, L and R, respectively.

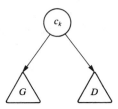

Figure 5.5.3.   A subtree of the optimal search tree is optimal.

To obtain a dynamic programming scheme, it remains to remark that the root k is chosen so as to minimize C_{ij}:

$$C_{ij} = m_{ij} + \min_{i \le k \le j} (C_{i,k-1} + C_{k+1,j}) \, . \tag{*}$$

In particular, $C_{ii} = p_i$.

Example 5.5.1.   To find the optimal search tree if the probabilities associated with five keys c_1 to c_5 are

$$\begin{array}{cccccc}
i & 1 & 2 & 3 & 4 & 5 \\
p_i & 0.30 & 0.05 & 0.08 & 0.45 & 0.12
\end{array}$$

we first calculate the matrix m.

$$m = \begin{pmatrix}
0.30 & 0.35 & 0.43 & 0.88 & 1.00 \\
 & 0.05 & 0.13 & 0.58 & 0.70 \\
 & & 0.08 & 0.53 & 0.65 \\
 & & & 0.45 & 0.57 \\
 & & & & 0.12
\end{pmatrix}$$

Now, we note that $C_{ii} = p_i$, $1 \le i \le 5$, and next, we use (*) to calculate the other values of C_{ij}.

$$C_{12} = m_{12} + \min(C_{10} + C_{22}, C_{11} + C_{32})$$
$$= 0.35 + \min(0.05, 0.30) = 0.40$$

Similarly

$$C_{23} = 0.18, \quad C_{34} = 0.61, \quad C_{45} = 0.69 .$$

Then

$$C_{13} = m_{13} + \min(C_{10} + C_{23}, C_{11} + C_{33}, C_{12} + C_{43})$$
$$= 0.43 + \min(0.18, 0.38, 0.40) = 0.61$$

$$C_{24} = m_{24} + \min(C_{21} + C_{34}, C_{22} + C_{44}, C_{23} + C_{54})$$
$$= 0.58 + \min(0.61, 0.50, 0.18) = 0.76$$

$$C_{35} = m_{35} + \min(C_{32} + C_{45}, C_{33} + C_{55}, C_{34} + C_{65})$$
$$= 0.65 + \min(0.69, 0.20, 0.61) = 0.85$$

$$C_{14} = m_{14} + \min(C_{10} + C_{24}, C_{11} + C_{34}, C_{12} + C_{44}, C_{13} + C_{54})$$
$$= 0.88 + \min(0.76, 0.91, 0.85, 0.61) = 1.49$$

$$C_{25} = m_{25} + \min(C_{21} + C_{35}, C_{22} + C_{45}, C_{23} + C_{55}, C_{24} + C_{65})$$
$$= 0.70 + \min(0.85, 0.74, 0.30, 0.76) = 1.00$$

$$C_{15} = m_{15} + \min(C_{10} + C_{25}, C_{11} + C_{35}, C_{12} + C_{45}, C_{13} + C_{55}, C_{14} + C_{65})$$
$$= 1.00 + \min(1.00, 1.15, 1.09, 0.73, 1.49)$$
$$= 1.73 .$$

The optimal search tree for these keys requires 1.73 comparisons on the average to find a key (see Figure 5.5.4).   □

Problem 5.5.6.   We know how to find the minimum number of comparisons necessary in the optimal tree, but how do we find the form of this tree ?   □

Problem 5.5.7. Write an algorithm that accepts the values n and p_i, $i = 1, 2, \ldots, n$, and that produces a description of the optimal search tree for these probabilities. (Provided the keys are sorted, we do not need their exact values.)   □

In this algorithm we calculate the values of C_{ij} first for $j - i = 1$, then for $j - i = 2$, and so on. When $j - i = m$, there are $n - m$ values of C_{ij} to calculate, each involving a choice among $m + 1$ possibilities. The required computation time is therefore in

$$\Theta\left(\sum_{m=1}^{n-1} (n-m)(m+1) \right) = \Theta(n^3) .$$

Problem 5.5.8. Prove this last equality.   □

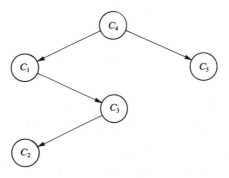

Figure 5.5.4. An optimal binary search tree.

*Problem 5.5.9. Generalize the preceding argument to take account of the possibility that a request may involve a key that is not in fact in the tree. Specifically, let p_i, $i = 1, 2, \ldots, n$, be the probability that a request concerns a key c_i that is in the tree, and let q_i, $i = 0, 1, 2, \ldots, n$, be the probability that it concerns a missing key situated between c_i and c_{i+1} (with the obvious interpretation for q_0 and q_n). We now have

$$\sum_{i=1}^{n} p_i + \sum_{i=0}^{n} q_i = 1 .$$

The optimal tree must minimize the average number of comparisons required to either find a key, if it is present in the tree, or to ascertain that it is missing.

Give an algorithm that can determine the optimal search tree in this context.   □

**Problem 5.5.10. For $1 \le i \le j \le n$, let

$$r_{ij} = \max\{k \mid i \le k \le j \ \text{and} \ C_{ij} = m_{ij} + C_{i,k-1} + C_{k+1,j}\}$$

be the root of an optimal subtree containing c_i, $c_{i+1}, \ldots, c_j$. Write also $r_{i,i-1} = i$. Prove that $r_{i,j-1} \leq r_{ij} \leq r_{i+1,j}$ for every $1 \leq i \leq j \leq n$.   □

Problem 5.5.11.   Use the result of Problem 5.5.10 to show how to calculate an optimal search tree in a time in $O(n^2)$. (Problems 5.5.10 and 5.5.11 generalize to the case discussed in Problem 5.5.9.)   □

Problem 5.5.12.   There is an obvious greedy approach to the problem of constructing an optimal search tree: place the most probable key, c_k, say, at the root, and construct the left- and right-hand subtrees for $c_1, c_2, \ldots, c_{k-1}$ and $c_{k+1}, c_{k+2}, \ldots, c_n$ recursively in the same way.

    i. How much time does this algorithm take in the worst case, assuming the keys are already sorted?

    ii. Show with the help of a simple, explicit example that this greedy algorithm does not always find the optimal search tree. Give an optimal search tree for your example, and calculate the average number of comparisons needed to find a key for both the optimal tree and the tree found by the greedy algorithm.   □

5.6 THE TRAVELLING SALESPERSON PROBLEM

We have already met this problem in Section 3.4.2. Given a graph with nonnegative lengths attached to the edges, we are required to find the shortest possible circuit that begins and ends at the same node, after having gone exactly once through each of the other nodes.

Let $G = \langle N, A \rangle$ be a directed graph. As usual, we take $N = \{1, 2, \ldots, n\}$, and the lengths of the edges are denoted by L_{ij}, with $L[i, i] = 0$, $L[i, j] \geq 0$ if $i \neq j$, and $L[i, j] = \infty$ if the edge (i, j) does not exist.

Suppose without loss of generality that the circuit begins and ends at node 1. It therefore consists of an edge $(1, j)$, $j \neq 1$, followed by a path from j to 1 that passes exactly once through each node in $N \setminus \{1, j\}$. If the circuit is optimal (as short as possible), then so is the path from j to 1: the principle of optimality holds.

Consider a set of nodes $S \subseteq N \setminus \{1\}$ and a node $i \in N \setminus S$, with $i = 1$ allowed only if $S = N \setminus \{1\}$. Define $g(i, S)$ as the length of the shortest path from node i to node 1 that passes exactly once through each node in S. Using this definition, $g(1, N \setminus \{1\})$ is the length of an optimal circuit. By the principle of optimality, we see that

$$g(1, N \setminus \{1\}) = \min_{2 \leq j \leq n} (L_{1j} + g(j, N \setminus \{1, j\})). \qquad (*)$$

More generally, if $i \neq 1$, $S \neq \varnothing$, $S \neq N \setminus \{1\}$, and $i \notin S$,

$$g(i, S) = \min_{j \in S} (L_{ij} + g(j, S \setminus \{j\})). \qquad (**)$$

Furthermore,

$$g(i, \varnothing) = L_{i1}, \quad i = 2, 3, \dots, n .$$

The values of $g(i, S)$ are therefore known when S is empty. We can apply (**) to calculate the function g for all the sets S that contain exactly one node (other than 1); then we can apply (**) again to calculate g for all the sets S that contain two nodes (other than 1), and so on. Once the value of $g(j, N \setminus \{1, j\})$ is known for all the nodes j except node 1, we can use (*) to calculate $g(1, N \setminus \{1\})$ and solve the problem.

Example 5.6.1.   Let G be the complete graph on four nodes given in Figure 5.6.1:

$$L = \begin{bmatrix} 0 & 10 & 15 & 20 \\ 5 & 0 & 9 & 10 \\ 6 & 13 & 0 & 12 \\ 8 & 8 & 9 & 0 \end{bmatrix} .$$

We initialize

$$g(2, \varnothing) = 5, \quad g(3, \varnothing) = 6, \quad g(4, \varnothing) = 8 .$$

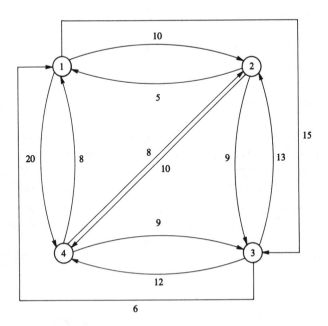

Figure 5.6.1.   A directed graph for the travelling salesperson problem.

Using (**), we obtain

$$g(2, \{3\}) = L_{23} + g(3, \varnothing) = 15$$
$$g(2, \{4\}) = L_{24} + g(4, \varnothing) = 18$$

and similarly

$$g(3, \{2\}) = 18, \quad g(3, \{4\}) = 20$$
$$g(4, \{2\}) = 13, \quad g(4, \{3\}) = 15.$$

Next, using (**) for sets of two nodes, we have

$$g(2, \{3,4\}) = \min(L_{23} + g(3, \{4\}), L_{24} + g(4, \{3\}))$$
$$= \min(29, 25) = 25$$

$$g(3, \{2,4\}) = \min(L_{32} + g(2, \{4\}), L_{34} + g(4, \{2\}))$$
$$= \min(31, 25) = 25$$

$$g(4, \{2,3\}) = \min(L_{42} + g(2, \{3\}), L_{43} + g(3, \{2\}))$$
$$= \min(23, 27) = 23 .$$

Finally we apply (*) to obtain

$$g(1, \{2,3,4\}) = \min(L_{12} + g(2, \{3,4\}), L_{13} + g(3, \{2,4\}), L_{14} + g(4, \{2,3\}))$$
$$= \min(35, 40, 43) = 35 .$$

The optimal circuit in Figure 5.6.1 has length 35.   □

To know where this circuit goes, we need an extra function: $J(i, S)$ is the value of j chosen to minimize g at the moment when we apply (*) or (**) to calculate $g(i, S)$.

Example 5.6.2.   (Continuation of Example 5.6.1.)   In this example we find

$$J(2, \{3,4\}) = 4$$
$$J(3, \{2,4\}) = 4$$
$$J(4, \{2,3\}) = 2$$
$$J(1, \{2,3,4\}) = 2$$

and the optimal circuit is

$$1 \rightarrow J(1, \{2,3,4\}) = 2$$
$$\rightarrow J(2, \{3,4\}) = 4$$
$$\rightarrow J(4, \{3\}) = 3$$
$$\rightarrow 1 .$$ □

The required computation time can be calculated as follows:

— to calculate $g(j, \emptyset)$: $n-1$ consultations of a table;

— to calculate all the $g(i, S)$ such that $1 \leq \#S = k \leq n-2$: $(n-1) \begin{bmatrix} n-2 \\ k \end{bmatrix} k$ additions in all;

— to calculate $g(1, N \setminus \{1\})$: $n-1$ additions.

These operations can be used as a barometer. The computation time is thus in

$$\Theta\left(2(n-1) + \sum_{k=1}^{n-2} (n-1)k \begin{bmatrix} n-2 \\ k \end{bmatrix}\right) = \Theta(n^2 2^n) \quad \text{since} \quad \sum_{k=1}^{r} k \begin{bmatrix} r \\ k \end{bmatrix} = r \, 2^{r-1}.$$

This is considerably better than having a time in $\Omega(n!)$, as would be the case if we simply tried all the possible circuits, but it is still far from offering a practical algorithm. What is more ...

Problem 5.6.1.   Verify that the space required to hold the values of g and J is in $\Omega(n \, 2^n)$, which is not very practical either.   □

TABLE 5.6.1.   SOLVING THE TRAVELLING SALESPERSON PROBLEM.

n	Time: Direct method $n!$	Time: Dynamic programming $n^2 2^n$	Space: Dynamic programming $n \, 2^n$
5	120	800	160
10	3,628,800	102,400	10,240
15	1.31×10^{12}	7,372,800	491,520
20	2.43×10^{18}	419,430,400	20,971,520

Problem 5.6.2.   The preceding analysis assumes that we can find in constant time a value of $g(j, S)$ that has already been calculated. Since S is a set, which data structure do you suggest to hold the values of g? With your suggested structure, how much time is needed to access one of the values of g?   □

Table 5.6.1 illustrates the dramatic increase in the time and space necessary as n goes up. For instance, $20^2 2^{20}$ microseconds is less than 7 minutes, whereas 20! microseconds exceeds 77 thousand years.

5.7 MEMORY FUNCTIONS

If we want to implement the method of Section 5.6 on a computer, it is easy to write a function that calculates g recursively. For example, consider

```
function g (i ,S )
  if S = ∅ then return L [i ,1]
  ans ← ∞
  for each j ∈ S do
    distviaj ← L [i , j ] + g ( j ,S \ { j })
    if distviaj < ans then ans ← distviaj
  return ans  .
```

Unfortunately, if we calculate g in this top-down way, we come up once more against the problem outlined at the beginning of this chapter: most values of g are recalculated many times and the program is very inefficient. (In fact, it ends up back in $\Omega((n-1)!)$.)

So how can we calculate g in the bottom-up way that characterizes dynamic programming? We need an auxiliary program that generates first the empty set, then all the sets containing just one element from $N \setminus \{1\}$, then all the sets containing two elements from $N \setminus \{1\}$, and so on. Although it is maybe not too hard to write such a generator, it is not immediately obvious how to set about it.

One easy way to take advantage of the simplicity of a recursive formulation without losing the efficiency offered by dynamic programming is to use a *memory function*. To the recursive function we add a table of the necessary size. Initially, all the entries in this table hold a special value to indicate that they have not yet been calculated. Thereafter, whenever we call the function, we first look in the table to see whether it has already been evaluated with the same set of parameters. If so, we return the value held in the table. If not, we go ahead and calculate the function. Before returning the calculated value, however, we save it at the appropriate place in the table. In this way it is never necessary to calculate the function twice for the same values of its parameters.

For the algorithm of Section 5.6 let *gtab* be a table all of whose entries are initialized to -1 (since a distance cannot be negative). Formulated in the following way:

```
function g (i ,S )
  if S = ∅ then return L [i ,1]
  if gtab [i ,S ] ≥ 0 then return gtab [i ,S ]
  ans ← ∞
  for each j ∈ S do
    distviaj ← L [i , j ] + g ( j ,S \ { j })
    if distviaj < ans then ans ← distviaj
  gtab [i ,S ] ← ans
  return ans  ,
```

the function g combines the clarity obtained from a recursive formulation and the efficiency of dynamic programming.

Problem 5.7.1.   Show how to calculate (i) a binomial coefficient and (ii) the function $series(n,p)$ of Section 5.2 using a memory function.   □

We sometimes have to pay a price for using this technique. We saw in Section 5.1, for instance, that we can calculate a binomial coefficient $\binom{n}{k}$ using a time in $O(nk)$ and space in $O(k)$. Implemented using a memory function, the calculation takes the same amount of time but needs space in $\Omega(nk)$.

* **Problem 5.7.2.**   If we are willing to use a little more space (the space needed is only multiplied by a constant factor, however), it is possible to avoid the initialization time needed to set all the entries of the table to some special value. This is particularly desirable when in fact only a few values of the function are to be calculated, but we do not know in advance which ones. (For an example, see Section 6.6.2.) Show how an array $T[1..n]$ can be *virtually initialized* with the help of two auxiliary arrays $B[1..n]$ and $P[1..n]$ and a few pointers. You should write three algorithms.

procedure *init*
    { virtually initializes $T[1..n]$ }

procedure *store*(i,v)
    { sets $T[i]$ to the value v }

function *val*(i)
    { returns the last value given to $T[i]$, if any ;
      returns a default value (such as -1) otherwise }

A call on any of these procedures or functions (including a call on *init* !) should take constant time in the worst case.   □

5.8 SUPPLEMENTARY PROBLEMS

* **Problem 5.8.1.**   Let u and v be two strings of characters. We want to transform u into v with the smallest possible number of operations of the following types :

- delete a character,
- add a character,
- change a character.

For instance, we can transform *abbac* into *abcbc* in three stages.

$$abbac \rightarrow abac \quad \text{(delete } b\text{)}$$
$$\rightarrow ababc \quad \text{(add } b\text{)}$$
$$\rightarrow abcbc \quad \text{(change } a \text{ into } c\text{)}.$$

Show that this transformation is not optimal.

Write a dynamic programming algorithm that finds the minimum number of operations needed to transform u into v, and that tells us what these operations are. How much time does your algorithm take as a function of the lengths of u and v?   □

Problem 5.8.2.   Consider the alphabet $\Sigma = \{a, b, c\}$. The elements of Σ have the multiplication table given in Table 5.8.1. Thus $ab = b$, $ba = c$, and so on. Note that the multiplication defined by this table is neither commutative nor associative.

TABLE 5.8.1.   AN ABSTRACT MULTIPLICATION TABLE

		Right-hand symbol		
		a	b	c
Left-hand	a	b	b	a
symbol	b	c	b	a
	c	a	c	c

Find an efficient algorithm that examines a string $x = x_1 x_2 \cdots x_n$ of characters of Σ and decides whether or not it is possible to parenthesize x in such a way that the value of the resulting expression is a. For instance, if $x = bbbba$, your algorithm should return "yes" because $(b(bb))(ba) = a$. (This expression is not unique. For example, $(b(b(b(ba)))) = a$ as well.)

In terms of n, the length of the string x, how much time does your algorithm take?   □

Problem 5.8.3.   Modify your algorithm from the previous problem so it returns the number of different ways of parenthesizing x to obtain a.   □

Problem 5.8.4.   There are N Hudson's Bay Company posts on the River Koksoak. At any of these posts you can rent a canoe to be returned at any other post downstream. (It is next to impossible to paddle against the current.) For each possible departure point i and each possible arrival point j the company's tariff gives the cost of a rental between i and j. However, it can happen that the cost of renting from i to j is higher than the total cost of a series of shorter rentals, in which case you can return the first canoe at some post k between i and j and continue the journey in a second canoe. There is no extra charge if you change canoes in this way.

Find an efficient algorithm to determine the minimum cost of a trip by canoe from each possible departure point i to each possible arrival point j. In terms of N, what is the computing time needed by your algorithm?   □

Problem 5.8.5.   In the introduction to Chapter 3 we saw a greedy algorithm for making change. This algorithm works correctly in a country where there are coins worth 1, 5, 10, and 25 units, but it does not always find an optimal solution if there

also exists a coin worth 12 units (see Problem 3.1.1). The general problem can be
solved exactly using dynamic programming. Let n be the number of different coins
that exist, and let $T [1 .. n]$ be an array giving the value of these coins. We suppose
that an unlimited number of coins of each value is available. Let L be a bound on the
sum we wish to obtain.

    i. For $1 \leq i \leq n$ and $1 \leq j \leq L$, let c_{ij} be the minimum number of coins required to
obtain the sum j if we may only use coins of types $T [1], T [2], \ldots, T [i]$, or
$c_{ij} = +\infty$ if the amount j cannot be obtained using just these coins. Give a
recurrence for c_{ij}, including the initial conditions.

    ii. Give a dynamic programming algorithm that calculates all the c_{nj}, $1 \leq j \leq L$.
Your algorithm may use only a single array of length L. As a function of n and
L, how much time does your algorithm take?

    iii. Give a greedy algorithm that can make change using the minimum number of
coins for any amount $M \leq L$ once the c_{nj} have been calculated. Your algorithm
should take a time in $O (n + c_{nM})$, provided $c_{nM} \neq +\infty$.   □

    *Problem 5.8.6.   You have n objects, which you wish to put in order using the
relations "<" and "=". For example, 13 different orderings are possible with three
objects.

$$
\begin{array}{lllll}
A = B = C & A = B < C & A < B = C & A < B < C & A < C < B \\
A = C < B & B < A = C & B < A < C & B < C < A & B = C < A \\
C < A = B & C < A < B & C < B < A
\end{array}
$$

Give a dynamic programming algorithm that can calculate, as a function of n, the
number of different possible orderings. Your algorithm should take a time in $O (n^2)$
and space in $O (n)$.   □

    **Problem 5.8.7.   Ackermann's function is defined recursively as follows:

$$
\begin{cases}
A (0,n) = n + 1 \\
A (m ,0) = A (m - 1, 1) & \text{if } m > 0 \\
A (m ,n) = A (m - 1, A (m ,n - 1)) & \text{if } m ,n > 0 .
\end{cases}
$$

This function grows extremely rapidly.

    i. Calculate $A (2, 5)$, $A (3, 3)$, and $A (4, 4)$.

    ii. Give a dynamic programming algorithm to calculate $A (m ,n)$. Your algorithm
must consist simply of two nested loops (recursion is not allowed). Moreover, it
is restricted to using a space in $O (m)$, although some of these memory words can

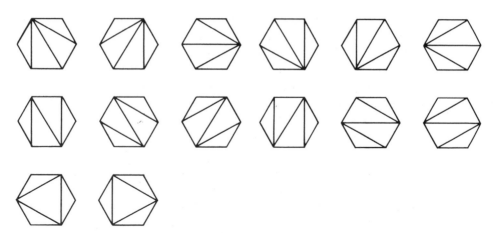

Figure 5.8.1.   Cutting a hexagon into triangles.

grow quite large. (*Hint :* use two arrays $val[0..m]$ and $ind[0..m]$ such that at every instant $val[i] = A(i, ind[i])$.)   □

Problem 5.8.8.   Prove that the number of ways to cut an n-sided convex polygon into $n-2$ triangles using diagonal lines that do not cross is $T(n-1)$, the $(n-1)$st Catalan number (see Section 5.3). For example, a hexagon can be cut in 14 different ways, as shown in Figure 5.8.1.   □

5.9 REFERENCES AND FURTHER READING

Several books are concerned with dynamic programming. We mention only Bellman 1957, Bellman and Dreyfus (1962), Nemhauser (1966), and Laurière (1979). The algorithm in Section 5.3 is described in Godbole (1973); a more efficient algorithm, able to solve the problem of chained matrix multiplications in a time in $O(n \log n)$, can be found in Hu and Shing (1982, 1984). Catalan's numbers are discussed in many places, including Sloane (1973) and Purdom and Brown (1985).

Floyd's algorithm for calculating all shortest paths is due to Floyd (1962). A theoretically more efficient algorithm is known: Fredman (1976) shows how to solve the problem in a time in $O(n^3(\log \log n / \log n)^{1/3})$. The solution to Problem 5.4.2 is supplied by the algorithm in Warshall (1962). Both Floyd's and Warshall's algorithms are essentially the same as the one in Kleene (1956) to determine the regular expression corresponding to a given finite automaton (Hopcroft and Ullman 1979). All these algorithms (with the exception of Fredman's) are unified in Tarjan (1981).

The algorithm of Section 5.5 for constructing optimal search trees, including the solution to Problem 5.5.9, comes from Gilbert and Moore (1959). The improvements suggested by Problems 5.5.10 and 5.5.11 come from Knuth (1971, 1973). A solution to Problem 5.5.10 that is both simpler and more general is given by Yao (1980); this paper gives a sufficient condition for certain dynamic programming algorithms that run in cubic time to be transformable automatically into quadratic algorithms. The optimal search tree for the 31 most common words in English is compared in Knuth (1973) with the tree obtained using the greedy algorithm suggested in Problem 5.5.12.

The algorithm for the travelling salesperson problem given in Section 5.6 comes from Held and Karp (1962). Memory functions are introduced in Michie (1968); for further details see Marsh (1970). Problem 5.7.2, which suggests how to avoid initializing a memory function, comes from Exercise 2.12 in Aho, Hopcroft and Ullman (1974).

A solution to Problem 5.8.1 is given in Wagner and Fischer (1974). Problem 5.8.5 is discussed in Wright (1975) and Chang and Korsh (1976). Problem 5.8.6 suggested itself to the authors one day when they set an exam including a question resembling Problem 2.1.11: we were curious to know what proportion of all the possible answers was represented by the 69 different answers suggested by the students (see also Lemma 10.1.2). Problem 5.8.7 is based on Ackermann (1928). Problem 5.8.8 is discussed in Sloane (1973). An important dynamic programming algorithm that we have not mentioned is the one in Kasimi (1965) and Younger (1967), which takes cubic time to carry out the syntactic analysis of any context-free language (Hopcroft and Ullman 1979).

6

Exploring Graphs

6.1 INTRODUCTION

A great many problems can be formulated in terms of graphs. We have seen, for instance, the shortest route problem and the problem of the minimal spanning tree. To solve such problems, we often need to look at all the nodes, or all the edges, of a graph. Sometimes, the structure of the problem is such that we need only visit some of the nodes or edges. Up to now, the algorithms we have seen have implicitly imposed an order on these visits: it was a case of visiting the nearest node, the shortest edge, and so on. In this chapter we introduce some general techniques that can be used when no particular order of visits is required.

We shall use the word "graph" in two different ways. A graph may be a data structure in the memory of a computer. In this case, the nodes are represented by a certain number of bytes, and the edges are represented by pointers. The operations to be carried out are quite concrete: to "mark a node" means to change a bit in memory, to "find a neighbouring node" means to follow a pointer, and so on.

At other times, the graph exists only implicitly. For instance, we often use abstract graphs to represent games: each node corresponds to a particular position of the pieces on the board, and the fact that an edge exists between two nodes means that it is possible to get from the first to the second of these positions by making a single legal move. When we explore such a graph, it does not really exist in the memory of the machine. Most of the time, all we have is a representation of the current position (that is, of the node we are in the process of visiting) and possibly representations of a few other positions. In this case to "mark a node" means to take any appropriate measures that enable us to recognize a position we have already seen, or to avoid arriving at the same position twice; to "find a neighbouring node" means to change the current position by making a single legal move; and so on.

However, whether the graph is a data structure or merely an abstraction, the techniques used to traverse it are essentially the same. In this chapter we therefore do not distinguish the two cases.

6.2 TRAVERSING TREES

We shall not spend long on detailed descriptions of how to explore a tree. We simply remind the reader that in the case of binary trees three techniques are often used. If at each node of the tree we visit first the node itself, then all the nodes in the left-hand subtree, and finally, all the nodes in the right-hand subtree, we are traversing the tree in *preorder*; if we visit first the left-hand subtree, then the node itself, and finally, the right-hand subtree, we are traversing the tree in *inorder*; and if we visit first the left-hand subtree, then the right-hand subtree, and lastly, the node itself, then we are visiting the tree in *postorder*. Preorder and postorder generalize in the obvious way to nonbinary trees.

These three techniques explore the tree from left to right. Three corresponding techniques explore the tree from right to left. It is obvious how to implement any of these techniques using recursion.

Lemma 6.2.1. For each of the six techniques mentioned, the time $T(n)$ needed to explore a binary tree containing n nodes is in $\Theta(n)$.

Proof. Suppose that visiting a node takes a time in $\Theta(1)$, that is, the time required is bounded above by some constant c. Without loss of generality, we may suppose that $c \geq T(0)$.

Suppose further that we are to explore a tree containing n nodes, $n > 0$, of which one node is the root, g nodes are situated in the left-hand subtree, and $n-g-1$ nodes are in the right-hand subtree. Then

$$T(n) \leq \max_{0 \leq g \leq n-1} (T(g) + T(n-g-1) + c) \quad n > 0 .$$

We prove by mathematical induction that $T(n) \leq dn + c$ where d is a constant such that $d \geq 2c$. By the choice of c the hypothesis is true for $n = 0$. Now suppose that it is true for all n, $0 \leq n < m$, for some $m > 0$. Then

$$T(m) \leq \max_{0 \leq g \leq m-1} (T(g) + T(m-g-1) + c)$$

$$\leq \max_{0 \leq g \leq m-1} (dg + c + d(m-g-1) + c + c)$$

$$\leq dm + 3c - d \leq dm + c$$

so the hypothesis is also true for $n = m$. This proves that $T(n) \leq dn + c$ for every $n \geq 0$, and hence $T(n)$ is in $O(n)$.

On the other hand, it is clear that $T(n)$ is in $\Omega(n)$ since each of the n nodes is visited. Therefore $T(n)$ is in $\Theta(n)$.   □

Problem 6.2.1.   Prove that for any of the techniques mentioned, a recursive implementation takes memory space in $\Omega(n)$ in the worst case.   □

*Problem 6.2.2.   Show how the preceding exploration techniques can be implemented so as to take only a time in $\Theta(n)$ and space in $\Theta(1)$, even when the nodes do not contain a pointer to their parents (otherwise the problem becomes trivial).   □

Problem 6.2.3.   Show how to generalize the concepts of preorder and postorder to arbitrary (nonbinary) trees. Assume the trees are represented as in Figure 1.9.5. Prove that both these techniques still run in a time in the order of the number of nodes in the tree to be traversed.   □

6.3 DEPTH-FIRST SEARCH : UNDIRECTED GRAPHS

Let $G = <N, A>$ be an undirected graph all of whose nodes we wish to visit. Suppose that it is somehow possible to mark a node to indicate that it has already been visited. Initially, no nodes are marked.

To carry out a *depth-first* traversal of the graph, choose any node $v \in N$ as the starting point. Mark this node to show that it has been visited. Next, if there is a node adjacent to v that has not yet been visited, choose this node as a new starting point and call the depth-first search procedure recursively. On return from the recursive call, if there is another node adjacent to v that has not been visited, choose this node as the next starting point, call the procedure recursively once again, and so on. When all the nodes adjacent to v have been marked, the search starting at v is finished.

If there remain any nodes of G that have not been visited, choose any one of them as a new starting point, and call the procedure yet again. Continue in this way until all the nodes of G have been marked. Here is the recursive algorithm.

```
procedure search (G )
    for each v∈N do mark [v] ← not-visited
    for each v∈N do
        if mark [v] ≠ visited then dfs (v )

procedure dfs (v : node )
    { node v has not been visited }
    mark [v] ← visited
    for each node w adjacent to v do
        if mark [w] ≠ visited then dfs (w )
```

The algorithm is called depth-first search since it tries to initiate as many recursive calls as possible before it ever returns from a call. The recursivity is only stopped when exploration of the graph is blocked and can go no further. At this point the recursion "unwinds" so that alternative possibilities at higher levels can be explored.

Example 6.3.1. If we suppose that the neighbours of a given node are examined in numerical order, and that node 1 is the first starting point, a depth-first search of the graph in Figure 6.3.1 progresses as follows:

1.	*dfs* (1)	initial call
2.	*dfs* (2)	recursive call
3.	*dfs* (3)	recursive call
4.	*dfs* (6)	recursive call
5.	*dfs* (5)	recursive call; progress is blocked
6.	*dfs* (4)	a neighbour of node 1 has not been visited
7.	*dfs* (7)	recursive call
8.	*dfs* (8)	recursive call; progress is blocked
9.	there are no more nodes to visit.	□

Problem 6.3.1. Show how a depth-first search progresses through the graph in Figure 6.3.1 if the neighbours of a given node are examined in numerical order but the initial starting point is node 6. □

How much time is needed to explore a graph with n nodes and a edges? Since each node is visited exactly once, there are n calls of the procedure *dfs*. When we visit a node, we look at the mark on each of its neighbouring nodes. If the graph is represented in such a way as to make the lists of adjacent nodes directly accessible (type *lisgraph* of Section 1.9.2), this work is proportional to a in total. The algorithm therefore takes a time in $O(n)$ for the procedure calls and a time in $O(a)$ to inspect the marks. The execution time is thus in $O(\max(a,n))$.

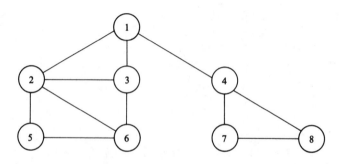

Figure 6.3.1. An undirected graph.

Problem 6.3.2.   What happens if the graph is represented by an adjacency matrix (type *adjgraph* of Section 1.9.2) rather than by lists of adjacent nodes ?   □

Problem 6.3.3.   Show how depth-first search can be used to find the connected components of an undirected graph.   □

A depth-first traversal of a connected graph associates a spanning tree to the graph. The edges of the tree correspond to the edges used to traverse the graph ; they are directed from the first node visited to the second. Edges that are not used in the traversal of the graph have no corresponding edge in the tree. The initial starting point of the exploration becomes the root of the tree.

Example 6.3.2. (Continuation of Example 6.3.1)   The edges used in the depth-first search of Example 6.3.1 are $\{1,2\}$, $\{2,3\}$, $\{3,6\}$, $\{6,5\}$, $\{1,4\}$, $\{4,7\}$ and $\{7,8\}$. The corresponding directed edges $(1,2)$, $(2,3)$, and so on, form a spanning tree for the graph in Figure 6.3.1. The root of the tree is node 1. See Figure 6.3.2.   □

If the graph being explored is not connected, a depth-first search associates to it not merely a single tree, but rather a forest of trees, one for each connected component of the graph.

A depth-first search also provides a way to number the nodes of the graph being visited : the first node visited (the root of the tree) is numbered 1, the second is numbered 2, and so on. In other words, the nodes of the associated tree are numbered in preorder. To implement this numbering, we need only add the following two statements at the beginning of the procedure *dfs* :

$pnum \leftarrow pnum + 1$
$prenum[v] \leftarrow pnum$

where *pnum* is a global variable initialized to zero.

Example 6.3.3. (Continuation of Example 6.3.1)   The depth-first search illustrated by Example 6.3.1 numbers the nodes as follows :

node	1	2	3	4	5	6	7	8
prenum	1	2	3	6	5	4	7	8

.   □

Of course, the tree and the numbering generated by a depth-first search in a graph are not unique, but depend on the chosen starting point and on the order in which neighbours are visited.

Problem 6.3.4.   Exhibit the tree and the numbering generated by the search of Problem 6.3.1.   □

6.3.1 Articulation Points

A node v of a connected graph is an *articulation point* if the subgraph obtained by deleting v and all the edges incident on v is no longer connected. For example, node 1 is an articulation point of the graph in Figure 6.3.1; if we delete it, there remain two connected components $\{2,3,5,6\}$ and $\{4,7,8\}$. A graph G is *biconnected* (or *unarticulated*) if it is connected and has no articulation points. It is *bicoherent* (or *isthmus-free*, or *2-edge-connected*) if each articulation point is joined by at least two edges to each component of the remaining sub-graph. These ideas are important in practice: if the graph G represents, say, a telecommunications network, then the fact that it is biconnected assures us that the rest of the network can continue to function even if the equipment in one of the nodes fails; if G is bicoherent, we can be sure that all the nodes will be able to communicate with one another even if one transmission line stops working.

The following algorithm finds the articulation points of a connected graph G.

a. Carry out a depth-first search in G, starting from any node. Let T be the tree generated by the depth-first search, and for each node v of the graph, let *prenum* $[v]$ be the number assigned by the search.

b. Traverse the tree T in postorder. For each node v visited, calculate *lowest* $[v]$ as the minimum of

  i. *prenum* $[v]$

  ii. *prenum* $[w]$ for each node w such that there exists an edge $\{v,w\}$ in G that has no corresponding edge in T

  iii. *lowest* $[x]$ for every child x of v in T.

c. Articulation points are now determined as follows:

  i. The root of T is an articulation point of G if and only if it has more than one child.

  ii. A node v other than the root of T is an articulation point of G if and only if v has a child x such that *lowest* $[x] \geq$ *prenum* $[v]$.

Example 6.3.4. (Continuation of Examples 6.3.1, 6.3.2, and 6.3.3)   The search described in Example 6.3.1 generates the tree illustrated in Figure 6.3.2. The edges of G that have no corresponding edge in T are represented by broken lines. The value of *prenum* $[v]$ appears to the left of each node v, and the value of *lowest* $[v]$ to the right. The values of *lowest* are calculated in postorder, that is, for nodes 5, 6, 3, 2, 8, 7, 4, and 1 successively. The articulation points of G are nodes 1 (by rule c(i)) and 4 (by rule c(ii)).   □

Problem 6.3.5.   Verify that the same articulation points are found if we start the search at node 6.   □

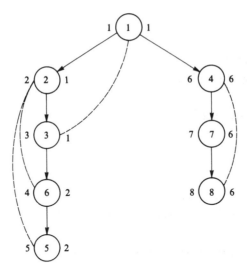

Figure 6.3.2.   A depth-first search tree; *prenum* on the left and *lowest* on the right.

Problem 6.3.6.   Prove that an edge of G that has no corresponding edge in T (a broken line in Figure 6.3.2) necessarily joins some node v to one of its ancestors in T.   □

Informally, we can define *lowest* [v] by

$$lowest\,[v] = \min\{\,prenum\,[w] \mid \text{you can get to } w \text{ from } v \text{ by following}$$
$$\text{down as many solid lines as you like and then}$$
$$\text{going up at most one broken line }\}\,.$$

If x is a child of v and if *lowest* [x] < *prenum* [v], there thus exists a chain of edges that joins x to the other nodes of the graph even if v is deleted. On the other hand, there is no chain joining x to the parent of v if v is not the root and if *lowest* [x] ≥ *prenum* [v].

Problem 6.3.7.   Complete the proof that the algorithm is correct.   □

Problem 6.3.8.   Show how to carry out the operations of steps (a) and (b) in parallel and write the corresponding algorithm.   □

* Problem 6.3.9.   Write an algorithm that decides whether or not a given con- nected graph is bicoherent.   □

* Problem 6.3.10.   Write an efficient algorithm that, given an undirected graph that is connected but not biconnected, finds a set of edges that could be added to make the graph biconnected. Your algorithm should find the smallest possible set of edges. Analyse the efficiency of your algorithm.   □

Problem 6.3.11.   Prove or give a counterexample:

i. If a graph is biconnected, then it is bicoherent.
ii. If a graph is bicoherent, then it is biconnected.   □

Problem 6.3.12.   Prove that a node v in a connected graph is an articulation point if and only if there exist two nodes a and b different from v such that every path joining a and b passes through v.   □

Problem 6.3.13.   Prove that for every pair of distinct nodes v and w in a biconnected graph, there exist at least two chains of edges joining v and w that have no nodes in common (except the starting and ending nodes).   □

6.4 DEPTH-FIRST SEARCH : DIRECTED GRAPHS

The algorithm is essentially the same as the one for undirected graphs, the difference being in the interpretation of the word "adjacent". In a directed graph, node w is adjacent to node v if the directed edge (v, w) exists. If (v, w) exists and (w, v) does not, then w is adjacent to v but v is not adjacent to w. With this change of interpretation the procedures *dfs* and *search* from Section 6.3 apply equally well in the case of a directed graph.

The algorithm behaves quite differently, however. Consider a depth-first search of the directed graph in Figure 6.4.1. If the neighbours of a given node are examined in numerical order, the algorithm progresses as follows:

1.	*dfs* (1)	initial call
2.	*dfs* (2)	recursive call
3.	*dfs* (3)	recursive call; progress is blocked
4.	*dfs* (4)	a neighbour of node 1 has not been visited
5.	*dfs* (8)	recursive call
6.	*dfs* (7)	recursive call; progress is blocked
7.	*dfs* (5)	new starting point
8.	*dfs* (6)	recursive call; progress is blocked
9.	there are no more nodes to visit.	

Problem 6.4.1.   Illustrate the progress of the algorithm if the neighbours of a given node are examined in decreasing numerical order, and if the starting point is node 1.   □

An argument identical with the one in Section 6.3 shows that the time taken by this algorithm is also in $O(\max(a, n))$. In this case, however, the edges used to visit all the nodes of a directed graph $G = <N, A>$ may form a forest of several trees even

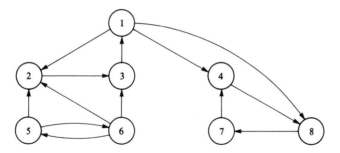

Figure 6.4.1.   A directed graph.

if G is connected. This happens in our example : the edges used, namely $(1,2)$, $(2,3)$, $(1,4)$, $(4,8)$, $(8,7)$, and $(5,6)$, form the forest shown by the solid lines in Figure 6.4.2. (The numbers to the left of each node are explained in Section 6.4.2.)

Let F be the set of edges in the forest. In the case of an undirected graph the edges of the graph with no corresponding edge in the forest necessarily join some node to one of its ancestors (Problem 6.3.6). In the case of a directed graph three kinds of edges can appear in $A \setminus F$ (these edges are shown by the broken lines in Figure 6.4.2).

i. Those like $(3,1)$ or $(7,4)$ that lead from a node to one of its ancestors ;

ii. those like $(1,8)$ that lead from a node to one of its descendants ; and

iii. those like $(5,2)$ or $(6,3)$ that join one node to another that is neither its ancestor nor its descendant. Edges of this type are necessarily directed from right to left.

Problem 6.4.2.   Prove that if (v,w) is an edge of the graph that has no corresponding edge in the forest, and if v is neither an ancestor nor a descendant of w in the forest, then *prenum* $[v] >$ *prenum* $[w]$, where the values of *prenum* are attributed as in Section 6.3. □

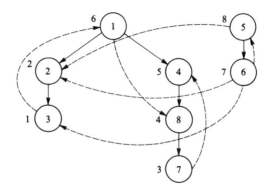

Figure 6.4.2.   A depth-first search forest.

6.4.1 Acyclic Graphs: Topological Sorting

Directed acyclic graphs can be used to represent a number of interesting relations. This class of structures includes trees, but is less general than the class of all directed graphs. For example, a directed acyclic graph can be used to represent the structure of an arithmetic expression that includes repeated subexpressions: thus Figure 6.4.3 represents the structure of the expression

$$(a+b)(c+d) + (a+b)(c-d) .$$

Such graphs also offer a natural representation for partial orderings (such as the relation "smaller than" defined on the integers and the set-inclusion relation). Figure 6.4.4 illustrates part of another partial ordering defined on the integers. (What is the partial ordering in question?) Finally, directed acyclic graphs are often used to represent the different stages of a complex project: the nodes are different states of the project, from the initial state to final completion, and the edges correspond to activities that have to be completed to pass from one state to another. Figure 6.4.5 gives an example of this type of diagram.

Depth-first search can be used to detect whether a given directed graph is acyclic.

Problem 6.4.3.   Let F be the forest generated by a depth-first search on a directed graph $G = <N, A>$. Prove that G is acyclic if and only if $A \setminus F$ includes no edge of type (i) (that is, from a node of G to one of its ancestors in the forest).   □

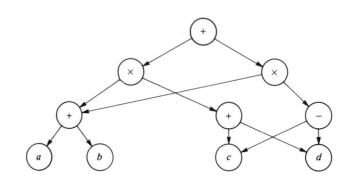

Figure 6.4.3   A directed acyclic graph.

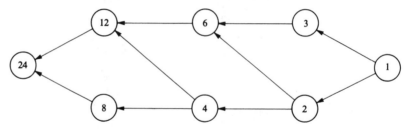

Figure 6.4.4.   Another directed acyclic graph.

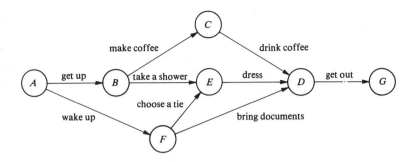

Figure 6.4.5. Yet another directed acyclic graph.

A *topological sort* of the nodes of a directed acyclic graph is the operation of arranging the nodes in order in such a way that if there exists an edge (i, j), then i precedes j in the list. For example, for the graph of Figure 6.4.4, the natural order 1, 2, 3, 4, 6, 8, 12, 24 is adequate; but the order 1, 3, 2, 6, 4, 12, 8, 24 is also acceptable, as are several others. For graphs as in Figure 6.4.5, a topological ordering of the states can be used to obtain a feasible schedule for the activities involved in the project; in our example, the order *A, B, F, C, E, D, G* will serve.

The necessary modification to the procedure *dfs* to make it into a topological sort is immediate: if we add a supplementary line

> **write** *v*

at the end of the procedure, the numbers of the nodes will be printed in *reverse* topological order.

Problem 6.4.4.   Prove this.   □

Problem 6.4.5.   For the graph of Figure 6.4.4, what is the topological order obtained if the neighbours of a node are visited in numerical order and if the depth-first search begins at node 1 ?   □

6.4.2 Strongly Connected Components

A directed graph is *strongly connected* if there exists a path from u to v and also a path from v to u for every distinct pair of nodes u and v. If a directed graph is not strongly connected, we are interested in the largest sets of nodes such that the corresponding subgraphs are strongly connected. Each of these subgraphs is called a *strongly connected component* of the original graph. In the graph of Figure 6.4.1, for instance, nodes $\{1, 2, 3\}$ and the corresponding edges form a strongly connected component. Another component corresponds to the nodes $\{4, 7, 8\}$. Despite the fact that there exist edges $(1, 4)$ and $(1, 8)$, it is not possible to merge these two strongly connected components into a single component because there exists no path from node 4 to node 1.

To detect the strongly connected components of a directed graph, we must first modify the procedure *dfs*. In Section 6.3 we number each node at the instant when exploration of the node begins. Here, we number each node at the moment when exploration of the node has been completed. In other words, the nodes of the tree produced are numbered in postorder. To do this, we need only add at the end of procedure *dfs* the following two statements:

$$nump \leftarrow nump + 1$$
$$postnum[v] \leftarrow nump \ ,$$

where *nump* is a global variable initialized to zero. Figure 6.4.2 shows to the left of each node the number thus assigned.

The following algorithm finds the strongly connected components of a directed graph G.

   i. Carry out a depth-first search of the graph starting from an arbitrary node. For each node v of the graph let $postnum[v]$ be the number assigned during the search.

   ii. Construct a new graph G' : G' is the same as G except that the direction of every edge is reversed.

   iii. Carry out a depth-first search in G'. Begin this search at the node w that has the highest value of *postnum*. (If G contains n nodes, it follows that $postnum[w] = n$.) If the search starting at w does not reach all the nodes, choose as the second starting point the node that has the highest value of *postnum* among all the unvisited nodes; and so on.

   iv. To each tree in the resulting forest there corresponds one strongly connected component of G.

Example 6.4.1.   On the graph of Figure 6.4.1, the first depth-first search assigns the values of *postnum* shown to the left of each node in Figure 6.4.2. The graph G' is illustrated in Figure 6.4.6, with the values of *postnum* shown to the left of each node. We carry out a depth-first search starting from node 5, since $postnum[5] = 8$; the search reaches nodes 5 and 6. For our second starting point, we choose node 1, with $postnum[1] = 6$; this time the search reaches nodes 1, 3, and 2. For the third starting point we take node 4, with $postnum[4] = 5$; this time the remaining nodes 4, 7, and 8 are all reached. The corresponding forest is illustrated in Figure 6.4.7. The strongly connected components of the original graph (Fig. 6.4.1) are the subgraphs corresponding to the sets of nodes $\{5,6\}$, $\{1,3,2\}$ and $\{4,7,8\}$.   □

*Problem 6.4.6.   Prove that if two nodes u and v are in the same strongly connected component of G, then they are in the same tree when we carry out the depth-first search of G'.   □

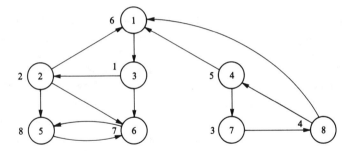

Figure 6.4.6.   Reversing the arrows in the graph of Figure 6.4.1.

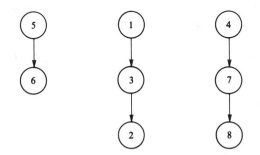

Figure 6.4.7.   The forest of strongly connected components.

It is harder to prove the result the other way. Let v be a node that is in the tree whose root is r when we carry out the search of G', and suppose $v \neq r$. This implies that there exists a path from r to v in G' ; thus there exists a path from v to r in G. When carrying out the search of G', we always choose as a new starting point (that is, as the root of a new tree) that node not yet visited with the highest value of *postnum*. Since we chose r rather than v to be the root of the tree in question, we have *postnum* $[r]$ > *postnum* $[v]$.

When we carried out the search in G, three possibilities seem to be open a priori:

- r was an ancestor of v ;
- r was a descendant of v ; or
- r was neither an ancestor nor a descendant of v.

The second possibility is ruled out by the fact that *postnum* $[r]$ > *postnum* $[v]$. In the third case it would be necessary for the same reason that r be to the right of v. However, there exists at least one path from v to r in G. Since in a depth-first search the edges not used by the search never go from left to right (Problem 6.4.2), any such path must go up the tree from v to a common ancestor (x, say) of v and r, and then go down the tree to r. But this is quite impossible. We should have *postnum* $[x]$ > *postnum* $[r]$ since x is an ancestor of r. Next, since there exists a path from v to x in G, there would exist a path from x to v in G'. Before choosing r as the root of a tree in the search of G', we would have already visited x (otherwise x rather than r would be chosen as the root of the new tree) and therefore also v. This contradicts the hypothesis that v is in the tree whose root is r when we carry out the search of G. Only the first possibility remains: r was an ancestor of v when we searched G. This implies that there exists a path from r to v in G.

We have thus proved that if node v is in the tree whose root is r when we carry out the search of G', then there exist in G both a path from v to r and a path from r to v. If two nodes u and v are in the same tree when we search G', they are therefore both in the same strongly connected component of G since there exist paths from u to v and from v to u in G via node r.

With the result of Problem 6.4.6, this completes the proof that the algorithm works correctly.

Problem 6.4.7.   Estimate the time and space needed by this algorithm.   □

6.5 BREADTH-FIRST SEARCH

When a depth-first search arrives at some node v, it next tries to visit some neighbour of v, then a neighbour of this neighbour, and so on. When a breadth-first search arrives at some node v, on the other hand, it first visits all the neighbours of v, and not until this has been done does it go on to look at nodes farther away. Unlike depth-first search, breadth-first search is not naturally recursive. To underline the similarities and the differences between the two methods, we begin by giving a nonrecursive formulation of the depth-first search algorithm. Let *stack* be a data type allowing two operations, *push* and *pop*. The type is intended to represent a list of elements that are to be handled in the order "last come, first served". The function *top* denotes the element at the top of the stack. Here is the modified depth-first search algorithm.

```
procedure dfs'(v : node )
    P ← empty-stack
    mark [v] ← visited
    push v on P
    while P is not empty do
        while there exists a node w adjacent to top (P)
            such that mark [w] ≠ visited do
                mark [w] ← visited
                push w on P  { w is now top (P) }
        pop top (P)
```

For the breadth-first search algorithm, by contrast, we need a type *queue* that allows two operations *enqueue* and *dequeue*. This type represents a list of elements that are to be handled in the order "first come, first served". The function *first* denotes the element at the front of the queue. Here now is the breadth-first search algorithm.

> **procedure** *bfs* (*v* : *node*)
> $Q \leftarrow$ *empty-queue*
> *mark* [*v*] $\leftarrow$ *visited*
> enqueue *v* into *Q*
> **while** *Q* is not empty **do**
>   *u* $\leftarrow$ *first* (*Q*)
>   dequeue *u* from *Q*
>   **for** each node *w* adjacent to *u* **do**
>   **if** *mark* [*w*] $\neq$ *visited* **then** *mark* [*w*] $\leftarrow$ *visited*
>   enqueue *w* into *Q*

In both cases we need a main program to start the search.

> **procedure** *search* (*G*)
>   **for** each *v* $\in$ *N* **do** *mark* [*v*] $\leftarrow$ *not-visited*
>   **for** each *v* $\in$ *N* **do**
>   **if** *mark* [*v*] $\neq$ *visited* **then** {*dfs'* or *bfs* }(*v*)

Example 6.5.1.   On the graph of Figure 6.3.1, if the neighbours of a node are visited in numerical order, and if node 1 is used as the starting point, breadth-first search proceeds as follows.

	Node Visited	*Q*
1.	1	2,3,4
2.	2	3,4,5,6
3.	3	4,5,6
4.	4	5,6,7,8
5.	5	6,7,8
6.	6	7,8
7.	7	8
8.	8	—

□

As for depth-first search, we can associate a tree with the breadth-first search. Figure 6.5.1 shows the tree generated by the search in Example 6.5.1. The edges of the graph that have no corresponding edge in the tree are represented by broken lines.

Problem 6.5.1.   After a breadth-first search in an undirected graph $G = <N, A>$, let F be the set of edges that have a corresponding edge in the forest of trees that is generated. Show that the edges $\{u, v\} \in A \setminus F$ are such that u and v are in the same tree, but that neither u nor v is an ancestor of the other.   □

It is easy to show that the time required by a breadth-first search is in the same order as that required by a depth-first search, namely $O(\max(a, n))$. If the appropriate

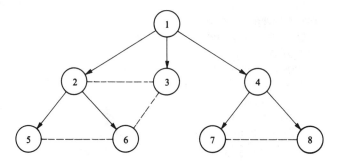

Figure 6.5.1.   A breadth-first search tree.

interpretation of the word "neighbouring" is used, the breadth-first search algorithm can be applied without modification to either directed or undirected graphs.

   Problem 6.5.2.   Show how a breadth-first search progresses through the graph of Figure 6.4.1, assuming that the neighbours of a node are always visited in numerical order, and that necessary starting points are also chosen in numerical order.   □

   Problem 6.5.3 (Continuation of Problem 6.5.2)   Sketch the corresponding forest and the remaining edges of the graph. How many kinds of "broken" edges are possible ? (see Section 6.4)   □

   Breadth-first search is most often used to carry out a partial exploration of certain infinite graphs or to find the shortest path from one point to another.

6.6 IMPLICIT GRAPHS AND TREES

   As mentioned at the outset of this chapter, various problems can be thought of in terms of abstract graphs. For instance, we can use the nodes of a graph to represent configurations in a game of chess and edges to represent legal moves (see Section 6.6.2). Often the original problem translates to searching for a specific node, path or pattern in the associated graph. If the graph contains a large number of nodes, it may be wasteful or infeasible to build it explicitly in computer memory before applying one of the search techniques we have encountered so far.

   An *implicit* graph is one for which we have available a description of its nodes and edges. Relevant portions of the graph can thus be built as the search progresses. Therefore computing time is saved whenever the search succeeds before the entire graph has been constructed. The economy in memory space is even more dramatic when nodes that have already been searched can be discarded, making room for subsequent nodes to be explored.

   Backtracking is a basic search technique on implicit graphs. One powerful application is in playing games of strategy by techniques known as minimax and alpha-beta pruning. Some optimization problems can be handled by the more sophisticated branch-and-bound technique. We now discuss these notions.

6.6.1 Backtracking

Backtracking algorithms use a special technique to explore implicit directed graphs. These graphs are usually trees, or at least they contain no cycles. A backtracking algorithm carries out a systematic search, looking for solutions to some problem. At least one application of this technique dates back to antiquity: it allows one to find the way through a labyrinth without danger of going round and round in circles. To illustrate the general principle, we shall, however, use a different example. Consider the classic problem of placing eight queens on a chess-board in such a way that none of them threatens any of the others. (A queen threatens the squares in the same row, in the same column, or on the same diagonals.)

Problem 6.6.1.   Solve this problem without using a computer.   □

The first obvious way to solve this problem consists of trying systematically all the ways of placing eight queens on a chess-board, checking each time to see whether a solution has been obtained. This approach is of no practical use, since the number of positions we have to check would be $\left\lceil \begin{smallmatrix} 64 \\ 8 \end{smallmatrix} \right\rceil = 4{,}426{,}165{,}368$. The first improvement we might try consists of never putting more than one queen in any given row. This reduces the computer representation of the chess-board to a simple vector of eight elements, each giving the position of the queen in the corresponding row. For instance, the vector $(3, 1, 6, 2, 8, 6, 4, 7)$ does not represent a solution since the queens in the third and the sixth rows are in the same column, and also two pairs of queens lie on the same diagonal. Using this representation, we can write the algorithm very simply using eight nested loops.

```
program Queens 1
  for i₁ ← 1 to 8 do
    for i₂ ← 1 to 8 do
      ⋅
      ⋅
      ⋅
        for i₈ ← 1 to 8 do
          try ← (i₁, i₂, . . . , i₈)
          if solution (try) then write try
                              stop
  write "there is no solution"
```

This time, the number of cases to be considered is reduced to $8^8 = 16{,}777{,}216$, although in fact the algorithm finds a solution and stops after considering only 1,299,852 cases.

Problem 6.6.2.   If you have not yet solved the previous problem, the information just given should be of considerable help !   □

Once we have realized that the chess-board can be represented by a vector, which prevents us from ever trying to put two queens in the same row, it is natural to be equally systematic in our use of the columns. Hence we now represent the board by a vector of eight *different* numbers between 1 and 8, that is, by a permutation of the first eight integers. The algorithm becomes

> **program** *Queens* 2
>   *try* ← *initial-permutation*
>   **while** *try* ≠ *final-permutation* **and not** *solution* (*try*) **do**
>   *try* ← *next-permutation*
>   **if** *solution* (*try*) **then write** *try*
>   **else write** "there is no solution" .

There are several natural ways to generate systematically all the permutations of the first n integers. For instance, we might put each value in turn in the leading position and generate recursively, for each of these leading values, all the permutations of the $n-1$ remaining elements.

> **procedure** *perm* (*i*)
>   **if** $i = n$ **then** *use* (*T*) { *T* is a new permutation }
>   **else for** $j ← i$ **to** n **do** exchange $T[i]$ and $T[j]$
>   *perm* (*i* +1)
>   exchange $T[i]$ and $T[j]$

Here $T[1..n]$ is a global array initialized to $[1, 2, \ldots, n]$ and the initial call is *perm* (1).

Problem 6.6.3.   If *use* (*T*) consists simply of printing the array T on a new line, show the result of calling *perm* (1) when $n = 4$.   □

Problem 6.6.4.   Assuming that *use* (*T*) takes constant time, how much time is needed, as a function of n, to execute the call *perm* (1)? Now rework the problem assuming that *use* (*T*) takes a time in $\Theta(n)$.   □

This approach reduces the number of possible cases to 8! = 40,320. If the preceding algorithm is used to generate the permutations, only 2,830 cases are in fact considered before the algorithm finds a solution. Although it is more complicated to generate permutations rather than all the possible vectors of eight integers between 1 and 8, it is, on the other hand, easier in this case to verify whether a given position is a solution. Since we already know that two queens can neither be in the same row nor in the same column, it suffices to verify that they are not in the same diagonal.

Starting from a crude method that tried to put the queens absolutely anywhere on the chess-board, we progressed first to a method that never puts two queens in the same row, and then to a better method still where the only positions considered are those where we know that two queens can neither be in the same row nor in the same column. However, all these algorithms share an important defect: they never test a position to see if it is a solution until all the queens have been placed on the board. For instance, even the best of these algorithms makes 720 useless attempts to put the last six queens on the board when it has started by putting the first two on the main diagonal, where of course they threaten one another!

Backtracking allows us to do better than this. As a first step, let us reformulate the eight queens problem as a tree searching problem. We say that a vector $V[1..k]$ of integers between 1 and 8 is *k-promising*, for $0 \le k \le 8$, if none of the k queens placed in positions $(1,V[1]), (2,V[2]), \ldots, (k,V[k])$ threatens any of the others. Mathematically, a vector V is k-promising if, for every $i \ne j$ between 1 and k, we have $V[i]-V[j] \notin \{i-j, 0, j-i\}$. For $k \le 1$, any vector V is k-promising. Solutions to the eight queens problem correspond to vectors that are 8-promising.

Let N be the set of k-promising vectors, $0 \le k \le 8$. Let $G = <N, A>$ be the directed graph such that $(U,V) \in A$ if and only if there exists an integer k, $0 \le k < 8$, such that U is k-promising, V is $(k+1)$-promising, and $U[i]=V[i]$ for every $i \in [1..k]$. This graph is a tree. Its root is the empty vector ($k=0$). Its leaves are either solutions ($k=8$) or else they are dead ends ($k<8$) such as $[1,4,2,5,8]$ where it is impossible to place a queen in the next row without her threatening at least one of the queens already on the board. The solutions to the eight queens problem can be obtained by exploring this tree. We do not generate the tree explicitly so as to explore it thereafter, however: rather, nodes are generated and abandoned during the course of the exploration. Depth-first search is the obvious method to use, particularly if we only require one solution.

This technique has two advantages over the previous algorithm that systematically tried each permutation. First, the number of nodes in the tree is less than $8! = 40,320$. Although it is not easy to calculate this number theoretically, it is straightforward to count the nodes using a computer: $\#N = 2057$. In fact, it suffices to explore 114 nodes to obtain a first solution. Secondly, in order to decide whether a vector is k-promising, knowing that it is an extension of a $(k-1)$-promising vector, we only need to check the last queen to be added. This check can be speeded up if we associate with each promising node the sets of columns, of positive diagonals (at 45 degrees), and of negative diagonals (at 135 degrees) controlled by the queens already placed. On the other hand, to decide if some given permutation represents a solution, it seems at first sight that we have to check each of the 28 pairs of queens on the board.

To print all the solutions to the eight queens problem, call *Queens* $(0, \emptyset, \emptyset, \emptyset)$, where $try[1..8]$ is a global array.

procedure *Queens* $(k, col, diag45, diag135)$
   $\{ try[1..k]$ is k-promising,
     $col = \{try[i] \mid 1 \le i \le k\}$,
     $diag45 = \{try[i] - i + 1 \mid 1 \le i \le k\}$, and
     $diag135 = \{try[i] + i - 1 \mid 1 \le i \le k\} \}$
   if $k = 8$
   then { an 8-promising vector is a solution }
       write *try*
   else { explore $(k+1)$-promising extensions of *try* }
      for $j \leftarrow 1$ **to** 8 **do**
        if $j \notin col$ **and** $j - k \notin diag45$ **and** $j + k \notin diag135$
        then $try[k+1] \leftarrow j$
           $\{ try[1..k+1]$ is $(k+1)$-promising }
           Queens $(k+1, col \cup \{j\}, diag45 \cup \{j-k\}, diag135 \cup \{j+k\})$

It is clear that the problem generalizes to an arbitrary number of queens: how can we place n queens on an $n \times n$ "chess-board" in such a way that none of them threatens any of the others?

Problem 6.6.5. Show that the problem for n queens may have no solution. Find a more interesting case than $n = 2$.   □

As we might expect, the advantage to be gained by using the backtracking algorithm instead of an exhaustive approach becomes more pronounced as n increases. For example, for $n = 12$ there are 479,001,600 possible permutations to be considered, and the first solution to be found (using the generator given previously) corresponds to the 4,546,044th position examined; on the other hand, the tree explored by the backtracking algorithm contains only 856,189 nodes, and a solution is obtained when the 262nd node is visited.

** Problem 6.6.6. Analyse mathematically, as a function of the number n of queens, the number of nodes in the tree of k-promising vectors. How does this number compare to $n!$?   □

Backtracking algorithms can also be used even when the solutions sought do not necessarily all have the same length. Here is the general scheme.

procedure *backtrack* $(v[1..k])$
   $\{ v$ is a k-promising vector }
   if v is a solution **then write** v
   { **otherwise** } **for** each $(k+1)$-promising vector w
               such that $w[1..k] = v[1..k]$ **do** *backtrack* $(w[1..k+1])$

The **otherwise** should be present if and only if it is impossible to have two different solutions such that one is a prefix of the other.

Problem 6.6.7.   *Instant Insanity* is a puzzle consisting of four coloured cubes. Each of the 24 faces is coloured blue, red, green, or white. The four cubes are to be placed side by side in such a way that each colour appears on one of the four top faces, on one of the four bottom faces, on one of the front faces, and on one of the rear faces. Show how to solve this problem using backtracking.   □

The n-queens problem was solved using depth-first search in the corresponding tree. Some problems that can be formulated in terms of exploring an implicit graph have the property that they correspond to an infinite graph. In this case, it may be necessary to use breadth-first search to avoid the interminable exploration of some fruitless infinite branch. Breadth-first search is also appropriate if we have to find a solution starting from some initial position and making as few changes as possible. (This last constraint does not apply to the eight queens problem where each solution involves exactly the same number of pieces.) The two following problems illustrate these ideas.

Problem 6.6.8.   Give an algorithm capable of transforming some initial integer n into a given final integer m by the application of the smallest possible number of transformations $f(i) = 3i$ and $g(i) = \lfloor i/2 \rfloor$. For instance, 15 can be transformed into 4 using four function applications: $4 = gfgg(15)$. What does your algorithm do if it is impossible to transform n into m in this way?   □

*Problem 6.6.9.   Give an algorithm that determines the shortest possible series of manipulations needed to change one configuration of Rubik's Cube into another. If the required change is impossible, your algorithm should say so rather than calculating forever.   □

Problem 6.6.10.   Give other applications of backtracking.   □

6.6.2 Graphs and Games: An Introduction

Most games of strategy can be represented in the form of directed graphs. A node of the graph corresponds to a particular position in the game, and an edge corresponds to a legal move between two positions. The graph is infinite if there is no a priori limit on the number of positions possible in the game. For simplicity, we assume that the game is played by two players, each of whom moves in turn, that the game is symmetric (the rules are the same for both players), and that chance plays no part in the outcome (the game is deterministic). The ideas we present can easily be adapted to more general contexts. We further suppose that no instance of the game can last forever and that no position in the game offers an infinite number of legal moves to the player whose turn it is. In particular, some positions in the game offer no legal moves, and hence some nodes in the graph have no successors: these are the *terminal positions*.

To determine a winning strategy for a game of this kind, we need only attach to each node of the graph a label chosen from the set *win, lose, draw*. The label corresponds to the situation of a player about to move in the corresponding position, assuming that neither player will make an error. The labels are assigned systematically in the following way.

i. The labels assigned to terminal positions depend on the game in question. For most games, if you find yourself in a terminal position, then there is no legal move you can make, and you have lost; but this is not necessarily the case (think of stalemate in chess).

ii. A nonterminal position is a winning position if *at least one* of its successors is a losing position.

iii. A nonterminal position is a losing position if *all* of its successors are winning positions.

iv. Any remaining positions lead to a draw.

Problem 6.6.11.   Grasp intuitively how these rules arise. Can a player who finds himself in a winning position lose if his opponent makes an "error"?   □

Problem 6.6.12.   In the case of an acyclic finite graph (corresponding to a game that cannot continue for an indefinite number of moves), find a relationship between this method of labelling the nodes and topological sorting (Section 6.4.1).   □

We illustrate these ideas with the help of a variant of Nim (also known as the Marienbad game). Initially, at least two matches are placed on the table between two players. The first player removes as many matches as he likes, except that he must take at least one and he must leave at least one. Thereafter, each player in turn must remove at least one match and at most twice the number of matches his opponent just took. The player who removes the last match wins. There are no draws.

Example 6.6.1.   There are seven matches on the table initially. If I take two of them, my opponent may take one, two, three, or four. If he takes more than one, I can remove all the matches that are left and win. If he takes only one match, leaving four matches on the table, I can in turn remove a single match, and he cannot prevent me from winning on my next turn. On the other hand, if at the outset I choose to remove a single match, or to remove more than two, then you may verify that my opponent has a winning strategy.

The player who has the first move in a game with seven matches is therefore certain to win provided that he does not make an error. On the other hand, you may verify that a player who has the first move in a game with eight matches cannot win unless his opponent makes an error.   □

Problem 6.6.13.   For $n \geq 2$, give a necessary and sufficient condition on n to ensure that the player who has the first move in a game involving n matches have a winning strategy. Your characterization of n should be as simple as possible. Prove your answer.   □

A position in this game is not specified merely by the number of matches that remain on the table. It is also necessary to know the upper limit on the number of matches that it is permissible to remove on the next move. The nodes of the graph corresponding to this game are therefore pairs $<i, j>$. In general, $<i, j>$, $1 \leq j \leq i$, indicates that i matches remain on the table and that any number of them between 1 and j may be removed in the next move. The edges leaving node $<i, j>$ go to the j nodes $<i-k, \min(2k, i-k)>$, $1 \leq k \leq j$. The node corresponding to the initial position in a game with n matches, $n \geq 2$, is $<n, n-1>$. All the nodes whose second component is zero correspond to terminal positions, but only $<0, 0>$ is interesting: the nodes $<i, 0>$ for $i > 0$ are inaccessible. Similarly, nodes $<i, j>$ with j odd and $j < i - 1$ cannot be reached starting from any initial position.

Figure 6.6.1 shows part of the graph corresponding to this game. The square nodes represent losing positions and the round nodes are winning positions. The heavy edges correspond to winning moves: in a winning position, choose one of the heavy edges in order to win. There are no heavy edges leaving a losing position, corresponding to the fact that such positions offer no winning move.

We observe that a player who has the first move in a game with two, three, or five matches has no winning strategy, whereas he does have such a strategy in the game with four matches.

Problem 6.6.14.   Add nodes $<8, 7>$, $<7, 6>$, $<6, 5>$ and their descendants to the graph of Figure 6.6.1.   □

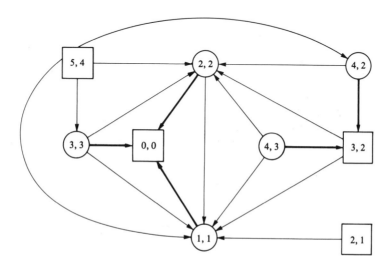

Figure 6.6.1.   Part of a game graph.

Problem 6.6.15.   Can a winning position have more than one losing position among its successors? In other words, are there positions in which several different winning moves are available? Can this happen in the case of a winning initial position $<n, n-1>$?   □

The obvious algorithm to determine whether a position is winning is the following.

> **function** $rec(i, j)$
> { returns *true* if and only if the node $<i, j>$ is winning;
>   we assume that $0 \le j \le i$ }
> **for** $k \leftarrow 1$ **to** j **do**
>   **if not** $rec(i-k, \min(2k, i-k))$
>   **then return** *true*
> **return** *false*

Problem 6.6.16.   Modify this algorithm so that it returns an integer k such that $k = 0$ if the position is a losing position and $1 \le k \le j$ if it is a winning move to take away k matches.   □

This algorithm suffers from the same defect as the algorithm *fib*1 in Section 1.7.5: it calculates the same value over and over. For instance, $rec(5, 4)$ returns *false* having called successively $rec(4, 2)$, $rec(3, 3)$, $rec(2, 2)$ and $rec(1, 1)$, but $rec(3, 3)$ also calls $rec(2, 2)$ and $rec(1, 1)$.

Problem 6.6.17.   Find two ways to remove this inefficiency. (If you want to work on this problem, do not read the following paragraphs yet!)   □

The first approach consists of using dynamic programming to create a Boolean array G such that $G[i, j] = true$ if and only if $<i, j>$ is a winning position. As usual with dynamic programming, we proceed in a bottom-up fashion, calculating all the values of $G[l, k]$ for $1 \le k \le l < i$, as well as all the values of $G[i, k]$ for $1 \le k < j$, before calculating $G[i, j]$.

> **procedure** $dyn(n)$
> { for each $1 \le j \le i \le n$, $G[i, j]$ is set to *true*
>   if and only if configuration $<i, j>$ is winning }
> $G[0, 0] \leftarrow false$
> **for** $i \leftarrow 1$ **to** n **do**
>   **for** $j \leftarrow 1$ **to** i **do**
> $k \leftarrow 1$
>   **while** $k < j$ **and** $G[i-k, \min(2k, i-k)]$ **do** $k \leftarrow k+1$
> $G[i, j] \leftarrow$ **not** $G[i-k, \min(2k, i-k)]$

Problem 6.6.18.   The preceding algorithm only uses $G[0,0]$ and the values of $G[l,k]$, $1 \le k \le l < i$, to calculate $G[i,j]$. Show how to improve its efficiency by also using the values of $G[i,k]$ for $1 \le k < j$.   □

In this context dynamic programming leads us to calculate wastefully some entries of the array G that are never needed. For instance, we know that $<15,14>$ is a winning position as soon as we discover that its second successor $<13,4>$ is a losing position. It is no longer of any interest to know whether the next successor $<12,6>$ is a winning or a losing position. In fact, only 27 nodes are really useful when we calculate $G[15,14]$, although the dynamic programming algorithm looks at 121 of them. About half this work can be avoided if we do not calculate $G[i,j]$ whenever j is odd and $j < i-1$, since these nodes are never of interest, but there is no "bottom-up" reason for not calculating $G[12,6]$.

The recursive algorithm given previously is inefficient because it recalculates the same value several times. Because of its top-down nature, however, it never calculates an unnecessary value. A solution that combines the advantages of both the algorithms consists of using a memory function (Section 5.7). This involves remembering which nodes have already been visited during the recursive computation using a global Boolean array $init[0..n,0..n]$, initialized to *false*, where n is an upper bound on the number of matches to be used.

```
function nim (i, j)
    if init [i,j] then return G [i,j]
    init [i,j] ← true
    for k ← 1 to j do
        if not nim (i−k, min(2k, i−k)) then G [i,j] ← true
                                             return true
    G [i,j] ← false
    return false
```

At first sight, there is no particular reason to favour this approach over dynamic programming, because in any case we have to take the time to initialize the whole array $init[0..n,0..n]$. Using the technique suggested in Problem 5.7.2 allows us, however, to avoid this initialization and to obtain a worthwhile gain in efficiency.

The game we have considered up to now is so simple that it can be solved without really using the associated graph. Here, without explanation, is an algorithm for determining a winning strategy that is more efficient than any of those given previously. In an initial position with n matches, first call *precond*(n). Thereafter a call on *whatnow*(i,j), $1 \le j \le i$, determines in a time in $\Theta(1)$ the move to make in a situation where i matches remain on the table and the next player has the right to take at most j of them. The array $T[0..n]$ is global. The initial call of *precond*(n) is an application of the preconditioning technique to be discussed in the next chapter.

```
procedure precond (n)
    T [0] ← ∞
    for i ← 1 to n do
        k ← 1
        while T [i−k] ≤ 2k do k ← k + 1
        T [i] ← k

function whatnow (i , j )
    if j < T [i] then { prolong the agony ! }
                            return 1
    return T [i]
```

*Problem 6.6.19. Prove that this algorithm works correctly and that *precond* (n) takes a time in $\Theta(n)$.   ☐

Consider now a more complex game, namely chess. At first sight, the graph associated with this game contains cycles, since if two positions u and v of the pieces differ only by the legal move of a rook, say, the king not being in check, then we can move equally well from u to v and from v to u. However, this problem disappears on closer examination. Remember first that in the game we just looked at, a position is defined not merely by the number of matches on the table, but also by an invisible item of information giving the number of matches that can be removed on the next move. Similarly, a position in chess is not defined simply by the positions of the pieces on the board. We also need to know whose turn it is to move, which rooks and kings have moved since the beginning of the game (to know if it is legal to castle), and whether some pawn has just been moved two squares forward (to know whether a capture *en passant* is possible). Furthermore, the International Chess Federation has rules that prevent games dragging on forever: for example, a game is declared to be a draw after 50 moves in which no irreversible action (movement of a pawn, or a capture) took place. Thus we must include in our notion of position the number of moves made since the last irreversible action. Thanks to such rules, there are no cycles in the graph corresponding to chess. (For simplicity we ignore exceptions to the 50-move rule, as well as the older rule that makes a game a draw if the pieces return three times to exactly the same positions on the board.)

Adapting the general rules given at the beginning of this section, we can therefore label each node as being a winning position for White, a winning position for Black, or a draw. Once constructed, this graph allows us to play a perfect game of chess, that is, to win whenever it is possible and to lose only when it is inevitable. Unfortunately (or perhaps fortunately for the game of chess), the graph contains so many nodes that it is quite out of the question to explore it completely, even with the fastest existing computers.

*Problem 6.6.20. Estimate the number of ways in which the pieces can be placed on a chess-board. For simplicity ignore the fact that certain positions are impossible, that is, they can never be obtained from the initial position by a legal series of moves (but take into account the fact that each bishop moves only on either white or black squares, and that both kings must be on the board). Ignore also the possibility of having promoted pawns.   □

Since a complete search of the graph associated with the game of chess is out of the question, it is not practical to use a dynamic programming approach. In this situation the recursive approach comes into its own. Although it does not allow us to be certain of winning, it underlies an important heuristic called *minimax*. This technique finds a move that may reasonably be expected to be among the best moves possible while exploring only a part of the graph starting from some given position. Exploration of the graph is usually stopped before the leaves are reached, using one of several possible criteria, and the positions thus reached are evaluated heuristically. Then we make the move that seems to cause our opponent the most trouble. This is in a sense merely a systematic version of the method used by some human players that consists of looking ahead a small number of moves. Here we give only an outline of the technique.

The minimax principle.   The first step is to define a static evaluation function *eval* that attributes some value to each possible position. Ideally, we want the value of *eval*(*u*) to increase as the position *u* becomes more favourable to White. It is customary to give values not too far from zero to positions where neither side has a marked advantage, and large negative values to positions that favour Black. This evaluation function must take account of many factors: the number and the type of pieces remaining on both sides, control of the centre, freedom of movement, and so on. A compromise must be made between the accuracy of this function and the time needed to calculate it. When applied to a terminal position, the evaluation function should return $+\infty$ if Black has been mated, $-\infty$ if White has been mated, and 0 if the game is a draw. For example, an evaluation function that takes good account of the static aspects of the position but that is too simplistic to be of real use might be the following: for nonterminal configurations, count 1 point for each white pawn, $3^{1}/_{4}$ points for each white bishop or knight, 5 points for each white rook, and 10 points for each white queen; subtract a similar number of points for each black piece.

If the static evaluation function were perfect, it would be easy to determine the best move to make. Suppose it is White's turn to move from position *u*. The best move would be to go to the position *v* that maximizes *eval*(*v*) among all the successors *w* of *u*.

$$val \leftarrow -\infty$$
for each configuration *w* that is a successor of *u* **do**
    if *eval*(*w*) $\geq$ *val* **then** *val* $\leftarrow$ *eval*(*w*)
                    v $\leftarrow$ *w*

It is clear that this simplistic approach would not be very successful using the evaluation function suggested earlier, since it would not hesitate to sacrifice a queen in order to take a pawn!

If the evaluation function is not perfect, a better strategy for White is to assume that Black will reply with the move that minimizes the function *eval*, since the smaller the value taken by this function, the better the position is supposed to be for him. (Ideally, he would like a large negative value.) We are now looking half a move ahead.

$$val \leftarrow -\infty$$
for each configuration w that is a successor of u **do**
    if w has no successor
    then $valw \leftarrow eval(w)$
    else $valw \leftarrow \min\{eval(x) \mid x$ is a successor of $w\}$
    if $valw \geq val$ **then** $val \leftarrow valw$
               $v \leftarrow w$

There is now no question of giving away a queen to take a pawn, which of course may be exactly the wrong rule to apply if it prevents White from finding the winning move : maybe if he looked further ahead the gambit would turn out to be profitable. On the other hand, we are sure to avoid moves that would allow Black to mate immediately (provided we *can* avoid this).

To add more dynamic aspects to the static evaluation provided by *eval*, it is preferable to look several moves ahead. To look n half-moves ahead from position u, White should move to the position v given by

$$val \leftarrow -\infty$$
for each configuration w that is a successor of u **do**
    if $Black(w,n) \geq val$ **then** $val \leftarrow Black(w,n)$
               $v \leftarrow w$

where the functions *Black* and *White* are the following :

function $Black(w,n)$
    if $n = 0$ **or** w has no successor
    then return $eval(w)$
    else return $\min\{White(x,n-1) \mid x$ is a successor of $w\}$

function $White(x,n)$
    if $n = 0$ **or** x has no successor
    then return $eval(x)$
    else return $\max\{Black(w,n-1) \mid w$ is a successor of $x\}$.

We see why the technique is called minimax : Black tries to minimize the advantage he allows to White, and White, on the other hand, tries to maximize the advantage he obtains from each move.

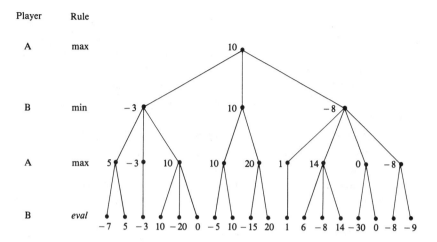

Figure 6.6.2.   The minimax principle.

Problem 6.6.21.   Let u correspond to the initial position of the pieces. What can you say about $White(u, 12800)$, besides the fact that it would take far too long to calculate in practice? Justify your answer.   □

Example 6.6.2.   Figure 6.6.2 shows part of the graph corresponding to some game. If the values attached to the leaves are obtained by applying the function *eval* to the corresponding positions, the values for the other nodes can be calculated using the minimax rule. In the example we suppose that player A is trying to maximize the evaluation function and that player B is trying to minimize it.

If A plays so as to maximize his advantage, he will choose the second of the three possible moves. This assures him of a value of at least 10.   □

Alpha-beta pruning.   The basic minimax technique can be improved in a number of ways. For example, it may be worthwhile to explore the most promising moves in greater depth. Similarly, the exploration of certain branches can be abandoned early if the information we have about them is already sufficient to show that they cannot possibly influence the values of nodes farther up the tree. This second type of improvement is generally known as *alpha-beta pruning*. We give just one simple example of the technique.

Example 6.6.3.   Look back at Figure 6.6.2. Let $<i,j>$ represent the jth node in the ith row of the tree. We want to calculate the value of the root $<1,1>$ starting from the values calculated by the function *eval* for the leaves $<4,j>$, $1 \leq j \leq 18$. To do this, we carry out a bounded depth-first search in the tree, visiting the successors of a given node from left to right.

If we want to abandon the exploration of certain branches because it is no longer useful, we have to transmit immediately to the higher levels of the tree any information obtained by evaluating a leaf. Thus as soon as the first leaf $<4,1>$ is evaluated, we know that $<4,1>$ has value -7 and that $<3,1>$ (a node that maximizes *eval*) has value at least -7. After evaluation of the second leaf $<4,2>$, we know that $<4,2>$ has value 5, $<3,1>$ has value 5, and $<2,1>$ (a node that minimizes *eval*) has value at most 5.

Continuing in this way, we arrive after evaluation of the leaf $<4,4>$ at the situation illustrated in Figure 6.6.3. Since node $<3,3>$ has value at least 10, whereas node $<2,1>$ has value at most -3, the exact value of node $<3,3>$ cannot have any influence on the value of node $<2,1>$. It is therefore unnecessary to evaluate the other descendants of node $<3,3>$; we say that the corresponding branches of the tree have been *pruned*.

Similarly, after evaluation of the leaf $<4,11>$, we are in the situation shown in Figure 6.6.4. Node $<2,3>$ has value at most 1. Since we already know that the value of $<1,1>$ is at least 10, there is no need to evaluate the other children of node $<2,3>$.

To establish that the value of the root $<1,1>$ is 10, we visit only 19 of the 31 nodes in the tree.   □

** Problem 6.6.22.   Write a program capable of playing brilliantly your favourite game of strategy.   □

** Problem 6.6.23.   Write a program that can beat the world backgammon champion. (This has already been done!)   □

** Problem 6.6.24.   What modifications should be made to the principles set out in this section to take account of those games of strategy in which chance plays a certain part? What about games with more than two players?   □

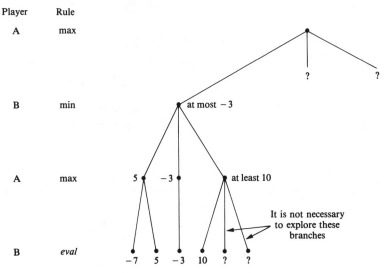

Figure 6.6.3. Alpha-beta pruning.

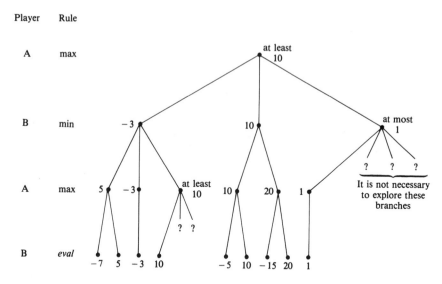

Player   Rule

Figure 6.6.4.   More alpha-beta pruning.

6.6.3 Branch-and-Bound

Like backtracking, branch-and-bound is a technique for exploring an implicit directed graph. Again, this graph is usually acyclic or even a tree. This time, we are looking for the optimal solution to some problem. At each node we calculate a bound on the possible value of any solutions that might happen to be farther on in the graph. If the bound shows that any such solution must necessarily be worse than the best solution we have found so far, then we do not need to go on exploring this part of the graph.

In the simplest version, calculation of these bounds is combined with a breadth-first or a depth-first search, and serves only, as we have just explained, to prune certain branches of a tree or to close certain paths in a graph. More often, however, the calculated bound is used not only to close off certain paths, but also to choose which of the open paths looks the most promising, so that it can be explored first.

In general terms we may say that a depth-first search finishes exploring nodes in inverse order of their creation, using a stack to hold those nodes that have been generated but not yet explored fully; a breadth-first search finishes exploring nodes in the order of their creation, using this time a queue to hold those that have been generated but not yet explored (see Section 6.5). Branch-and-bound uses auxiliary computations to decide at each instant which node should be explored next, and a priority list to hold those nodes that have been generated but not yet explored.

An example illustrates the technique.

Example 6.6.4.   We return to the travelling salesperson problem (see Sections 3.4.2 and 5.6).

Let G be the complete graph on five points with the following distance matrix:

$$\begin{bmatrix} 0 & 14 & 4 & 10 & 20 \\ 14 & 0 & 7 & 8 & 7 \\ 4 & 5 & 0 & 7 & 16 \\ 11 & 7 & 9 & 0 & 2 \\ 18 & 7 & 17 & 4 & 0 \end{bmatrix} .$$

We are looking for the shortest tour starting from node 1 that passes exactly once through each other node before finally returning to node 1.

The nodes in the implicit graph correspond to partially specified paths. For instance, node $(1,4,3)$ corresponds to two complete tours: $(1,4,3,2,5,1)$ and $(1,4,3,5,2,1)$. The successors of a given node correspond to paths in which one additional node has been specified. At each node we calculate a lower bound on the length of the corresponding complete tours.

To calculate this bound, suppose that half the distance between two points i and j is counted at the moment we leave i, and the other half when we arrive at j. For instance, leaving node 1 costs us at least 2, namely the lowest of the values 14/2, 4/2, 10/2, and 20/2. Similarly, visiting node 2 costs us at least 6 (at least 5/2 when we arrive and at least 7/2 when we leave). Returning to node 1 costs at least 2, the minimum of 14/2, 4/2, 11/2, and 18/2. To obtain a bound on the length of a path, it suffices to add elements of this kind. For instance, a complete tour must include a departure from node 1, a visit to each of the nodes 2, 3, 4, and 5 (not necessarily in this order) and a return to 1. Its length is therefore at least

$$2 + 6 + 4 + 3 + 3 + 2 = 20 .$$

Notice that this calculation does not imply the existence of a solution that costs only 20.

In Figure 6.6.5 the root of the tree specifies that the starting point for our tour is node 1. Obviously, this arbitrary choice of a starting point does not alter the length of the shortest tour. We have just calculated the lower bound shown for this node. (This bound on the root of the implicit tree serves no purpose in the algorithm; it was computed here for the sake of illustration.) Our search begins by generating (as though for a breadth-first search) the four possible successors of the root, namely, nodes $(1,2)$, $(1,3)$, $(1,4)$, and $(1,5)$. The bound for node $(1,2)$, for example, is calculated as follows. A tour that begins with $(1,2)$ must include

- The trip $1-2$: 14 (formally, leaving 1 for 2 and arriving at 2 from 1: $7+7$)
- A departure from 2 toward 3, 4, or 5: minimum 7/2
- A visit to 3 that neither comes from 1 nor leaves for 2: minimum 11/2
- A similar visit to 4: minimum 3
- A similar visit to 5: minimum 3
- A return to 1 from 3, 4, or 5: minimum 2 .

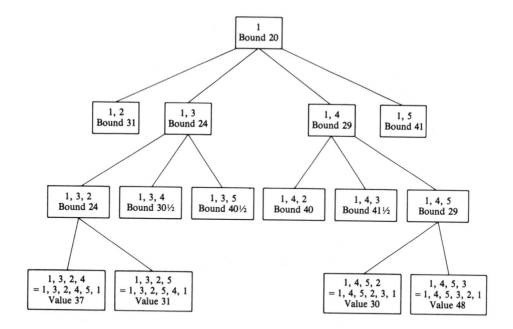

Figure 6.6.5. Branch-and-bound.

The length of such a tour is therefore at least 31. The other bounds are calculated similarly.

Next, the most promising node seems to be $(1,3)$, whose bound is 24. The three children $(1,3,2)$, $(1,3,4)$, and $(1,3,5)$ of this node are therefore generated. To give just one example, we calculate the bound for node $(1,3,2)$ as follows:

- The trip $1-3-2$: 9
- A departure from 2 toward 4 or 5: minimum 7/2
- A visit to 4 that comes from neither 1 nor 3 and that leaves for neither 2 nor 3: minimum 3
- A similar visit to 5: minimum 3
- A return to 1 from 4 or 5: minimum 11/2,

which gives a total length of at least 24.

The most promising node is now $(1,3,2)$. Its two children $(1,3,2,4)$ and $(1,3,2,5)$ are generated. This time, as node $(1,3,2,4)$, for instance, corresponds to exactly one complete tour $(1,3,2,4,5,1)$, we do not need to calculate a lower bound since we may calculate immediately its length 37.

We find that the length of the tour $(1,3,2,5,4,1)$ is 31. If we are only concerned to find one optimal solution, we do not need to continue exploration of the nodes $(1,2)$,

$(1,5)$ and $(1,3,5)$, which cannot possibly lead to a better solution. Even exploration of the node $(1,3,4)$ is pointless. (Why?) There remains only node $(1,4)$ to explore. The only child to offer interesting possibilities is $(1,4,5)$. After looking at the two complete tours $(1,4,5,2,3,1)$ and $(1,4,5,3,2,1)$, we find that the tour $(1,4,5,2,3,1)$ of length 30 is optimal. This example illustrates the fact that although at one point $(1,3)$ was the most promising node, the optimal solution does not come from there.

To obtain our answer, we have looked at merely 15 of the 41 nodes that are present in a complete tree of the type illustrated in Figure 6.6.5.   □

Problem 6.6.25.   Solve the same problem using the method of Section 5.6.   □

* Problem 6.6.26.   Implement this algorithm on a computer and test it on our example.   □

Problem 6.6.27.   Show how to solve the same problem using a backtracking algorithm that calculates a bound as shown earlier to decide whether or not a partially defined path is promising.   □

The need to keep a list of nodes that have been generated but not yet completely explored, situated in all the levels of the tree and preferably sorted in order of the corresponding bounds, makes branch-and-bound quite hard to program. The heap is an ideal data structure for holding this list. Unlike depth-first search and its related techniques, no elegant recursive formulation of branch-and-bound is available to the programmer. Nevertheless, the technique is sufficiently powerful that it is often used in practical applications.

It is next to impossible to give any idea of how well the technique will perform on a given problem using a given bound. There is always a compromise to be made concerning the quality of the bound to be calculated: with a better bound we look at less nodes, but on the other hand, we shall most likely spend more time at each one calculating the corresponding bound. In the worst case it may turn out that even an excellent bound does not allow us to cut any branches off the tree, and all the extra work we have done is wasted. In practice, however, for problems of the size encountered in applications, it almost always pays to invest the necessary time in calculating the best possible bound (within reason). For instance, one finds applications such as integer programming handled by branch-and-bound, the bound at each node being obtained by solving a related problem in linear programming with continuous variables.

6.7 SUPPLEMENTARY PROBLEMS

Problem 6.7.1.   Write algorithms to determine whether a given undirected graph is in fact a tree (i) using a depth-first search; (ii) using a breadth-first search. How much time do your algorithms take?   □

Problem 6.7.2.   Write an algorithm to determine whether a given directed graph is in fact a rooted tree, and if so, to find the root. How much time does your algorithm take?   □

*Problem 6.7.3.   A node p of a directed graph $G = < N, A >$ is called a *sink* if for every node $v \in N$, $v \neq p$, the edge (v, p) exists, whereas the edge (p, v) does not exist. Write an algorithm that can detect the presence of a sink in G in a time in $O(n)$. Your algorithm should accept the graph represented by its adjacency matrix (type *adjgraph* of Section 1.9.2). Notice that a running time in $O(n)$ for this problem is remarkable given that the instance takes a space in $\Omega(n^2)$ merely to write down.   □

*Problem 6.7.4. Euler's problem.   An *Euler path* in a finite undirected graph is a path such that every edge appears in it exactly once. Write an algorithm that determines whether or not a given graph has an Euler path, and prints the path if so. How much time does your algorithm take?   □

*Problem 6.7.5.   Repeat Problem 6.7.4 for a directed graph.   □

Problem 6.7.6.   The value 1 is available. To construct other values, you have available the two operations $\times 2$ (multiplication by 2) and $/3$ (division by 3, any resulting fraction being dropped). Operations are executed from left to right. For instance

$$10 = 1 \times 2 \times 2 \times 2 \times 2 / 3 \times 2 .$$

We want to express 13 in this way. Show how the problem can be expressed in terms of exploring a graph and find a minimum-length solution.   □

*Problem 6.7.7.   Show how the problem of carrying out a syntactic analysis of a programming language can be solved in top-down fashion using a backtracking algorithm. (This approach is used in a number of compilers.)   □

Problem 6.7.8.   A Boolean array $M[1..n, 1..n]$ represents a square maze. In general, starting from a given point, it is permissible to go to adjacent points in the same row or in the same column. If $M[i, j]$ is *true*, then you may pass through point (i, j); if $M[i, j]$ is *false*, then you may not pass through point (i, j). Figure 6.7.1 gives an example.

i. Give a backtracking algorithm that finds a path, if one exists, from $(1, 1)$ to (n, n). Without being completely formal (for instance, you may use statements such as "**for** each point v that is a neighbour of x **do** $\cdots$"), your algorithm must be clear and precise.

ii. Without giving all the details of the algorithm, indicate how to solve this problem by branch-and-bound.   □

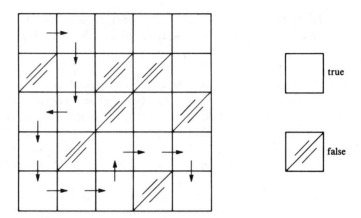

Figure 6.7.1.   A maze.

6.8 REFERENCES AND FURTHER READING

There exist a number of books concerning graph algorithms or combinatorial problems that are often posed in terms of graphs. We mention in chronological order Christofides (1975), Lawler (1976), Reingold, Nievergelt, and Deo (1977), Gondran and Minoux (1979), Even (1980), Papadimitriou and Steiglitz (1982), and Tarjan (1983). The mathematical notion of a graph is treated at length in Berge (1958, 1970).

A solution of problem 6.2.2 is given in Robson (1973).

Several applications of depth-first search are taken from Tarjan (1972) and Hopcroft and Tarjan (1973). Problem 6.3.10 is solved in Rosenthal and Goldner (1977). A linear time algorithm for testing the planarity of a graph is given in Hopcroft and Tarjan (1974). Other algorithms based on depth-first search appear in Aho, Hopcroft, and Ullman (1974, 1983).

Backtracking is described in Golomb and Baumert (1965) and techniques for analysing its efficiency are given in Knuth (1975a). Some algorithms for playing chess appear in Good (1968). The book by Nilsson (1971) is a gold mine of ideas concerning graphs and games, the minimax technique, and alpha-beta pruning. The latter is analysed in Knuth (1975b). A lively account of the first time a computer program beat the world backgammon champion (Problem 6.6.23) is given in Deyong (1977). For a more technical description of this feat, consult Berliner (1980). The branch-and-bound technique is explained in Lawler and Wood (1966). The use of this technique to solve the travelling salesperson problem is described in Bellmore and Nemhauser (1968).

7

Preconditioning

and Precomputation

If we know that we shall have to solve several similar instances of the same problem, it is sometimes worthwhile to invest some time in calculating auxiliary results that can thereafter be used to speed up the solution of each instance. This is *preconditioning*. Even when there is only one instance to be solved, *precomputation* of auxiliary tables may lead to a more efficient algorithm.

7.1 PRECONDITIONING

7.1.1 Introduction

Let I be the set of instances of a given problem. Suppose each instance $i \in I$ can be separated into two components $j \in J$ and $k \in K$ (that is, $I \subseteq J \times K$).

A preconditioning algorithm for this problem is an algorithm A that accepts as input some element j of J and produces as output a new algorithm B_j. This algorithm B_j must be such that if $k \in K$ and $<j, k> \in I$, then the application of B_j on k gives the solution to the instance $<j, k>$ of the original problem.

Example 7.1.1. Let J be a set of grammars for a family of programming languages. For example, J might be a set of grammars in Backus-Naur form for such languages as Algol, Pascal, Simula, and so on. Let K be a set of programs. The general problem is to know whether a given program is syntactically correct with respect to some given language. In this case I is the set of instances of the type "Is $k \in K$ a valid program in the language defined by the grammar $j \in J$?".

One possible preconditioning algorithm for this example is a compiler generator: applied to the grammar $j \in J$, it generates a compiler B_j for the language in question. Thereafter, to know whether $k \in K$ is a program in language j, we simply apply the compiler B_j to k. □

Let

$a(j) =$ the time required to produce B_j given j

$b_j(k) =$ the time required to apply B_j to k

$t(j,k) =$ the time required to solve $<j,k>$ directly .

It is usually the case that $b_j(k) \leq t(j,k) \leq a(j) + b_j(k)$. Obviously, we are wasting our time using preconditioning if $b_j(k) > t(j,k)$, and on the other hand, one way of solving $<j,k>$ from scratch consists of producing B_j from j and then applying it on k. Preconditioning can be useful in two situations.

a. We need to be able to solve any instance $i \in I$ very rapidly, for example to ensure a sufficiently fast response time for a real-time application. In this case it is sometimes impractical to calculate and store ahead of time the $\#I$ solutions to all the relevant instances. It may, on the other hand, be possible to calculate and store ahead of time $\#J$ preconditioned algorithms. Such an application of preconditioning may be of practical importance even if only one crucial instance is solved in the whole lifetime of the system: this may be just the instance that enables us, for example, to stop a runaway reactor. The time you spend studying before an exam may also be considered as an example of this kind of preconditioning.

b. We have to solve a series of instances $<j,k_1>, <j,k_2>, \dots, <j,k_n>$ with the same j. In this case the time taken to solve all the instances is

$$t_1 = \sum_{i=1}^{n} t(j,k_i)$$

if we work without preconditioning, and

$$t_2 = a(j) + \sum_{i=1}^{n} b_j(k_i)$$

with preconditioning. Whenever n is sufficiently large, it often happens that t_2 is much smaller than t_1.

Example 7.1.2.   Let J be a set of sets of keywords, for example

$J = \{$ {**if, then, else, endif**}, {**si, alors, sinon, finsi**},

{**for, to, by**}, {**pour, jusqu'à, pas**} $\}$.

Let K be a set of keywords, for example

$$K = \{\text{si}, \text{begin}, \text{jusqu'à}, \text{function}\}.$$

We have to solve a large number of instances of the type "Is the keyword $k \in K$ a member of the set $j \in J$?". If we solve each instance directly, we have

$$t(j,k) \in \Theta(n_j)$$

in the worst case, where n_j is the number of elements in the set j. On the other hand, if we start by sorting j (this is the preconditioning), then we can subsequently solve $<j,k>$ by a binary search algorithm.

$$a(j) \in \Theta(n_j \log n_j) \qquad \text{for the sort}$$
$$b_j(k) \in \Theta(\log n_j) \qquad \text{for the search}$$

If there are many instances to be solved for the same j, then the second technique is clearly preferable.   □

Example 7.1.3.   We are to solve the system of equations $Ax = b$, where A is a non-singular square matrix and b is a column vector. If we expect to have several systems of equations to solve with the same matrix A but different vectors b, then it is probably worthwhile to calculate the inverse of A once and for all, and to multiply this inverse by each b.   □

Example 7.1.4.   Problem 5.8.5 suggests how to obtain an efficient greedy algorithm for making change. Calculating the necessary values c_{nj} is an example of preconditioning that allows us subsequently to make change quickly every time this is required.   □

Example 7.1.5.   Creating an optimal search tree (Section 5.5) is a further example of preconditioning.   □

7.1.2 Ancestry in a rooted tree

Let J be the set of all rooted trees, and let K be the set of pairs $<v,w>$ of nodes. For a given pair $k = <v,w>$ and a given rooted tree j we want to know whether node v is an ancestor of node w in tree j. (By definition, every node is its own ancestor and, recursively, the ancestor of all the nodes of which its children are ancestors.)

If the tree j contains n nodes, any direct solution of this instance takes a time in $\Omega(n)$ in the worst case.

Problem 7.1.1.   Why?   □

It is, however, possible to precondition the tree in a time in $\Theta(n)$, so that we can subsequently solve any particular instance in a time in $\Theta(1)$.

We illustrate this approach using the tree in Figure 7.1.1. It contains 13 nodes. To precondition the tree, we traverse it first in preorder and then in postorder (see Section 6.2), numbering the nodes sequentially as we visit them. For a node v, let *prenum* [v] be the number assigned to the node when we traverse the tree in preorder, and let *postnum* [v] be the number assigned during the traversal in postorder. In Figure 7.1.1 these two numbers appear to the left and the right of the node, respectively.

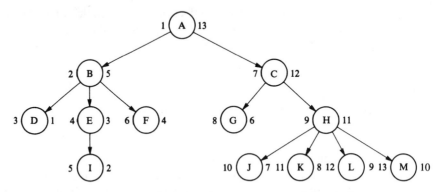

Figure 7.1.1. A rooted tree with preorder and postorder numberings.

Let v and w be two nodes in the tree. In preorder we first number a node and then we number its subtrees from left to right. Thus

$$prenum\,[v] \le prenum\,[w] \iff v \text{ is an ancestor of } w \textbf{ or}$$
$$v \text{ is to the left of } w \text{ in the tree.}$$

In postorder we first number the subtrees of a node from left to right, and then we number the node itself. Thus

$$postnum\,[v] \ge postnum\,[w] \iff v \text{ is an ancestor of } w \textbf{ or}$$
$$v \text{ is to the right of } w \text{ in the tree.}$$

It follows that

$$prenum\,[v] \le prenum\,[w] \textbf{ and } postnum\,[v] \ge postnum\,[w]$$
$$\iff v \text{ is an ancestor of } w.$$

Once all the values of *prenum* and *postnum* have been calculated in a time in $\Theta(n)$, the required condition can be checked in a time in $\Theta(1)$.

Problem 7.1.2.   There exist several similar ways of preconditioning a tree so as to be able thereafter to verify rapidly whether one node is an ancestor of another. Show, for example, that this can be done using a traversal in preorder followed by a traversal in inverted preorder, which visits first a node and then its subtrees from right to left. (Notice, however, that traversal of a rooted tree in inverted preorder requires more work than traversal in postorder if the representation of rooted trees suggested in Figure 1.9.5 is used.)   □

7.1.3 Repeated Evaluation of a Polynomial

Let J be the set of polynomials in one variable x, and let K be the set of values this variable can take. The problem consists of evaluating a given polynomial at a given point.

For simplicity, we restrict ourselves to polynomials with integer coefficients, evaluated at integer values of x. We use the number of integer multiplications that have to be carried out as a barometer to measure the efficiency of an algorithm, taking no account of the size of the operands involved nor of the number of additions and subtractions.

Initially, we restrict ourselves even further and consider only monic polynomials (the leading coefficient is 1) of degree $n = 2^k - 1$ for some integer $k \geq 1$.

Example 7.1.6.   Consider the polynomial

$$p(x) = x^7 - 5x^6 + 4x^5 - 13x^4 + 3x^3 - 10x^2 + 5x - 17.$$

A naive method for evaluating this polynomial is to calculate first the series of values $x^2, x^3, \ldots, x^7$, from which we can obtain $5x, 10x^2, \ldots, 5x^6$ and finally $p(x)$. This method requires 12 multiplications and 7 additions (counting the subtractions as additions).

It is easy to do better. If we evaluate $p(x)$ as

$$p(x) = ((((((x-5)x+4)x-13)x+3)x-10)x+5)x-17$$

we need only 6 multiplications and 7 additions. Better still, we can calculate

$$p(x) = (x^4+2)[(x^2+3)(x-5) + (x+2)] + [(x^2-4)x + (x+9)]$$

using only 5 multiplications (of which two are to calculate x^2 and x^4) plus 9 additions. (A typical 7th degree monic polynomial would require 5 multiplications as is the case here, but 10 additions.)   □

In general, if $p(x)$ is a monic polynomial of degree $n = 2^k - 1$, we first express it in the form

$$p(x) = (x^{(n+1)/2} + a)q(x) + r(x),$$

where a is a constant and $q(x)$ and $r(x)$ are monic polynomials of degree $2^{k-1} - 1$. Next, we apply the same procedure recursively to $q(x)$ and $r(x)$. Finally, $p(x)$ is expressed entirely in terms of polynomials of the form $x^i + c$, where i is a power of 2.

In the preceding example we first express $p(x)$ in the form

$$(x^4+a)(x^3+q_2x^2+q_1x+q_0) + (x^3+r_2x^2+r_1x+r_0).$$

Equating the coefficients of $x^6, x^5, \ldots, 1$, we obtain successively $q_2 = -5$, $q_1 = 4$, $q_0 = -13$, $a = 2$, $r_2 = 0$, $r_1 = -3$, and $r_0 = 9$. Thus

$$p(x) = (x^4+2)(x^3-5x^2+4x-13) + (x^3-3x+9).$$

Similarly, we find

$$x^3 - 5x^2 + 4x - 13 = (x^2+3)(x-5) + (x+2)$$

$$x^3 - 3x + 9 = (x^2-4)x + (x+9)$$

to arrive finally at the expression for $p(x)$ given in Example 7.1.6. This expression is the *preconditioned form* of the polynomial.

Problem 7.1.3.   Express $p(x) = x^7 + 2x^6 - 5x^4 + 2x^3 - 6x^2 + 6x - 32$ in preconditioned form.   □

Problem 7.1.4.   Express $p(x) = x^7$ in preconditioned form.   □

Analysis of the method.   Let $M(k)$ be the number of multiplications required to evaluate $p(x)$, a preconditioned monic polynomial of degree $n = 2^k - 1$. Let $\hat{M}(k) = M(k) - k + 1$ be the number of multiplications required if we do not count those used in the calculation of $x^2, x^4, \ldots, x^{(n+1)/2}$. We obtain the recurrence equation

$$\hat{M}(k) = \begin{cases} 0 & k = 1 \\ 2\hat{M}(k-1) + 1 & k \geq 2. \end{cases}$$

Consequently, $\hat{M}(k) = 2^{k-1}-1$ for $k \geq 1$, and hence $M(k) = 2^{k-1}+k-2$. In other words, $(n-3)/2+\lg(n+1)$ multiplications are sufficient to evaluate a preconditioned polynomial of degree $n = 2^k - 1$.

*Problem 7.1.5.   Prove that if the monic polynomial $p(x)$ is given by its coefficients, there does not exist an algorithm that can calculate $p(x)$ using less than $n-1$ multiplications in the worst case. In other words, the time invested in preconditioning the polynomial allows us to evaluate it subsequently using essentially half the number of multiplications otherwise required.   □

Problem 7.1.6.   Show that evaluation of a preconditioned polynomial of degree $n = 2^k - 1$ requires $(3n-1)/2$ additions in the worst case.   □

Problem 7.1.7.   Generalize this method of preconditioning to polynomials that are not monic. Your generalization must give an exact answer, with no risk of rounding error due to the use of floating-point arithmetic.   □

Problem 7.1.8. (Continuation of Problem 7.1.7)   Further generalize the method to polynomials of any degree.   □

Problem 7.1.9.   Show using an explicit example that the method described here does not necessarily give an optimal solution (that is, it does not necessarily minimize the required number of multiplications) even in the case of monic polynomials of degree $n = 2^k - 1$.   □

Problem 7.1.10.   Is the method appropriate for polynomials involving real coefficients and real variables? Justify your answer.   □

7.2 PRECOMPUTATION FOR STRING-SEARCHING PROBLEMS

The following problem occurs frequently in the design of text-processing systems (editors, macroprocessors, information retrieval systems, etc.). Given a *target string* consisting of n characters, $S = s_1 s_2 \cdots s_n$, and a *pattern* consisting of m characters, $P = p_1 p_2 \cdots p_m$, we want to know whether P is a substring of S, and if so, whereabouts in S it occurs. Suppose without loss of generality that $n \geq m$. In the analyses that follow, we use the number of comparisons between pairs of characters as a barometer to measure the efficiency of our algorithms.

The following naive algorithm springs immediately to mind. It returns r if the first occurrence of P in S begins at position r (that is, r is the smallest integer such that $s_{r+i-1} = p_i$, $i = 1, 2, \ldots, m$), and it returns 0 if P is not a substring of S.

```
for i ← 0 to n − m do
    ok ← true
    j ← 1
    while ok and j ≤ m do
        if p[j] ≠ s[i+j] then ok ← false
                          else j ← j + 1
    if ok then return i + 1
return 0
```

The algorithm tries to find the pattern P at every position in S. In the worst case it makes m comparisons at each position to see whether or not P occurs there. (Think of $S =$ "aaa $\cdots$ aab", $P =$ "aaaab".) The total number of comparisons to be made is therefore in $\Omega(m(n-m))$, which is in $\Omega(mn)$ if n is much larger than m. Can we do better?

7.2.1 Signatures

Suppose that the target string S can be decomposed in a natural way into substrings, $S = S_1 S_2 \cdots S_t$, and that the pattern P, if it occurs in S, must occur entirely within one of these substrings (thus we exclude the possibility that P might straddle several consecutive substrings). This situation occurs, for example, if the S_i are the lines in a text file S and we are searching for the lines in the file that contain P.

The basic idea is to use a Boolean function $T(P, S_i)$ that can be calculated rapidly to make a preliminary test. If $T(P, S_i)$ is false, then P cannot be a substring of S_i; if $T(P, S_i)$ is true, however, it is possible that P might be a substring of S_i, but we have to carry out a detailed check to verify this (for instance, using the naive algorithm given earlier). *Signatures* offer a simple way of implementing such a function.

Suppose that the character set used for the strings S and P is $\{a, b, c, \ldots, x, y, z, other\}$, where we have lumped all the non-alphabetic characters together. Suppose too that we are working on a computer with 32-bit words. Here is one common way of defining a signature.

i. Define $val("a") = 0$, $val("b") = 1, \ldots,$ $val("z") = 25$, $val(other) = 26$.

ii. If c_1 and c_2 are characters, define

$$B(c_1, c_2) = (27\, val(c_1) + val(c_2)) \bmod 32.$$

iii. Define the signature $sig(C)$ of a string $C = c_1 c_2 \cdots c_r$ as a 32-bit word where the bits numbered $B(c_1, c_2)$, $B(c_2, c_3), \ldots,$ $B(c_{r-1}, c_r)$ are set to 1 and the other bits are 0.

Example 7.2.1.   If C is the string "computers", we calculate $B("c", o") = 27 \times 2 + 14 \bmod 32 = 4$, $B("o", m") = 27 \times 14 + 12 \bmod 32 = 6, \ldots,$ $B("r", s") = 27 \times 17 + 18 \bmod 32 = 29$. If the bits of a word are numbered from 0 (on the left) to 31 (on the right), the signature of this string is the word

$$0000\ 1110\ 0100\ 0001\ 0001\ 0000\ 0000\ 0100\ .$$

Only seven bits are set to 1 in the signature because $B("e", r") = B("r", s") = 29$.   □

We calculate a signature for each substring S_i and for the pattern P. If S_i contains the pattern P, then all the bits that are set to 1 in the signature of P are also set to 1 in the signature of S_i. This gives us the function T we need:

$$T(P, S_i) = [(sig(P) \text{ and } sig(S_i)) = sig(P)],$$

where the **and** operator represents the bitwise conjunction of two whole words. T can be computed very rapidly once all the signatures have been calculated.

This is yet another example of preconditioning. Calculating the signatures for S takes a time in $O(n)$. For each pattern P we are given we need a further time in $O(m)$ to calculate its signature, but from then on we hope that the preliminary test will allow us to speed up the search for P. The improvement actually obtained in practice depends on the judicious choice of a method for calculating signatures.

* **Problem 7.2.1.**   If signatures are calculated as described, and if the characters $a, b, \ldots, z$ and *other* are equiprobable, what is the probability that the signature of a random string of n characters contains a 1 in all the bit positions that contain a 1 in the signature of another random string of m characters? Calculate the numerical value of this probability for some plausible values of m and n (for instance, $n = 40$, $m = 5$).   □

Problem 7.2.2.   Is the method illustrated of interest if the target string is very long and if it cannot be divided into substrings?   □

Problem 7.2.3.   If $T(P, S_i)$ is true with probability $\varepsilon > 0$ even if S_i does not contain P, what is the order of the number of operations required in the worst case to find P in S or to confirm that it is absent?   □

Many variations on this theme are possible. In the preceding example the function B takes two consecutive characters of the string as parameters. It is easy to invent such functions based on three consecutive characters, and so on. The number of bits in the signature can also be changed.

Problem 7.2.4.   Can we define a function B based on a single character? If this is possible, is it useful?   □

Problem 7.2.5.   If the character set contains the 128 characters of the ASCII code, and if the computer in use has 32-bit words, we might define B by

$$B(c_1, c_2) = (128\, val(c_1) + val(c_2)) \bmod 32.$$

Is this to be recommended? If not, what do you suggest instead?   □

7.2.2 The Knuth-Morris-Pratt Algorithm

We confine ourselves to giving an informal description of this algorithm (henceforth: the KMP algorithm), which finds the occurrences of P in S in a time in $O(n)$.

Example 7.2.2.   Let S = "babcbabcabcaabcabcabcacabc" and P = "abcabcacab". To find P in S we slide P along S from left to right, looking at the characters that are opposite one another. Initially, we try the following configuration:

```
S        b a b c b a b c a b c a a b c a b c a b c a c a b c
P        a b c a b c a c a b
         ↑
```

We check the characters of P from left to right. The arrows show the comparisons carried out before we find a character that does not match. In this case there is only one comparison. After this failure we try

```
S        b a b c b a b c a b c a a b c a b c a b c a c a b c
P          a b c a b c a c a b
           ↑↑↑↑
```

This time the first three characters of P are the same as the characters opposite them in S, but the fourth does not match. Up to now, we have proceeded exactly as in the naive algorithm. However we now know that the last four characters examined in S are abc$x$ where $x \neq$ "a". Without making any more comparisons with S, we can conclude that it is useless to slide P one, two, or three characters along: such an alignment cannot be correct. So let us try sliding P four characters along.

$$S \qquad \text{b a b c b a b c a b c a a b c a b c a b c a c a b c}$$
$$P \qquad \qquad \qquad \text{a b c a b c a c a b}$$
$$\qquad \qquad \qquad \qquad \uparrow\uparrow\uparrow\uparrow\uparrow\uparrow\uparrow\uparrow$$

Following this mismatch, we know that the last eight characters examined in S are abcabca$x$ where $x \neq$ "c". Sliding P one or two places along cannot be right; however moving it three places might work.

$$S \qquad \text{b a b c b a b c a b c a a b c a b c a b c a c a b c}$$
$$P \qquad \qquad \qquad \qquad \text{a b c a b c a c a b}$$
$$\qquad \qquad \qquad \qquad \qquad \qquad \uparrow$$

There is no need to recheck the first four characters of P: we chose the movement of P in such a way as to ensure that they necessarily match. It suffices to start checking at the current position of the pointer. In this case we have a second mismatch in the same position. This time, sliding P four places along might work. (A three-place movement is not enough: we know that the last characters examined in S are ax, where x is not a "b".)

$$S \qquad \text{b a b c b a b c a b c a a b c a b c a b c a c a b c}$$
$$P \qquad \qquad \qquad \qquad \qquad \text{a b c a b c a c a b}$$
$$\qquad \qquad \qquad \qquad \qquad \qquad \uparrow\uparrow\uparrow\uparrow\uparrow\uparrow\uparrow\uparrow$$

Yet again we have a mismatch, and this time a three-place movement is necessary.

$$S \qquad \text{b a b c b a b c a b c a a b c a b c a b c a c a b c}$$
$$P \qquad \qquad \qquad \qquad \qquad \qquad \qquad \text{a b c a b c a c a b}$$
$$\qquad \qquad \qquad \qquad \qquad \qquad \qquad \qquad \uparrow\uparrow\uparrow\uparrow\uparrow\uparrow$$

We complete the verification starting at the current position of the pointer, and this time the correspondence between the target string and the pattern is complete.   □

To implement this algorithm, we need an array $next[1..m]$. This array tells us what to do when a mismatch occurs at position j in the pattern. If $next[j] = 0$, it is useless to compare further characters of the pattern to the target string at the current position. We must instead line up P with the first character of S

that has not yet been examined and start checking again at the beginning of P. If $next[j] = i > 0$, we should align the ith character of P on the current character of S and start checking again at this position. In both cases we slide P along $j - next[j]$ characters to the right with respect to S. In the preceding example we have

j	1	2	3	4	5	6	7	8	9	10
$p[j]$	a	b	c	a	b	c	a	c	a	b
$next[j]$	0	1	1	0	1	1	0	5	0	1

Once this array has been calculated, here is the algorithm for finding P in S.

> **function** *KMP*
> $j, k \leftarrow 1$
> **while** $j \le m$ **and** $k \le n$ **do**
>   **while** $j > 0$ **and** $s[k] \ne p[j]$ **do**
> $j \leftarrow next[j]$
> $k \leftarrow k + 1$
> $j \leftarrow j + 1$
> **if** $j > m$ **then return** $k - m$
>   **else return** 0

It returns either the position of the first occurrence of P in S, or else 0 if P is not a substring of S.

Problem 7.2.6.   Follow the execution of this algorithm step by step using the strings from Example 7.2.2.   □

After each comparison of two characters, we move either the pointer (the arrow in the diagrams, or the variable k in the algorithm) or the pattern P. The pointer and P can each be moved a maximum of n times. The time required by the algorithm is therefore in $O(n)$. Precomputation of the array $next[1..m]$ can be carried out in a time in $O(m)$, which can be neglected since $m \le n$. Overall, the execution time is thus in $O(n)$.

It is correct to talk of preconditioning in this case only if the same pattern is sought in several distinct target strings, which does happen in some applications. On the other hand, preconditioning does not apply if several distinct patterns are sought in a given target string. In all cases, including the search for a single pattern in a single target, it is correct to talk of precomputation.

*Problem 7.2.7.   Find a way to compute the array $next[1..m]$ in a time in $O(m)$.   □

Problem 7.2.8.   Modify the KMP algorithm so that it finds all the occurrences of P in S in a total time in $O(n)$.   □

7.2.3 The Boyer-Moore Algorithm

Like the KMP algorithm, the algorithm due to Boyer and Moore (henceforth: the BM algorithm) finds the occurrences of P in S in a time in $O(n)$ in the worst case. However, since the KMP algorithm examines every character of the string S at least once in the case when P is absent, it makes at least n comparisons. The BM algorithm, on the other hand, is often sublinear: it does not necessarily examine every character of S, and the number of comparisons carried out can be less than n. Furthermore, the BM algorithm tends to become more efficient as m, the number of characters in the pattern P, increases. In the best case the BM algorithm finds all the occurrences of P in S in a time in $O(m+n/m)$.

As with the KMP algorithm, we slide P along S from left to right, checking corresponding characters. This time, however, the characters of P are checked from right to left after each movement of the pattern. We use two rules to decide how far we should move P after a mismatch.

i. If we have a mismatch immediately after moving P, let c be the character opposite $p[m]$. We know that $c \neq p[m]$. If c appears elsewhere in the pattern, we slide the latter along in such a way as to align the last occurrence of c in the pattern with the character c in the target string. If c does not appear in the pattern, we align the latter just after the occurrence of c in the target string.

ii. If a certain number of characters at the end of P correspond to the characters in S, then we use this partial knowledge of S (just as in the KMP algorithm) to slide P along to a new position compatible with the information we possess.

Example 7.2.3. Let S = "This is a delicate topic" and P = "cat".

The target string and the pattern are initially aligned as follows:

```
S       This is a delicate topic
P       cat
          ↑
```

We examine P from right to left. There is an immediate mismatch in the position shown by the arrow. The character opposite $p[m]$ is "i". Since the pattern does not include this character, we slide the pattern just to the right of the arrow.

```
S       This is a delicate topic
P          cat
             ↑
```

Again we examine P from right to left, and again there is an immediate mismatch. Since "i" does not appear in the pattern, we try

```
S        This  is  a  delicate  topic
P                 cat
                    ↑
```

There is once again an immediate mismatch, but this time the character "a" that appears opposite $p[m]$ also appears in P. We slide P one place along to align the occurrences of the letter "a", and start checking again (at the right-hand end of P).

```
S        This  is  a  delicate  topic
P                 cat
                   ↑
```

After two more immediate mismatches we are in this situation.

```
S        This  is  a  delicate  topic
P                      cat
                        ↑
```

Now, when we slide P along one position to align the "a" in the target string with the "a" in the pattern, P is correctly aligned. A final check, always from right to left, will confirm this. In this example we have found P without ever using rule (ii). We have made only 9 comparisons between a character of P and a character of S.   □

Example 7.2.4.   Consider the same strings as in Example 7.2.2:

```
S        babcbabcabcaabcabcabcacabc
P        abcabcacab
                ↑↑↑↑
```

We examine P from right to left. The left-hand arrow shows the position of the first mismatch. We know that starting at this position S contains the characters $x$cab where $x \neq$ "a". If we slide P five places right, this information is not contradicted. (Underscores show which characters were aligned.)

```
S        babcbabcabcaabcabcabcacabc
P             abcabcacab
                   ↑
```

Unlike the KMP algorithm, we check all the positions of P after moving the pattern. Some unnecessary checks (corresponding to the underscored characters in P) may be made at times. In our example when we start over checking P from right to left, there is an immediate mismatch. We slide P along to align the "c" found in S with the last "c" in P.

S   b a b c b a b c a b c a a b c a b c a b c a c a b c
P   a b c a b c a c a b
                                        ↑↑↑↑

After four comparisons between P and S (of which one is unnecessary), carried out as usual from right to left, we again have a mismatch. A second application of rule (ii) gives us

S   b a b c b a b c a b c a a b c a b c a b c a c a b c
P   a b c a b c a c a b
                                              ↑

We apply rule (i) once to align the letters "a":

S   b a b c b a b c a b c a a b c a b c a b c a c a b c
P   a b c a b c a c a b
                                                ↑

and one last time to align the letters "c":

S   b a b c b a b c a b c a a b c a b c a b c a c a b c
P   a b c a b c a c a b
                                  ↑↑↑↑↑↑↑↑↑↑

We have made 21 comparisons in all to find P.   □

To implement the algorithm, we need two arrays $d_1[\{character\ set\}]$ and $d_2[1..m-1]$, the former to implement rule (i) and the latter for rule (ii).

The array d_1, indexed by the character set we are using, is easy to compute. For every character c

$$d_1[c] \leftarrow \textbf{if } c \text{ does not appear in } p[1..m] \textbf{ then } m$$
$$\textbf{else } m - \max\{i \mid p[i] = c\} \ .$$

This is the distance to move P according to rule (i) when we have an immediate mismatch.

It is more complicated to compute d_2. We shall not give the details here, but only an example. The interpretation of d_2 is the following: after a mismatch at position i of the pattern, begin checking again at position m (that is, at the right-hand end) of the pattern and $d_2[i]$ characters further along S.

Example 7.2.5.   Suppose the pattern is P = "assesses". Suppose further that at some moment during our search for the string P in S we have a mismatch in position $p[7]$. Since we always examine the characters of P from right to left, we know that starting at the position of the mismatch the characters of S are xs, where $x \neq$ "e":

```
S        ? ? ? ? ? ? x s ? ? ? ? ? ? ? ?      x ≠ " e "
P        a s s e s s e s
                      ↑↑
```

The fact that $x \neq$ "e" does not rule out the possibility of aligning the "s" in p [6] with the "s" we have found in S. It may therefore be possible to align P as follows:

```
S        ? ? ? ? ? ? x s ? ? ? ? ? ? ? ?      x ≠ " e "
P            a s s e s s e s
                      ↑
```

We start checking again at the end of P, that is, 3 characters further on in S than the previous comparison: thus $d_2[7] = 3$.

Similarly, suppose now that we have a mismatch at position p [6]. Starting from the position of the mismatch, the characters of S are $x$es, where $x \neq$ "s":

```
S        ? ? ? ? ? x e s ? ? ? ? ? ? ? ?      x ≠ " s "
P        a s s e s s e s
                    ↑↑↑
```

The fact that $x \neq$ "s" rules out the possibility of aligning the "e" and the "s" in p [4] and p [5] with the "e" and the "s" found in S. It is therefore impossible to align P under these characters, and we must slide P all the way to the right under the characters of S that we have not yet examined:

```
S        ? ? ? ? ? x e s ? ? ? ? ? ? ? ?      x ≠ " s "
P                    a s s e s s e s
                              ↑
```

We start checking again at the end of P, that is, 10 characters further on in S than the previous comparison: thus $d_2[6] = 10$.

As a third instance, suppose we have a mismatch at position p [4]. Starting from the position of the mismatch, the characters of S are $x$sses, where $x \neq$ "e":

```
S        ? ? ? x s s e s ? ? ? ? ? ? ? ?      x ≠ " e "
P        a s s e s s e s
                ↑↑↑↑↑
```

In this case it may be possible to align P with S by sliding it three places right:

```
S        ? ? ? x s s e s ? ? ? ? ? ? ? ?      x ≠ " e "
P              a s s e s s e s
                      ↑
```

Now we start checking at the end of P, 7 characters further on in S than the previous comparison, so $d_2[4] = 7$.

For this example we find

i	1	2	3	4	5	6	7	8
$p[i]$	a	s	s	e	s	s	e	s
$d_2[i]$	15	14	13	7	11	10	3	

We also have $d_1["s"]=0$, $d_1["e"]=1$, $d_1["a"]=7$ and $d_1[\textit{any other character}]=8$. Note that $d_1["s"]$ has no significance, because an immediate mismatch is impossible at a position where S contains "s".   □

Problem 7.2.9.   Calculate d_1 and d_2 for the pattern in Example 7.2.4.   □

Problem 7.2.10.   Calculate d_1 and d_2 for P = "abracadabraaa".   □

Here finally is the BM algorithm.

function BM
   $j, k \leftarrow m$
   while $k \leq n$ **and** $j > 0$ **do**
     while $j > 0$ **and** $s[k] = p[j]$ **do**
      $k \leftarrow k - 1$
      $j \leftarrow j - 1$
     if $j \neq 0$ **then**
      if $j = m$ **then** $k \leftarrow k + d_1[s[k]]$
              else $k \leftarrow k + d_2[j]$
                $j \leftarrow m$
   if $j = 0$ **then return** $k + 1$
           else return 0

It returns either the position of the first occurrence of P in S, or else 0 if P is not a substring of S.

Problem 7.2.11.   In this algorithm, the choice between using rule (i) and rule (ii) depends on the test "$j = m$?". However, even if $j < m$, it is possible that rule (i) might allow k to advance more than rule (ii). Continuing from Example 7.2.5, consider the following situation:

   S ?????? t s ?????????
   P a s s e s s e s
                 ↑↑

The failure of the match between "t" and "e" is of the second kind, so k is increased by $d_2[7] = 3$ to obtain

$$S \quad\quad ? ? ? ? ? ? \, t \, s \, ? ? ? ? ? ? ? ?$$
$$P \quad\quad\quad\quad a \, s \, s \, e \, s \, s \, e \, s$$
$$\uparrow$$

However, the fact that "t" does not appear in P should have allowed us to increase k directly by $d_1["t"] = 8$ positions.

$$S \quad\quad ? ? ? ? ? ? \, t \, e \, ? ? ? ? ? ? ? ?$$
$$P \quad\quad\quad\quad\quad a \, s \, s \, e \, s \, s \, e \, s$$
$$\uparrow$$

Show that the algorithm is still correct if we replace

> **if** $j = m$ **then** $k \leftarrow k + d_1[s[k]]$
>   **else** $k \leftarrow k + d_2[j]$
> $j \leftarrow m$

by

> $k \leftarrow k + \max(d_1[s[k]], d_2[j])$
> $j \leftarrow m$,

provided we define $d_2[m] = 1$ and $d_1[p[m]] = 0$.

This modification corresponds to the algorithm usually known by the name Boyer-Moore (although these authors also suggest other improvements).   □

Problem 7.2.12.   Show the progress of the algorithm if we search (unsuccessfully, of course) for the pattern $P =$ "assesses" in $S =$ "I guess you possess a dress fit for a princess".

How many comparisons are made altogether before the failure to match is discovered, and how many of these comparisons are redundant (that is, they repeat comparisons previously made)?   □

* Problem 7.2.13.   Find a way to calculate the array d_2 in a time in $O(m)$.   □

** Problem 7.2.14.   Prove that the total execution time of the algorithm (computation of d_1 and d_2 and search for P) is in $O(n)$.   □

Problem 7.2.15.   Modify the BM algorithm so that it will find all the occurrences of P in S in a time in $O(n)$.   □

It is easy to see intuitively why the algorithm is often more efficient for longer patterns. For a character set of reasonable size (say, 52 letters if we count upper- and lowercase separately, ten figures and about a dozen other characters) and a pattern that is not too long, $d_1[c]$ is equal to m for most characters c. Thus we look at approximately one character out of every m in the target string. As long as m stays small compared to the size of the character set, the number of characters examined goes down as m goes up. Boyer and Moore give some empirical results: if the target string S is a text in English, about 20% of the characters are examined when $m = 6$; when $m = 12$, only 15% of the characters in S are examined.

7.3 REFERENCES AND FURTHER READING

Preconditioning polynomials for repeated evaluation is suggested in Belaga (1961). Signatures are discussed in Harrison (1971). The KMP and BM algorithms of Sections 7.2.2 and 7.2.3 come from Knuth, Morris, and Pratt (1977) and Boyer and Moore (1977). Rytter (1980) corrects the algorithm given in Knuth, Morris, and Pratt (1977) for calculating the array d_2 to be used in the Boyer-Moore algorithm. Finite automata, as described for instance in Hopcroft and Ullman (1979), can be used to introduce the KMP algorithm in an intuitively appealing way; see, for example, Baase (1978). For an efficient algorithm capable of finding all the occurrences of a finite set of patterns in a target string, consult Aho and Corasick (1975). For a probabilistic string-searching algorithm (see Chapter 8), read Karp and Rabin (1987).

8

Probabilistic Algorithms

8.1 INTRODUCTION

Imagine that you are the hero (or the heroine) of a fairy tale. A treasure is hidden at a place described by a map that you cannot quite decipher. You have managed to reduce the search to two possible hiding-places, which are, however, a considerable distance apart. If you were at one or the other of these two places, you would immediately know whether it was the right one. It takes five days to get to either of the possible hiding-places, or to travel from one of them to the other. The problem is complicated by the fact that a dragon visits the treasure every night and carries part of it away to an inaccessible den in the mountains. You estimate that it will take four more days' computation to solve the mystery of the map and thus to know with certainty where the treasure is hidden, but if you set out on a journey you will no longer have access to your computer. An elf offers to show you how to decipher the map if you pay him the equivalent of the treasure that the dragon can carry away in three nights.

Problem 8.1.1. Leaving out of consideration the possible risks and costs of setting off on a treasure-hunting expedition, should you accept the elf's offer? □

Obviously it is preferable to give three nights' worth of treasure to the elf rather than allow the dragon four extra nights of plunder. If you are willing to take a calculated risk, however, you can do better. Suppose that x is the value of the treasure remaining today, and that y is the value of the treasure carried off every night by the dragon. Suppose further that $x > 9y$. Remembering that it will take you five days to reach the hiding-place, you can expect to come home with $x - 9y$ if you wait four days to finish deciphering the map. If you accept the elf's offer, you can set out

immediately and bring back $x - 5y$, of which $3y$ will go to pay the elf; you will thus
have $x - 8y$ left. A better strategy is to toss a coin to decide which possible hiding-
place to visit first, journeying on to the other if you find you have decided wrong. This
gives you one chance out of two of coming home with $x - 5y$, and one chance out of
two of coming home with $x - 10y$. Your expected profit is therefore $x - 7.5y$. This is
like buying a ticket for a lottery that has a positive expected return.

   This fable can be translated into the context of algorithmics as follows: when an
algorithm is confronted by a choice, it is sometimes preferable to choose a course of
action at random, rather than to spend time working out which alternative is the best.
Such a situation arises when the time required to determine the optimal choice is prohibitive, compared to the time that will be saved on the average by making this optimal
choice. Clearly, the probabilistic algorithm can only be more efficient with respect to
its expected execution time. It is always possible that bad luck will force the algorithm
to explore many unfruitful possibilities.

   We make an important distinction between the words "average" and "expected".
The *average* execution time of a deterministic algorithm was discussed in section 1.4.
It refers to the average time taken by the algorithm when each possible instance of a
given size is considered equally likely. By contrast, the *expected* execution time of a
probabilistic algorithm is defined on each individual instance: it refers to the mean
time that it would take to solve the same instance over and over again. This makes it
meaningful to talk about the average expected time and the worst-case expected time
of a probabilistic algorithm. The latter, for instance, refers to the expected time taken
by the worst possible instance of a given size, *not* the time incurred if the worst possible probabilistic choices are unfortunately taken.

   Example 8.1.1.   Section 4.6 describes an algorithm that can find the kth smallest of an array of n elements in linear time in the worst case. Recall that this algorithm begins by partitioning the elements of the array on either side of a pivot, and that
it then calls itself recursively on the appropriate section of the array if need be. One
fundamental principle of the divide-and-conquer technique suggests that the nearer the
pivot is to the median of the elements, the more efficient the algorithm will be.
Despite this, there is no question of choosing the exact median as the pivot because
this would cause an infinite recursion (see Problem 4.6.3). Thus we choose a suboptimal so-called pseudomedian. This avoids the infinite recursion, but choosing the
pseudomedian still takes quite some time. On the other hand, we saw another algorithm that is much faster on the average, but at the price of a quadratic worst case: it
simply decides to use the first element of the array as the pivot. We shall see in Section 8.4.1 that choosing the pivot *randomly* gives a substantial improvement in the
expected execution time as compared to the algorithm using the pseudomedian, without
making the algorithm catastrophically bad for the worst-case instances.

   We once asked the students in an algorithmics course to implement the selection
algorithm of their choice. The only algorithms they had seen were those in Sec-

tion 4.6. Since the students did not know which instances would be used to test their programs (and suspecting the worst of their professors), none of them took the risk of using a deterministic algorithm with a quadratic worst case. Three students, however, thought of using a probabilistic approach. This idea allowed them to beat their colleagues hands down: their programs took an average of 300 milliseconds to solve the trial instance, whereas the majority of the deterministic algorithms took between 1500 and 2600 milliseconds.   □

Example 8.1.2.   Section 6.6.1 describes a systematic way of exploring an implicit tree to solve the eight queens problem. If we are content with finding one solution rather than all of them, we can improve the backtracking technique by placing the first few queens at random. Section 8.5.1 goes into this more thoroughly.   □

Example 8.1.3.   No known deterministic algorithm can decide in a reasonable time whether a given integer with several hundred decimal digits is prime or composite. Nevertheless, Section 8.6.2 describes an efficient probabilistic algorithm to solve this problem provided that one is willing to accept an arbitrarily small probability of error. This problem has important applications in cryptology (Section 4.8).   □

Example 8.1.2 raises an important consideration concerning probabilistic algorithms. They are sometimes used to solve problems that allow several correct solutions. Using the same probabilistic algorithm on the same instance, we may obtain different correct solutions on different occasions. For another example of the same phenomenon consider the problem: "Find a nontrivial factor of a given composite integer." Of course, such problems can also be handled by deterministic algorithms, but in this case the choice of algorithm determines uniquely which solution will be obtained whenever the algorithm is applied on any given instance.

The analysis of probabilistic algorithms is often complex, requiring an acquaintance with results in probability, statistics, and number theory beyond the scope of this book. For this reason, a number of results are cited without proof in the following sections. For more details, consult the references suggested in the last section.

Throughout this chapter we suppose that we have available a random number generator that can be called at unit cost. Let a and b, $a < b$, be real numbers. A call on $uniform(a,b)$ returns a real number x chosen randomly in the interval $a \leq x < b$. The distribution of x is uniform on the interval, and successive calls on the generator yield independent values of x. To generate random integers, we extend the notation to include $uniform(i..j)$, where i and j are integers, $i \leq j$, and the function returns an integer k chosen randomly, uniformly, and independently in the interval $i \leq k \leq j$. Similarly, $uniform(X)$, where X is a nonempty finite set, returns an element chosen randomly, uniformly, and independently among the elements of X.

Problem 8.1.2.   Show how the effect of $uniform(i..j)$ can be obtained if only $uniform(a,b)$ is available.   □

Example 8.1.4.   Let p be a prime number, and let a be an integer such that $1 \le a < p$. The *index* of a modulo p is the smallest strictly positive integer i such that $a^i \equiv 1 \pmod{p}$. It is thus the cardinality of $X = \{a^j \bmod p \mid j \ge 1\}$. For example, the index of 2 modulo 31 is 5, that of 3 is 30, and that of 5 is 3. By Fermat's theorem, an index modulo p always divides $p-1$ exactly. This suggests one way of making a random, uniform, independent choice of an element of X.

> **function** *draw* (a, p)
> $j \leftarrow uniform\,(1 .. p - 1)$
>   **return** *dexpoiter* (a, j, p) { Section 4.8 }   □

Problem 8.1.3.   Give other examples of sets in which there is an efficient way to choose an element randomly, uniformly, and independently.   □

Truly random generators are not usually available in practice. Most of the time *pseudorandom* generators are used instead: these are deterministic procedures that are able to generate long sequences of values that appear to have the properties of a random sequence. To start a sequence, we must supply an initial value called a *seed*. The same seed always gives rise to the same sequence, so to obtain different sequences, we may choose, for example, a seed that depends on the date or time. Most programming languages include such a generator, although some implementations should be used with caution. Using a good pseudorandom generator, the theoretical results obtained in this chapter concerning the efficiency of different algorithms can generally be expected to hold. However, the impractical hypothesis that a genuinely random generator is available is crucial when we carry out the analysis.

The theory of pseudorandom generators is complex, but a simple example will illustrate the general idea. Most generators are based on a pair of functions $S : X \rightarrow X$ and $R : X \rightarrow Y$, where X is a sufficiently large set and Y is the domain of pseudorandom values to be generated. Let $g \in X$ be a seed. Using the function S, this seed defines a sequence: $x_0 = g$ and $x_i = S(x_{i-1})$ for $i > 0$. Finally, the function R allows us to obtain the pseudorandom sequence $y_0, y_1, y_2, \ldots$ defined by $y_i = R(x_i)$, $i \ge 0$. This sequence is necessarily periodic, with a period that cannot exceed $\#X$. However, if S and R (and sometimes g) are chosen properly, the period can be made very long, and the sequence may be for most practical purposes statistically indistinguishable from a truly random sequence of elements of Y. Suggestions for further reading are given at the end of the chapter.

8.2 CLASSIFICATION OF PROBABILISTIC ALGORITHMS

By definition, a probabilistic algorithm leaves some of its decisions to chance. We shall not worry about the fact that such a concept conflicts with the definition of "algorithm" given at the beginning of the first chapter. The fundamental characteristic of these algorithms is that they may react differently if they are applied twice to the same instance. The execution time, and even the result obtained, may vary considerably from one use to the next. Probabilistic algorithms can be divided into four major classes: numerical, Monte Carlo, Las Vegas, and Sherwood. Some authors use the term "Monte Carlo" for any probabilistic algorithm, and in particular for those we call "numerical".

Randomness was first used in algorithmics for the approximate solution of *numerical* problems. Simulation can be used, for example, to estimate the mean length of a queue in a system so complex that it is impossible to get closed-form solutions or to get numerical answers by deterministic methods. The answer obtained by such a probabilistic algorithm is always approximate, but its expected precision improves as the time available to the algorithm increases. (The error is usually inversely proportional to the square root of the amount of work performed.) For certain real-life problems, computation of an exact solution is not possible even in principle, perhaps because of uncertainties in the experimental data to be used, or maybe because a digital computer can only handle binary or decimal values while the answer to be computed is irrational. For other problems, a precise answer exists but it would take too long to figure it out exactly. Sometimes the answer is given in the form of a confidence interval.

Monte Carlo algorithms, on the other hand, are used when there is no question of accepting an approximate answer, and only an exact solution will do. In the case of a decision problem, for example, it is hard to see what an "approximation" might be, since only two answers are possible. Similarly, if we are trying to factorize an integer, it is of little interest to know that such-and-such a value is "almost a factor". A way to put down seven queens on the chess-board is little help in solving the eight queens problem. A Monte Carlo algorithm always gives an answer, but the answer is not necessarily right; the *probability* of success (that is, of getting a correct answer) increases as the time available to the algorithm goes up. The principal disadvantage of such algorithms is that it is not in general possible to decide efficiently whether or not the answer given is correct. Thus a certain doubt will always exist.

Las Vegas algorithms never return an incorrect answer, but sometimes they do not find an answer at all. As with Monte Carlo algorithms, the probability of success increases as the time available to the algorithm goes up. However, any answer that is obtained is necessarily correct. Whatever the instance to be solved, the probability of failure can be made arbitrarily small by repeating the same algorithm enough times on this instance. These algorithms should not be confused with those, such as the simplex algorithm for linear programming, that are extremely efficient for the great majority of instances to be handled, but catastrophic for a few instances.

Finally, *Sherwood* algorithms always give an answer, and the answer is always correct. They are used when some known deterministic algorithm to solve a particular problem runs much faster on the average than in the worst case. Incorporating an element of randomness allows a Sherwood algorithm to reduce, and sometimes even to eliminate, this difference between good and bad instances. It is not a case of preventing the occasional occurrence of the algorithm's worst-case behaviour, but rather of breaking the link between the occurrence of such behaviour and the particular instance to be solved. Since it reacts more uniformly than the deterministic algorithm, a Sherwood algorithm is less vulnerable to an unexpected probability distribution of the instances that some particular application might give it to solve (see the end of Section 1.4).

Problem 8.2.1.   A problem is *well-characterized* if it is always possible to verify efficiently the correctness of a proposed solution for any given instance. Show that the problem of finding a nontrivial factor of a composite integer (Section 8.5.3) is well-characterized. You should realize this in no way implies that the problem is easy to solve. Intuitively, do you think the problem of finding the *smallest* nontrivial factor of a composite integer is well-characterized?   ☐

Problem 8.2.2.   Show how to obtain a Las Vegas algorithm to solve a well-characterized problem given that you already have a Monte Carlo algorithm for the same problem. Contrariwise, show how to obtain a Monte Carlo algorithm for any problem whatsoever given that you already have a Las Vegas algorithm for the same problem.   ☐

Problem 8.2.3.   Why "Sherwood", do you think?   ☐

8.3 NUMERICAL PROBABILISTIC ALGORITHMS

Remember that it is a question of finding an approximate answer for a numerical problem.

8.3.1 Buffon's Needle

You spill a box of toothpicks onto a wooden floor. The toothpicks spread out on the ground in random positions and at random angles, each one independently of all the others. If you know that there were 355 toothpicks in the box, and that each one is exactly half as long as the planks in the floor are wide (we realize that this gets unlikelier every minute!), how many toothpicks will fall across a crack between two planks?

Clearly any answer between 0 and 355 is possible, and this uncertainty is typical of probabilistic algorithms. However, as Georges Louis Leclerc showed, the average number of toothpicks expected to fall across a crack can be calculated: it is almost exactly 113.

Problem 8.3.1.   Why 113? Prove it. Why Buffon?   ☐

In fact, each toothpick has one chance in π of falling across a crack. This suggests a probabilistic "algorithm" for estimating the value of π by spilling a sufficiently large number of toothpicks onto the floor. Needless to say, this method is not used in practice since better methods of calculating the decimal expansion of π are known.

Furthermore, the precision of your estimate of π would be limited by the precision of the ratio of the length of the toothpicks to the width of the planks.

* Problem 8.3.2.   Supposing that the width of the planks is exactly twice the length of the toothpicks, how many of the latter should you drop in order to obtain with probability at least 90% an estimate of π whose absolute error does not exceed 0.001 ?   □

Problem 8.3.3.   Supposing that you have available a random generator of the type discussed previously, give an algorithm *Buffon (n)* that simulates the experiment of dropping n toothpicks. Your algorithm should count the number k of toothpicks that fall across a crack, and return n/k as its estimate of π. Try your algorithm on a computer with $n = 1000$ and $n = 10,000$, using a pseudorandom generator. What are your estimates of π? (It is likely that you will need the value of π during the simulation to generate the random angle — in radians — of each toothpick that falls. But then nobody said this was a practical method !)   □

Consider next the experiment that consists of throwing n darts at a square target and counting the number k that fall inside a circle inscribed in this square. We suppose that every point in the square has exactly the same probability of being hit by a dart. (It is much easier to simulate this experiment on a computer than to find a darts-player with exactly the degree of expertise — or of incompetence — required.) If the radius of the inscribed circle is r, then its area is πr^2, whereas that of the square target is $4r^2$, so the average proportion of the darts that fall inside the circle is $\pi r^2/4r^2 = \pi/4$. This allows us to estimate $\pi \approx 4k/n$. Figure 8.2.1 illustrates the experiment. In our example, where 28 darts have been thrown, we are not surprised to find 21 of them inside the circle, where we expect to see on average $28\pi/4 \approx 22$.

The following algorithm simulates this experiment, except that it only throws darts into the upper right quadrant of the target.

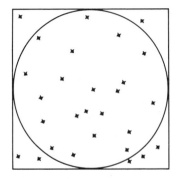

Figure 8.2.1.   Throwing darts to compute π.

```
function darts (n)
  k ← 0
  for i ← 1 to n do
    x ← uniform (0, 1)
    y ← uniform (0, 1)
    if x² + y² ≤ 1 then k ← k + 1
  return 4k / n
```

Problem 8.3.4.   What value is estimated if we replace "$x \leftarrow uniform(0,1)$; $y \leftarrow uniform(0,1)$" by "$x \leftarrow uniform(0,1)$; $y \leftarrow x$" in this algorithm?   □

8.3.2 Numerical Integration

This brings us to the best known of the numerical probabilistic algorithms: Monte Carlo integration. (This name is unfortunate, because in our terminology it is *not* an example of a Monte Carlo algorithm.) Recall that if $f : [0,1] \rightarrow [0,1]$ is a continuous function, then the area of the surface bounded by the curve $y = f(x)$, the x-axis, the y-axis, and the line $x = 1$ is given by

$$\int_0^1 f(x)\, dx \ .$$

To estimate this integral, we could throw a sufficient number of darts at the unit square and count how many of them fall below the curve.

```
function hitormiss (f, n)
  k ← 0
  for i ← 1 to n do
    x ← uniform (0, 1)
    y ← uniform (0, 1)
    if y ≤ f (x) then k ← k + 1
  return k / n
```

Thus the algorithm using darts to estimate π is equivalent to the evaluation of

$$4\int_0^1 (1-x^2)^{\frac{1}{2}}\, dx$$

by hit-or-miss Monte Carlo.

*Problem 8.3.5.   Consider two real constants ε and δ strictly between 0 and 1. Prove that if I is the correct value of the integral and if h is the value returned by the preceding algorithm, then $\text{Prob}[|h - I| < \varepsilon] \geq 1 - \delta$ whenever the number n of iterations is at least $I(1-I)/\varepsilon^2\delta$. Therefore it is sufficient to use $n = \lceil 1/4\varepsilon^2\delta \rceil$ (because $I(1-I) \leq \frac{1}{4}$) to reduce below δ the probability of an absolute error exceeding ε.

Notice that this is not very good: one more decimal digit of precision requires one hundred times more computation.   □

Problem 8.3.6.   Let a, b, c, and d be four real numbers such that $a \le b$ and $c \le d$, and let $f : [a,b] \to [c,d]$ be a continuous function. Generalize the preceding algorithm to estimate

$$\int_a^b f(x)\,dx \ .$$

Your algorithm should accept as parameters, besides f, a, and b, the number n of iterations to make and the values of c and d.   □

Usually more efficient probabilistic methods of estimating the value of a definite integral exist. The simplest consists of generating a number of points randomly and uniformly inside the interval concerned. The estimated value of the integral is obtained by multiplying the width of the interval by the arithmetic mean of the values of the function at these points.

```
function crude (f, n, a, b)
    sum ← 0
    for i ← 1 to n do
        x ← uniform (a, b)
        sum ← sum + f (x)
    return (b − a)×(sum/n)
```

Provided $\int_a^b f(x)\,dx$ and $\int_a^b f^2(x)\,dx$ exist, the variance of the estimate calculated by this algorithm is inversely proportional to the number of points generated randomly, and the distribution of the estimate is approximately normal when n is large. Moreover, for any fixed number of iterations, its variance is never worse than that of the *hit-or-miss* algorithm. One should not immediately conclude, however, that *crude* always outperforms *hit-or-miss*, because *hit-or-miss* can sometimes make more iterations than *crude* in a given amount of time. If both are used to compute π as previously described, for instance, each iteration of *crude* requires the computation of a square root, which *hit-or-miss* can do without by proceeding as in *darts*.

As presented thus far, Monte Carlo integration is of little practical use. A better estimate of the integral can generally be obtained by various deterministic methods, one of the simplest of which is the trapezoidal algorithm.

```
function trapezoidal (f, n, a, b)
    { we assume n ≥ 2 }
    delta ← (b − a)/(n − 1)
    sum ← (f (a)+f (b))/2
    for x ← a + delta step delta to b − delta do
        sum ← sum + f (x)
    return sum × delta
```

Problem 8.3.7. Try to grasp intuitively why this algorithm works. Why is it called the trapezoidal algorithm?   □

Problem 8.3.8. Compare experimentally the trapezoidal algorithm and the two probabilistic algorithms we have seen. In each case, estimate the value of π by calculating $\int_0^1 4(1-x^2)^{\frac{1}{2}} \, dx$.   □

In general, the trapezoidal algorithm needs many less iterations than does Monte Carlo integration to obtain a comparable degree of precision. This is typical of most of the natural functions that we may wish to integrate. However, to every deterministic integration algorithm, even the most sophisticated, there correspond continuous functions that can be constructed expressly to fool the algorithm. Consider for example the function $f(x) = \sin^2((100!)\pi x)$. Any call on *trapezoidal*$(f, n, 0, 1)$ with $2 \leq n \leq 101$ returns the value zero, even though the true value of this integral is $\frac{1}{2}$. No function can play this kind of trick on the Monte Carlo integration algorithm (although there is an extremely small probability that the algorithm might manage to make a similar kind of error, even when f is a thoroughly ordinary function).

In practice, Monte Carlo integration is of interest when we have to evaluate a multiple integral. If a deterministic integration algorithm using some systematic method to sample the function is generalized to several dimensions, the number of sample points needed to achieve a given precision grows exponentially with the dimension of the integral to be evaluated; If 100 points are needed to evaluate a simple integral, then it will probably be necessary to use all the points of a 100×100 grid, that is, 10,000 points, to achieve the same precision when a double integral is evaluated; one million points will be needed for a triple integral, and so on. In Monte Carlo integration, on the other hand, the dimension of the integral generally has little effect on the precision obtained, although the amount of work for each iteration is likely to increase slightly with the dimension. In practice, Monte Carlo integration is used to evaluate integrals of dimension four or higher. The precision of the answer can be improved using hybrid techniques that are partly systematic and partly probabilistic. If the dimension is fixed, it may even be preferable to use *quasi* Monte Carlo integration, a technique not discussed here (but Section 8.7 gives a reference for further reading).

8.3.3 Probabilistic Counting

In the preceding examples, numeric probabilistic algorithms are used to approximate a real number. The same technique can also be used to estimate the value of an integer. Let X be a finite set. We would like to know the cardinality of X, but the number of elements is too large for it to be practical simply to count them one by one. Suppose, on the other hand, that we are able to choose an element from X randomly, uniformly, and independently (see Example 8.1.4). A classic brain-teaser helps to explain how this ability to choose random elements from X allows us to estimate its cardinality.

Problem 8.3.9.   A room contains 25 randomly chosen people. Would you be willing to bet that at least two of them share the same birthday? (Do not read the following paragraphs if you wish to think about this.)   □

The intuitive answer to the preceding question is almost invariably "of course not". Nevertheless, the probability that you would win your bet is greater than 56%. More generally, there are $n!/(n-k)!$ different ways of choosing k distinct objects from among n objects, taking into account the order in which they are chosen. Since there are n^k different ways of choosing k objects if repetitions are allowed, the probability that k objects chosen randomly and uniformly from n (with repetition allowed) are all distinct is $n!/(n-k)!n^k$.

Problem 8.3.10.   Calculate $365!/340!365^{25}$ to four significant figures.   □

Problem 8.3.11.   The calculation in problem 8.3.10 does not correspond exactly to the puzzle in problem 8.3.9, because births are not uniformly distributed through the year. Does this make it more or less likely that you would win your bet? Justify your answer intuitively. What about leap years?   □

Stirling's approximation, $n! \approx \sqrt{2\pi n}\,(n/e)^n$, and the approximation $\ln(1+x) \approx x - x^2/2$ when x is near zero, allow us to estimate this probability.

Problem 8.3.12.   Show that $n!/(n-k)!n^k \approx e^{-k^2/2n}$.   □

*Problem 8.3.13.   Use the more accurate formulas

$$n! \in \sqrt{2\pi n}\,(n/e)^n\,[1 + 1/12n + \Theta(n^{-2})]$$

$$\ln(1+x) \in x - x^2/2 + x^3/3 - \Theta(x^4) \quad \text{when} \quad -1 < x < 1$$

to conclude that

$$n!/(n-k)!n^k \in e^{-k(k-1)/2n - k^3/6n^2 \pm O(\max(k^2/n^2,\, k^4/n^3))}\,,$$

provided that $1 \ll k \ll n$.   □

In particular, it is when $k \approx \alpha\sqrt{n}$, where $\alpha = \sqrt{2\ln 2} \approx 1.177$, that the probability of having a repetition exceeds 50%. It is harder to determine the average value of k corresponding to the *first* repetition.

**Problem 8.3.14.   Let X be a set of n elements from which we randomly, uniformly and independently choose elements with replacement. Let k be the number of choices before the occurrence of the first repetition. When n is large, show that the expected value of k tends to $\beta\sqrt{n}$, where $\beta = \sqrt{\pi/2} \approx 1.253$.   □

This suggests the following probabilistic algorithm for estimating the number of elements in a set X.

```
function count (X : set )
    k ← 0
    S ← ∅
    a ← uniform (X)
    repeat
        k ← k + 1
        S ← S ∪ {a}
        a ← uniform (X)
    until a∈S
    return 2k²/π
```

** Problem 8.3.15.   Carry out a statistical analysis of the random variable k^2 as a function of the value of n. Does the function *count* provide an unbiased estimator of n? Give an algorithm $count\,2(X,\varepsilon,\delta)$ that returns an estimate N of the number n of elements in X such that $\text{Prob}[\,|\,1-N/n\,|<\varepsilon\,]\geq 1-\delta$.   □

The algorithm $count\,(X)$ estimates the number n of elements in X in an expected time and space that are both in $\Theta(\sqrt{n}\,)$, provided operations on the set S are counted at unit cost. This quantity of space can be prohibitive if n is large. The space can be reduced to a constant with only a linear increase in the execution time by using a pseudorandom generator. This is one of the rare instances where using a truly random generator would be a hindrance rather than a help. We have not merely to choose an element at random from X, but also to step through X in a pseudorandom, and hence deterministic, way.

Let $f:X\to X$ be a pseudorandom function and let $x_0\in X$ be a randomly chosen starting point. This defines a walk $x_0,\ x_1,\ x_2,\ \ldots$ through X, where $x_i=f(x_{i-1})$, $i>0$. Because X is a finite set, the sequence $\{x_i\}_{i\geq 0}$ must eventually repeat itself. Let q be the smallest integer such that x_q appears more than once in the sequence, and let k be the smallest integer larger than q such that $x_q=x_k$. Let p stand for $k-q$. Because the walk is pseudorandom, we also have $x_{q+i}=x_{k+i}$ for every nonnegative integer i, and more generally, $x_i=x_j$ whenever $j\geq i\geq q$ and $j-i\equiv 0\ (\text{mod } p)$. For this reason $\{x_i\}_{i=0}^{q-1}$ is called the *tail* of the walk and $\{x_i\}_{i=q}^{q+p}$ is its *period*.

We are interested in computing the value of k, since this corresponds precisely to the first repetition, and thus $2k^2/\pi$ is our estimate on $\#X$. The following exercise shows how to obtain both q and p in constant space and in a time in $O(k)$, hence in an expected time in $O(\sqrt{\#X}\,)$.

* Problem 8.3.16.   Consider the sequence $\{y_i\}_{i\geq 0}$ defined by $y_i=x_{2i}$. Show that the smallest integer $t>0$ such that $y_t=x_t$ is such that $q\leq t\leq q+p$, with $t=q+p$ only possible if $q=0$. Show also that $t\equiv 0\ (\text{mod } p)$, and deduce that the smallest integer j such that $x_j=x_{t+j}$ is precisely q, the length of the tail. Incorporate all these ideas into a simple algorithm that is capable of finding the values of q and p, hence $k=q+p$, in a time in $O(k)$ and in constant space.   □

Problem 8.3.17. (Continuation of Problem 8.3.11)   The probabilistic counting algorithm no longer works if the generation of elements in X is not uniform, that is, if some elements of X are favoured to the detriment of others. Show that it can, nevertheless, be used unchanged to estimate a *lower bound* on the number of elements in X.   □

The variance of the estimate obtained from this algorithm is unfortunately too high for most practical applications (unless the solution to Problem 8.3.15 is used). The following example shows that it can nonetheless be useful if we simply need to know whether X contains less than a elements or more than b, where $a \ll b$.

Example 8.3.1.   An *endomorphic cryptosystem* consists of a finite set K of keys, a finite set M of messages and two permutations $E_k : M \rightarrow M$ and $D_k : M \rightarrow M$ for each key $k \in K$ such that $D_k(E_k(m)) = m$ for every $m \in M$ and $k \in K$. Such a system is *closed* (an undesirable property) if

$$(\forall k_1, k_2 \in K)(\exists k_3 \in K)(\forall m \in M)[E_{k_1}(E_{k_2}(m)) = E_{k_3}(m)] .$$

For every $m \in M$ consider the set $X_m = \{E_{k_1}(E_{k_2}(m)) \mid k_1, k_2 \in K \}$. It is clear that $\#X_m \leq \#K$ if the system is closed. On the other hand, if the system is not closed, it is reasonable to hope that $\#X_m \gg \#K$ provided that $\#M \gg \#K$. It suffices to choose k_1 and k_2 randomly in K and to calculate $E_{k_1}(E_{k_2}(m))$ in order to choose a random element from X_m, although this may not imply a uniform distribution on X_m.

All this suggests a probabilistic approach to testing whether or not a cryptosystem is closed. Let m be chosen randomly from M. Probabilistic counting is used to estimate a lower bound on the cardinality of X_m. (We can only estimate a lower bound since there is no reason to believe that elements are chosen uniformly from X_m; see Problem 8.3.17.) It is improbable that the system is closed if this estimate is significantly greater than the cardinality of K.

A similar approach was used to demonstrate that the American Data Encryption Standard is almost certainly not closed. In this application $\#K = 2^{56}$ and $\#M = 2^{64}$, which rules out any possibility of an exhaustive verification of the hypothesis that $\#X_m > 2^{56}$. (Even 2^{56} microseconds is more than two millennia.) Implemented using specialized hardware, the probabilistic algorithm was able to arrive at this conclusion in less than a day.   □

8.3.4 More Probabilistic Counting

You have the complete works of Shakespeare on a magnetic tape. How can you determine the number of different words he used, counting different forms of the same word (plurals, possessives, and so on) as distinct items?

Two obvious solutions to this problem are based on the techniques of sorting and searching. Let N be the total number of words on the tape, and let n be the number of different words. The first approach might be to sort the words on the tape so as to bring identical forms together, and then to make a sequential pass over the sorted tape to count the number of different words. This method takes a time in $\Theta(N \log N)$ but requires a relatively modest amount of space in central memory if a suitable external sorting technique is used. (Such techniques are not covered in this book.) The second approach consists of making a single pass over the tape and constructing in central memory a hash table (see Section 8.4.4) holding a single occurrence of each form so far encountered. The required time is thus in $O(N)$ on the average, but it is in $\Omega(Nn)$ in the worst case. Moreover, this second method requires a quantity of central memory in $\Omega(n)$, which will most likely prove prohibitive.

If we are willing to tolerate some imprecision in the estimate of n, and if we already know an upper bound M on the value of n (or failing this, on the value of N), then there exists a probabilistic algorithm for solving this problem that is efficient with respect to both time and space. We must first define what sequences of characters are to be considered as words. (This may depend not only on the character set we are using, but also whether we want to count such sequences as "jack-rabbit", "jack-o'lantern", and "jack-in-the-box" as one, two, three, or four words.) Let U be the set of such sequences. Let m be a parameter somewhat larger than $\lg M$ (a more detailed analysis shows that $m = 5 + \lceil \lg M \rceil$ suffices). Let $h : U \to \{0,1\}^m$ be a hash function that transforms a sequence from U in a pseudorandom way into a string of bits of length m. If y is a string of bits of length k, denote by $y[i]$ the ith bit of y, $1 \le i \le k$; denote by $\pi(y,b)$, $b \in \{0,1\}$, the smallest i such that $y[i] = b$, or $k+1$ if none of the bits of y is equal to b. Consider the following algorithm.

> **function** *wordcnt*
>   { initialization }
> $y \leftarrow$ string of $(m+1)$ bits set to zero
>   { sequential passage through the tape }
>   **for** each word x on the tape **do**
> $i \leftarrow \pi(h(x),1)$
> $y[i] \leftarrow 1$
>   **return** $\pi(y,0)$

Suppose, for example, that the value returned by this algorithm is 4. This means that the final y begins with 1110. Consequently, there are words x_1, x_2 and x_3 on the tape such that $h(x_i)$ begins with 1, 01, and 001, respectively, but there is no word x_4 such that $h(x_4)$ begins with 0001. Let k be the value returned by a call on *wordcnt*. Since the probability that a random binary string begins with 0001 is 2^{-4}, it is unlikely that there could be more than 16 distinct words on the tape. (The probability that $\pi(h(x_i),1) \ne 4$ for 16 different values of x_i is $(15/16)^{16} \approx 35.6\% \approx e^{-1}$, assuming h has sufficiently random behaviour; in fact, $\text{Prob}[k=4 \mid n=16] \approx 31\frac{3}{4}\%$.) Conversely, since the probability that a random binary string begins with 001 is 2^{-3}, it is unlikely that there could be less than 4 distinct words on the tape. (The probability that $\pi(h(x_i),1) = 3$ for at least one value of x_i among 4 different values is $1 - (7/8)^4 \approx 41.4\% \approx 1 - e^{-1/2}$; in fact, $\text{Prob}[k=4 \mid n=4] = 18\frac{3}{4}\%$.) This crude reasoning indicates that it is plausible to expect that the number of distinct words on the tape should lie between 2^{k-2} and 2^k. It is far from obvious how to carry out a more precise analysis of the unbiased estimate of n given by k.

** Problem 8.3.18.   Let R_n be the random variable returned by this algorithm when the tape contains n different words and the function $h : U \rightarrow \{0, 1\}^m$ is randomly chosen with uniform distribution among all such functions (this last assumption is not reasonable in practice). Prove that the expected value of R_n is in $\lg n + \Theta(1)$, where the hidden constant in $\Theta(1)$ fluctuates around 0.62950 when n is sufficiently large. Prove further that the standard deviation of R_n fluctuates around 1.12127.   □

This offers a first approach for estimating the number of different words: calculate k using the algorithm *wordcnt* and estimate n as $2^k / 1.54703$. Unfortunately, the standard deviation of R_n shows that this estimate may be in error by a factor of 2, which is unacceptable.

** Problem 8.3.19.   Show how to obtain an arbitrarily precise estimate by using a little more space but with no appreciable increase in execution time, provided n is sufficiently large. (*Hint*: by using t strings $y_1, y_2, \ldots, y_t$ of m bits, you can obtain a relative precision of about $0.78/\sqrt{t}$ provided t is sufficiently large ($t \geq 64$); your hash function should produce strings of $m + \lg t$ bits.)   □

Notice that this approximate counting algorithm is completely insensitive to the order in which the words appear on the tape and to the number of repetitions of each of them.

8.3.5 Numerical Problems in Linear Algebra

Many classic problems in linear algebra can be handled by numerical probabilistic algorithms. Among those are matrix multiplication, the solution of a system of simultaneous linear equations, matrix inversion, and the computation of eigenvalues and eigenvectors. We do not discuss any of them in detail here because it is only for very specialized applications that they perform better than the obvious deterministic algorithms. The reader is referred to the literature for further discussion.

An intriguing feature of these probabilistic algorithms is their ability to compute independently the various entries in the result. Consider, for instance, an $n \times n$ nonsingular matrix A. Classic deterministic inversion algorithms compute its inverse B as a whole or perhaps column by column. By contrast, there are probabilistic algorithms that are capable of estimating the value of B_{ij}, for any given $1 \leq i \leq n$ and $1 \leq j \leq n$, in about $1/n^2$ of the time they would require to compute the whole inverse. These algorithms are only applicable if the matrices concerned are well conditioned, typically requiring that $I - A$ have only small eigenvalues, where I stands for the identity matrix.

8.4 SHERWOOD ALGORITHMS

Section 1.4 mentions that analysing the average efficiency of an algorithm may sometimes give misleading results. The reason is that any analysis of the average case must
be based on a hypothesis about the probability distribution of the instances to be handled. A hypothesis that is correct for a given application of the algorithm may prove
disastrously wrong for a different application. Suppose, for example, that quicksort
(Section 4.5) is used as a subalgorithm inside a more complex algorithm. Analysis of
this sorting method shows that it takes an average time in $\Theta(n \log n)$ to sort n items
provided that the instances to be sorted are chosen randomly. This analysis no longer
bears any relation to reality if in fact we tend to give the algorithm only instances that
are already almost sorted. Sherwood algorithms free us from the necessity of worrying
about such situations by evening out the time required on different instances of a given
size.

Let A be a deterministic algorithm and let $t_A(x)$ be the time it takes to solve some
instance x. For every integer n let X_n be the set of instances of size n. Supposing that
every instance of a given size is equiprobable, the average time taken by the algorithm
to solve an instance of size n is

$$\bar{t}_A(n) = \sum_{x \in X_n} t_A(x)/\#X_n .$$

This in no way rules out the possibility that there exists an instance x of size n such
that $t_A(x) \gg \bar{t}_A(n)$. We wish to obtain a probabilistic algorithm B such that
$t_B(x) \approx \bar{t}_A(n)+s(n)$ for *every* instance x of size n, where $t_B(x)$ is the expected time
taken by algorithm B on instance x and $s(n)$ is the cost we have to pay for this unifor-
mity.

Algorithm B may occasionally take more time than $\bar{t}_A(n)+s(n)$ on an instance x
of size n, but this fortuitous behaviour is only due to the probabilistic choices made by
the algorithm, independently of the specific instance x to be solved. Thus there are no
longer worst-case instances, but only worst-case executions. If we define

$$\bar{t}_B(n) = \sum_{x \in X_n} t_B(x)/\#X_n$$

the average expected time taken by algorithm B on a random instance of size n, it is
clear that $\bar{t}_B(n) \approx \bar{t}_A(n)+s(n)$. The Sherwood algorithm thus involves only a small
increase in the average execution time if $s(n)$ is negligible compared to $\bar{t}_A(n)$.

8.4.1 Selection and Sorting

We return to the problem of finding the kth smallest element in an array T of n elements (Section 4.6 and Example 8.1.1). The heart of this algorithm is the choice of a

pivot around which the other elements of the array are partitioned. Using the pseudo-median as the pivot assures us of a linear execution time in the worst case, even though finding this pivot is a relatively costly operation. On the other hand, using the first element of the array as the pivot assures us of a linear execution time on the average, with the risk that the algorithm will take quadratic time in the worst case (Problems 4.6.5 and 4.6.6). Despite this prohibitive worst case, the simpler algorithm has the advantage of a much smaller hidden constant on account of the time that is saved by not calculating the pseudomedian. The decision whether it is more important to have efficient execution in the worst case or on the average must be taken in the light of the particular application. If we decide to aim for speed on the average thanks to the simpler deterministic algorithm, we must make sure that the instances to be solved are indeed chosen randomly and uniformly.

Suppose that the elements of T are distinct, and that we are looking for the median. The execution times of the algorithms in Section 4.6 do not depend on the values of the elements of the array, but only on their relative order. Rather than express this time as a function solely of n, which forces us to distinguish between the worst case and an average case, we can express it as a function of both n and σ, the permutation of the first n integers that corresponds to the relative order of the elements of the array.

Let $t_p(n,\sigma)$ and $t_s(n,\sigma)$ be the times taken by the algorithm that uses the pseudomedian and by the simplified algorithm, respectively. The simplified algorithm is generally faster: for every n, $t_s(n,\sigma) < t_p(n,\sigma)$ for most values of σ. On the other hand, the simplified algorithm is sometimes disastrous: $t_s(n,\sigma)$ is occasionally much greater than $t_p(n,\sigma)$. More precisely, let S_n be the set of $n!$ permutations of the first n integers. Define $\bar{t}_s(n) = \sum_{\sigma \in S_n} t_s(n,\sigma)/n!$. We have the following equations:

$$(\exists c_p)(\exists n_1 \in \mathbb{N})(\forall n \geq n_1)(\forall \sigma \in S_n)[t_p(n,\sigma) \leq c_p n]$$

$$(\exists c_s \ll c_p)(\exists n_2 \in \mathbb{N})(\forall n \geq n_2)[\bar{t}_s(n) \leq c_s n]$$

but

$$(\exists \hat{c}_s)(\exists n_3 \in \mathbb{N})(\forall n \geq n_3)(\exists \sigma \in S_n)[t_s(n,\sigma) \geq \hat{c}_s n^2 \gg c_p n \geq t_p(n,\sigma)].$$

For the execution time to be independent of the permutation σ, it suffices to choose the pivot randomly among the n elements of the array T. The fact that we no longer calculate a pseudomedian simplifies the algorithm and avoids recursive calls. The resulting algorithm resembles the iterative binary search of Section 4.3.

function *selectionRH* $(T[1..n],k)$
{ finds the kth smallest element in array T ;
we assume that $1 \leq k \leq n$ }
$i \leftarrow 1; j \leftarrow n$

```
while i < j do
    m ← T [uniform (i .. j)]
    partition (T, i, j, m, u, v)
    if k < u then j ← u − 1
    else if k > v then i ← v + 1
        else i, j ← k
return T [i]
```

Here $partition(T, i, j, m, \textbf{var } u, \textbf{var } v)$ pivots the elements of $T[i..j]$ around the value m; after this operation the elements of $T[i..u-1]$ are less than m, those of $T[u..v]$ are equal to m, and those of $T[v+1..j]$ are greater than m. The values of u and v are calculated and returned by the pivoting algorithm (see Problem 4.6.1).

A similar analysis to that of Problem 4.6.5 shows that the expected time taken by this probabilistic selection algorithm is linear, independently of the instance to be solved. Thus its efficiency is not affected by the peculiarities of the application in which the algorithm is used. It is always possible that some particular execution of the algorithm will take quadratic time, but the probability that this will happen becomes increasingly negligible as n gets larger, and, to repeat, this probability is independent of the instance concerned. Let $t_{RH}(n, \sigma)$ be the average time taken by the Sherwood algorithm to determine the median of an array of n elements arranged in the order specified by σ. The probabilistic nature of the algorithm ensures that $t_{RH}(n, \sigma)$ is independent of σ. Its simplicity ensures that

$$(\exists n_0 \in \mathbb{N})(\forall n \geq n_0)(\forall \sigma \in S_n)[t_{RH}(n, \sigma) < t_p(n, \sigma)] \ .$$

To sum up, we started with an algorithm that is excellent when we consider its average execution time on all the instances of some particular size but that is very inefficient on certain specific instances. Using the probabilistic approach, we have transformed this algorithm into a Sherwood algorithm that is efficient (with high probability) whatever the instance considered.

Problem 8.4.1.   Show how to apply the Sherwood style of probabilistic approach to quicksort (Section 4.5). Notice that quicksort must first be modified along the lines of Problem 4.5.4.   □

8.4.2 Stochastic Preconditioning

The modifications we made to the deterministic algorithms for sorting and for selection in order to obtain Sherwood algorithms are simple. There are, however, occasions when we are given a deterministic algorithm efficient on the average but that we cannot reasonably expect to modify. This happens, for instance, if the algorithm is part of a complicated, badly documented software package. Stochastic preconditioning allows us to obtain a Sherwood algorithm without changing the deterministic algorithm. The trick is to transform the instance to be solved into a random instance, to use the given deterministic algorithm to solve this random instance, and then to deduce the solution to the original instance.

Suppose the problem to be solved consists of the computation of some function $f : X \rightarrow Y$ for which we already have an algorithm that is efficient on the average. For every integer n, let X_n be the set of instances of size n, and let A_n be a set with the same number of elements. Assume random sampling with uniform distribution is possible efficiently within A_n. Let A be the union of all the A_n. *Stochastic preconditioning* consists of a pair of functions $u : X \times A \rightarrow X$ and $v : A \times Y \rightarrow Y$ such that

   i. $(\forall n \in \mathbb{N})(\forall x, y \in X_n)(\exists! r \in A_n)[u(x,r) = y]$;

   ii. $(\forall n \in \mathbb{N})(\forall x \in X_n)(\forall r \in A_n)[f(x) = v(r, f(u(x,r)))]$; and

   iii. the functions u and v can be calculated efficiently in the worst case.

We thus obtain the following Sherwood algorithm.

   function *RH* (x)
     { computation of $f(x)$ by Sherwood algorithm }
     let n be the size of x
     $r \leftarrow uniform(A_n)$
     $y \leftarrow u(x,r)$ { random instance of size n }
     $s \leftarrow f(y)$ { solved by the deterministic algorithm }
     return $v(r,s)$

Whatever the instance x to be solved, the first property ensures that this instance is transformed into an instance y chosen randomly and uniformly from all those of the same size. Thanks to the second property, the solution to this random instance allows us to recover the solution of the original instance x.

   Example 8.4.1.   The stochastic preconditioning required for selection or for sorting is the same: it is simply a question of randomly shuffling the elements of the array in question. No posttreatment (function v) is needed to recover the solution in these cases. Simply call the following procedure before the deterministic sorting or selection algorithm.

   procedure *shuffle* $(T[1..n])$
     for $i \leftarrow 1$ **to** $n-1$ **do**
       $j \leftarrow uniform(i..n)$
       interchange $T[i]$ and $T[j]$ $\square$

   Example 8.4.2.   Recall that no efficient algorithm is known for calculating discrete logarithms (Section 4.8). Suppose for the purposes of illustration that someone were to discover an algorithm efficient on the average but prohibitively slow in the worst case. Denote the discrete logarithm of x modulo p to the base g by

$\log_{g,p} x$. The following equations allow us to transform our hypothetical algorithm into a Sherwood algorithm:

    i. $\log_{g,p} (xy \bmod p) = (\log_{g,p} x + \log_{g,p} y) \bmod (p - 1)$;
    ii. $\log_{g,p} (g^r \bmod p) = r$ for $0 \le r \le p - 2$.

Here is the Sherwood algorithm.

       function $dlogRH(g, x, p)$
         $r \leftarrow uniform(0 .. p-2)$
         $b \leftarrow dexpoiter(g, r, p)$ {Section 4.8}
         $a \leftarrow bx \bmod p$
         $s \leftarrow \log_{g,p} a$ { using the assumed algorithm }
         return $(s - r) \bmod (p - 1)$ □

Problem 8.4.2.   Why does the algorithm $dlogRH$ work? Point out the functions corresponding to u and v.   □

Problem 8.4.3.   Find other problems that can benefit from stochastic preconditioning.   □

Stochastic preconditioning offers an intriguing possibility: *computing with an encrypted instance.* Assume that you would like to compute $f(x)$ for some instance x but that you lack the computing power or the efficient algorithm to do so. Assume, furthermore, that some other party is capable of carrying out this computation and willing to do so for you, perhaps for a fee. What should you do if you are unwilling to divulge your actual request x? The solution is easy if stochastic preconditioning applies to the computation of f: use the function u to encrypt x into some random y, have $f(y)$ computed for you, and then use the function v to deduce $f(x)$. This process yields *no* information on your actual request, except for its size, because the probability distribution of $u(x,r)$ is independent of x as long as r is chosen randomly with uniform probability.

8.4.3 Searching an Ordered List

A list of n keys sorted into ascending order is implemented using two arrays $val[1..n]$ and $ptr[1..n]$ and an integer *head*. The smallest key is in $val[head]$, the next smallest is in $val[ptr[head]]$, and so on. In general, if $val[i]$ is not the largest key, then $ptr[i]$ gives the index of the following key. The end of the list is marked by $ptr[i] = 0$. The *rank* of a key is the number of keys in the list that are less than or equal to the given key. For instance, here is one way to represent the list 1, 2, 3, 5, 8, 13, 21.

i	1	2	3	4	5	6	7
$val[i]$	2	3	13	1	5	21	8
$ptr[i]$	2	5	6	1	7	0	3

In this example *head* = 4 and the rank of 13 is 6.

We can use binary search to find a key efficiently in a sorted *array*. Here, however, there is no obvious way to select the middle of the list, which would correspond to the first step in binary search. In fact, any deterministic algorithm takes a time in $\Omega(n)$ in the worst case to find a key in this kind of list.

Problem 8.4.4.   Prove the preceding assertion. (*Hint :* show how a worst-case instance can be constructed systematically from the probes made into the list by any given deterministic algorithm.)   □

Despite this inevitable worst case, there exists a deterministic algorithm that is capable of carrying out such a search in an average time in $O(\sqrt{n})$. From this we can obtain a Sherwood algorithm whose expected execution time is in $O(\sqrt{n})$ whatever the instance to be solved. As usual, the Sherwood algorithm is no faster on the average than the corresponding deterministic algorithm, but it does not have such a thing as a worst-case instance.

Suppose for the moment that the required key is always in fact present in the list, and that all the elements of the list are distinct. Given a key x, the problem is thus to find that index i, $1 \le i \le n$, such that $val[i] = x$. Any instance can be characterized by a permutation σ of the first n integers and by the rank k of the key we are looking for. Let S_n be the set of all $n!$ permutations. If A is any deterministic algorithm, $t_A(n,k,\sigma)$ denotes the time taken by this algorithm to find the key of rank k among the n keys in the list when the order of the latter in the array *val* is specified by the permutation σ. In the case of a probabilistic algorithm, $t_A(n,k,\sigma)$ denotes the expected value of this time. Whether the algorithm is deterministic or probabilistic, $w_A(n)$ and $m_A(n)$ denote its worst-case and its mean time, respectively. Thus

$$w_A(n) = \max\{t_A(n,k,\sigma) \mid 1 \le k \le n \text{ and } \sigma \in S_n\},$$

and

$$m_A(n) = \frac{1}{n \times n!} \sum_{\sigma \in S_n} \sum_{k=1}^{n} t_A(n,k,\sigma).$$

Problem 8.4.4 implies that $w_A(n) \in \Omega(n)$ for every deterministic algorithm A. We want a deterministic algorithm B such that $m_B(n) \in O(\sqrt{n})$ and a Sherwood algorithm C such that $w_C(n) \approx m_B(n)$.

The following algorithm finds a key x starting from some position i in the list, provided that $x \ge val[i]$ and that x is indeed present.

```
function search (x, i)
    while x > val [i] do i ← ptr [i]
    return i
```

Here is the obvious deterministic search.

```
function A (x)
    return search (x, head)
```

Problem 8.4.5.   Let $\hat{t}_A(n,k)$ be the exact number of references to the array *val* made by the algorithm A to find the key of rank k in a list of n keys. (The order σ of the keys is irrelevant for this algorithm.) Define $\hat{w}_A(n)$ and $\hat{m}_A(n)$ similarly. Determine $\hat{t}_A(n,k)$ for every integer n and for every k between 1 and n. Determine $\hat{w}_A(n)$ and $\hat{m}_A(n)$ for every integer n. $\square$

Here is a first probabilistic algorithm.

```
function D (x)
    i ← uniform (1 .. n)
    y ← val [i]
    case x < y : return search (x, head)
         x > y : return search (x, ptr [i])
         otherwise : return i
```

Problem 8.4.6.   Determine $\hat{t}_D(n,k)$ for every integer n and for every k between 1 and n. Determine $\hat{w}_D(n)$ and $\hat{m}_D(n)$ for every integer n. As a function of n, what values of k maximize $\hat{t}_D(n,k)$? Compare $\hat{w}_D(n)$ and $\hat{m}_A(n)$. Give explicitly a function $f(n)$ such that $\hat{t}_D(n,k) < \hat{t}_A(n,k)$ if and only if $k > f(n)$. (See Problem 8.4.5 for the definition of $\hat{t}$, $\hat{w}$, and $\hat{m}$.) $\square$

Problem 8.4.7.   The quantities $\hat{t}$, $\hat{w}$, and $\hat{m}$ introduced in the previous problems facilitate our analysis. Show, however, that they do not tell the whole story by exhibiting a deterministic algorithm E such that $\hat{w}_E(n) \in O(\log n)$, thus apparently contradicting Problem 8.4.4. $\square$

The following deterministic algorithm is efficient on the average.

```
function B (x)
    i ← head
    max ← val [i]
    for j ← 1 to ⌊√n⌋ do
        y ← val [j]
        if max < y ≤ x then i ← j, max ← y
    return search (x, i)
```

Problem 8.4.8.   Intuitively, why should we choose to execute the **for** loop $\sqrt{n}$ times in algorithm B? $\square$

∗ Problem 8.4.9.   Prove that $m_B(n) \in O(\sqrt{n})$. [*Hint*: Let $M_{l,n}$ be the random variable that corresponds to the minimum of l integers chosen randomly, uniformly, and independently with replacement from the set $\{1, 2, \ldots, n\}$. Find a link between this random variable and the average-case analysis of algorithm B. Show that the expected value of $M_{l,n}$ is about $n/(l+1)$ when l is a constant and about $\sqrt{n}$ when $l = \lfloor \sqrt{n} \rfloor$.]   □

Problem 8.4.10.   Show, however, that $w_B(n) \in \Omega(n)$, which is unavoidable from Problem 8.4.4. To do this, give explicitly a permutation σ and a rank k such that $t_B(n, k, \sigma) \in \Omega(n)$.   □

Problem 8.4.11.   Starting from the deterministic algorithm B, give a Sherwood algorithm C such that $w_C(n) \in O(\sqrt{n})$.   □

∗ Problem 8.4.12.   (Continuation of Problem 8.4.11)   Show more precisely that $\hat{w}_C(n) \in 2\sqrt{n} + \Theta(1)$, where the meaning of $\hat{w}$ is given in Problem 8.4.5.   □

Problem 8.4.13.   Give an efficient Sherwood algorithm that takes into account the possibility that the key we are seeking may not be in the list and that the keys may not all be distinct. Analyse your algorithm.   □

Problem 8.4.14.   Use the structure and the algorithms we have just seen to obtain a Sherwood sorting algorithm that is able to sort n elements in a worst-case expected time in $O(n^{3/2})$. Is this better than $O(n \log n)$? Justify your answer.   □

8.4.4 Universal Hashing

Hash coding, or simply hashing, is used in just about every compiler to implement the symbol table. Let X be the set of possible identifiers in the language to be compiled, and let N be a parameter chosen to obtain an efficient system. A *hash function* is a function $h : X \to \{1, 2, \ldots, N\}$. Such a function is a good choice if it efficiently disperses all the probable identifiers, that is, if $h(x) \neq h(y)$ for most of the pairs $x \neq y$ that are likely to be found in the same program. When $x \neq y$ but $h(x) = h(y)$, we say that there is a *collision* between x and y. The *hash table* is an array $T[1..N]$ of lists in which $T[i]$ is the list of those identifiers x found in the program such that $h(x) = i$. The *load factor* of the table is the ratio $\alpha = n/N$, where n is the number of distinct identifiers in the table. (The ratio α may well be greater than 1.) If we suppose that every identifier and every pointer occupies a constant amount of space, the table takes space in $\Theta(N + n)$ and the average length of the lists is α. Thus we see that increasing the value of N reduces the average search time but increases the space occupied by the table.

Problem 8.4.15.   Other ways to handle collisions, besides the use of a table of lists as outlined here, are legion. Suggest a few.   □

Problem 8.4.16.   What do you think of the "solution" that consists of ignoring the problem? If we are given an a priori upper bound on the number of identifiers a program may contain, does it suffice to choose N rather larger than this bound to ensure that the probability of a collision is negligible? (*Hint*: solve Problem 8.3.9 before answering.)   □

Problem 8.4.17.   Show that n calls on the symbol table can take a total time in $\Omega(n^2)$ in the worst case.   □

This technique is very efficient provided that the function h disperses the identifiers properly. If we suppose, however, that $\#X$ is very much greater than N, it is inevitable that certain programs will cause a large number of collisions. These programs will compile slowly every time they are submitted. In a sense they are paying the price for all other programs to compile quickly. A Sherwood approach allows us to retain the efficiency of hashing on the average, without arbitrarily favouring some programs at the expense of others. (If you have not yet solved Problem 8.2.3, now is the time to think some more about it!)

The basic idea is to choose the hash function randomly at the beginning of each compilation. A program that causes a large number of collisions during one compilation will therefore probably be luckier next time it is compiled. Unfortunately, there are far too many functions from X into $\{1, 2, \ldots, N\}$ for it to be reasonable to choose one at random.

Problem 8.4.18.   How many functions $f: A \rightarrow B$ are there if the cardinalities of A and B are a and b, respectively?   □

This difficulty is solved by universal hashing. By definition a class H of functions from A to B is *universal$_2$* if $\#\{h \in H \mid h(x) = h(y)\} \leq \#H/\#B$ for every $x, y \in A$ such that $x \neq y$. Let H be a universal$_2$ class of functions from X to $\{1, 2, \ldots, N\}$, and let x and y be any two distinct identifiers. If we choose a hash function h randomly and uniformly in H, the probability of a collision between x and y is therefore at most $1/N$. The following problem generalizes this situation.

*Problem 8.4.19.   Let $S \subseteq X$ be a set of n identifiers already in the table. Let $x \in X \setminus S$ be a new identifier. Prove that the average number of collisions between x and the elements of S (that is, the average length of the list $T[h(x)]$) is less than or equal to the load factor α. Prove further that the probability that the number of collisions will be greater than $t\alpha$ is less than $1/t$ for all $t \geq 1$.   □

Several efficient universal$_2$ classes of functions are known. We content ourselves with mentioning just one.

*Problem 8.4.20.   Let X be $\{0, 1, 2, \ldots, a-1\}$, and let p be a prime number not less than N. Let m and n be two integers. Define $h_{m,n} : X \rightarrow \{0, 1, \ldots, N-1\}$ by $h_{m,n}(x) = ((mx + n) \bmod p) \bmod N$.

Prove that $H = \{h_{m,n} \mid 1 \le m < p \text{ and } 0 \le n < p\}$ is a universal$_2$ class of functions. (*Remarks*: In practice we take N to be a power of 2 so that the second **mod** operation can be executed efficiently. It is also more efficient to carry out all the computations in a Galois field whose cardinality is a power of 2.)   □

*Problem 8.4.21.   Find applications of universal hashing that have nothing to do with compilation, nor even with management of a symbol table.   □

8.5 LAS VEGAS ALGORITHMS

Although its behaviour is more uniform, a Sherwood algorithm is no faster on the average than the deterministic algorithm from which it arose. A *Las Vegas* algorithm, on the other hand, allows us to obtain an increase in efficiency, sometimes for *every* instance. It may be able to solve in practice certain problems for which no efficient deterministic algorithm is known even on the average. However, there is no upper bound on the time that may be required to obtain a solution, even though the expected time required for each instance may be small and the probability of encountering an excessive execution time is negligible. Contrast this to a Sherwood algorithm, where we are able to predict the maximum time needed to solve a given instance. For example, the Sherwood version of quicksort (Problem 8.4.1) never takes a time in excess of $O(n^2)$ to sort n elements, whatever happens.

The distinguishing characteristic of Las Vegas algorithms is that now and again they take the risk of making a random decision that renders it impossible to find a solution. Thus these algorithms react by either returning a correct solution or admitting that their random decisions have led to an impasse. In the latter case it suffices to resubmit the same instance to the same algorithm to have a second, independent chance of arriving at a solution. The overall probability of success therefore increases with the amount of time we have available.

Las Vegas algorithms usually have a return parameter *success*, which is set to *true* if a solution is obtained and *false* otherwise. The typical call to solve instance x is $LV(x, y, success)$, where y is a return parameter used to receive the solution thus obtained whenever *success* is set to *true*. Let $p(x)$ be the probability of success of the algorithm each time that it is asked to solve the instance x. For an algorithm to be correct, we require that $p(x) > 0$ for every instance x. Better still is the existence of some constant $\delta > 0$ such that $p(x) \ge \delta$ for every instance x. Let $s(x)$ and $e(x)$ be the expected times taken by the algorithm on instance x in the case of success and of failure, respectively. Now consider the following algorithm.

> **function** *obstinate* (*x*)
>   **repeat**
>   *LV* (*x*, *y*, *success*)
>   **until** *success*
>   **return** *y*

Let $t(x)$ be the expected time taken by the algorithm *obstinate* to find an exact solution to the instance x. Neglecting the time taken by the control of the **repeat** loop, we obtain the following recurrence:

$$t(x) = p(x)s(x) + (1-p(x))(e(x)+t(x)).$$

This follows because the algorithm succeeds at the first attempt with probability $p(x)$, thus taking an expected time $s(x)$. With probability $1-p(x)$ it first makes an unsuccessful attempt to solve the instance, taking an expected time $e(x)$, before starting all over again to solve the instance, which still takes an expected time $t(x)$. The recurrence is easily solved to yield

$$t(x) = s(x) + \frac{1-p(x)}{p(x)} e(x).$$

There is a compromise to be made between $p(x)$, $s(x)$, and $e(x)$ if we want to minimize $t(x)$. For example, it may be preferable to accept a smaller probability of success if this also decreases the time required to know that a failure has occurred.

Problem 8.5.1.   Suppose that $s(x)$ and $e(x)$ are not just expected times, but that they are in fact the *exact* times taken by a call on $LV(x, \ldots)$ in the case of success and of failure, respectively. What is the probability that the algorithm *obstinate* will find a correct solution in a time not greater than t, for any $t \geq s(x)$? Give your answer as a function of t, $s(x)$, $e(x)$ and $p(x)$.   □

8.5.1 The Eight Queens Problem Revisited

The eight queens problem (Section 6.6.1) provides a nice example of this kind of algorithm. Recall that the backtracking technique used involves systematically exploring the nodes of the implicit tree formed by the k-promising vectors. Using this technique, we obtain the first solution after examining only 114 of the 2,057 nodes in the tree. This is not bad, but the algorithm does not take into account one important fact: there is nothing systematic about the positions of the queens in most of the solutions. On the contrary, the queens seem more to have been positioned haphazardly. This observation suggests a greedy Las Vegas algorithm that places queens randomly on successive rows, taking care, however, that the queens placed on the board do not threaten one another. The algorithm ends either successfully if it manages to place all the queens on the board or in failure if there is no square in which the next queen can be added. The resulting algorithm is not recursive.

procedure *QueensLV* (**var** *success*)
  { if *success* = *true* at the end, then *try* [1 .. 8] (a global array)
   contains a solution to the eight queens problem }
  col, *diag*45, *diag*135 ← ∅
  $k ← 0$
  repeat
    { *try* [1 .. *k*] is *k*-promising }
    nb ← 0
    for $i ← 1$ **to** 8 **do**
      if $i ∉ col$ **and** $i-k ∉ diag45$ **and** $i+k ∉ diag135$
      then { column *i* is available for the $(k+1)$st queen }
          nb ← *nb* + 1
          if *uniform* (1 .. *nb*) = 1
          then { maybe try column *i* }
             $j ← i$
    if *nb* > 0
    then { amongst all *nb* possibilities for the $(k+1)$st queen,
        it is column *j* that has been chosen (with probability 1 / *nb*) }
        try [*k* + 1] ← *j*
        col ← *col* ∪ { *j* }
        *diag*45 ← *diag*45 ∪ { *j* − *k* }
        *diag*135 ← *diag*135 ∪ { *j* + *k* }
        { *try* [1 .. *k* + 1] is $(k+1)$-promising }
        $k ← k + 1$
  until *nb* = 0 **or** *k* = 8
  success ← (*nb* > 0)

To analyse the efficiency of this algorithm, we need to determine its probability p of success, the average number s of nodes that it explores in the case of success, and the average number e of nodes that it explores in the case of failure. Clearly $s = 9$ (counting the 0-promising empty vector). Using a computer we can calculate $p = 0.1293 \cdots$ and $e = 6.971 \cdots$ A solution is therefore obtained more than one time out of eight by proceeding in a completely random fashion! The expected number of nodes explored if we repeat the algorithm until a success is finally obtained is given by the general formula $s + (1-p)e/p = 55.927 \cdots$, less than half the number of nodes explored by the systematic backtracking technique.

Problem 8.5.2. When there is more than one position open for the $(k+1)$st queen, the algorithm *QueensLV* chooses one at random without first counting the number *nb* of possibilities. Show that each position has, nevertheless, the same probability of being chosen. □

We can do better still. The Las Vegas algorithm is too defeatist: as soon as it detects a failure it starts all over again from the beginning. The backtracking algorithm, on the other hand, makes a systematic search for a solution that we know has nothing systematic about it. A judicious combination of these two algorithms first places a number of queens on the board in a random way, and then uses backtracking to try and add the remaining queens, without, however, reconsidering the positions of the queens that were placed randomly.

An unfortunate random choice of the positions of the first few queens can make it impossible to add the others. This happens, for instance, if the first two queens are placed in positions 1 and 3, respectively. The more queens we place randomly, the smaller is the average time needed by the subsequent backtracking stage, but the greater is the probability of a failure.

The resulting algorithm is similar to *QueensLV*, except that the last two lines are replaced by

> **until** $nb = 0$ **or** $k = stopVegas$
> **if** $nb > 0$ **then** $backtrack(k, col, diag45, diag135, success)$
>   **else** $success \leftarrow false$,

where $1 \le stopVegas \le 8$ indicates how many queens are to be placed randomly before moving on to the backtracking phase. The latter looks like the algorithm *Queens* of Section 6.6.1 except that it has an extra parameter *success* and that it returns immediately after finding the first solution if there is one.

The following table gives for each value of *stopVegas* the probability p of success, the expected number s of nodes explored in the case of success, the expected number e of nodes explored in the case of failure, and the expected number $t = s + (1-p)e/p$ of nodes explored if the algorithm is repeated until it eventually finds a solution. The case *stopVegas* $= 0$ corresponds to using the deterministic algorithm directly.

stopVegas	p	s	e	t
0	1.0000	114.00	—	114.00
1	1.0000	39.63	—	39.63
2	0.8750	22.53	39.67	28.20
3	0.4931	13.48	15.10	29.01
4	0.2618	10.31	8.79	35.10
5	0.1624	9.33	7.29	46.92
6	0.1357	9.05	6.98	53.50
7	0.1293	9.00	6.97	55.93
8	0.1293	9.00	6.97	55.93

We tried these different algorithms on a CYBER 835. The pure backtracking algorithm finds the first solution in 40 milliseconds, whereas an average of 10 milliseconds is all that is needed if the first two or three queens are placed at random. The original greedy algorithm *QueensLV*, which places all the queens in a random way, takes on the average 23 milliseconds to find a solution. This is a fraction more than half the time taken by the backtracking algorithm because we must also take into account the time required to make the necessary pseudorandom choices of position.

Problem 8.5.3.   If you are still not convinced of the value of this technique, we suggest you try to solve the twelve queens problem without using a computer. First, try to solve the problem systematically, and then try again, this time placing the first five queens randomly.   □

For the eight queens problem, a systematic search for a solution beginning with the first queen in the first column takes quite some time. First the trees below the 2-promising nodes $[1,3]$ and $[1,4]$ are explored to no effect. Even when the search starting from node $[1,5]$ begins, we waste time with $[1,5,2]$ and $[1,5,7]$. This is one reason why it is more efficient to place the first queen at random rather than to begin the systematic search immediately. On the other hand, a systematic search that begins with the first queen in the fifth column is astonishingly quick. (Try it!) This unlucky characteristic of the upper left-hand corner is nothing more than a meaningless accident. For instance, the same corner is a better than average starting point for the problems with five or twelve queens. What *is* significant, however, is that a solution can be obtained more rapidly on the average if *several* queens are positioned randomly before embarking on the backtracking phase. Once again, this can be understood intuitively in terms of the lack of regularity in the solutions (at least when the number of queens is not $4k+2$ for some integer k).

Here are the values of p, s, e, and t for a few values of *stopVegas* in the case of the twelve queens problem.

stopVegas	p	s	e	t
0	1.0000	262.00	—	262.00
5	0.5039	33.88	47.23	80.39
12	0.0465	13.00	10.20	222.11

On the CYBER 835 the Las Vegas algorithm that places the first five queens randomly before starting backtracking requires only 37 milliseconds on the average to find a solution, whereas the pure backtracking algorithm takes 125 milliseconds. As for the greedy Las Vegas algorithm, it wastes so much time making its pseudorandom choices of position that it requires essentially the same amount of time as the pure backtracking algorithm.

An empirical study of the twenty queens problem was also carried out using an Apple II personal computer. The deterministic backtracking algorithm took more than 2 hours to find a first solution. Using the probabilistic approach and placing the first ten queens at random, 36 different solutions were found in about five and a half minutes. Thus the probabilistic algorithm turned out to be almost 1,000 times faster per solution than the deterministic algorithm.

** Problem 8.5.4.   If we want a solution to the general n queens problem, it is obviously silly to analyse exhaustively all the possibilities so as to discover the optimal value of *stopVegas*, and then to apply the corresponding Las Vegas algorithm. In fact, determining the optimal value of *stopVegas* takes longer than a straightforward search for a solution using backtracking. (We needed more than 50 minutes computation on the CYBER to establish that *stopVegas* = 5 is the optimal choice for the twelve queens problem!) Find an analytic method that enables a good, but not necessarily optimal, value of *stopVegas* to be determined rapidly as a function of n.   □

∗∗ Problem 8.5.5.   Technically, the general algorithm obtained using the previous problem (first determine *stopVegas* as a function of n, the number of queens, and then try to place the queens on the board) can only be considered to be a Las Vegas algorithm if its probability of success is strictly positive for every n. This is the case if and only if there exists at least one solution. If no solution exists, the obstinate probabilistic algorithm will loop forever without realizing what is happening. Prove or disprove: the n queens problem can be solved for every $n \geq 4$. Combining this with Problem 8.5.4, can you find a constant $\delta > 0$ such that the probability of success of the Las Vegas algorithm to solve the n queens problem is at least δ for every n?   □

8.5.2 Square Roots Modulo p

Let p be an odd prime. An integer x is a *quadratic residue* modulo p if $1 \leq x \leq p - 1$ and if there exists an integer y such that $x \equiv y^2 \pmod{p}$. Such a y is a *square root* of x modulo p provided $1 \leq y \leq p - 1$. For instance, 63 is a square root of 55 modulo 103. An integer z is a *quadratic nonresidue* modulo p if $1 \leq z \leq p - 1$ and z is not a quadratic residue modulo p. Any quadratic residue has at least two distinct square roots since $(p - y)^2 = p^2 - 2py + y^2 \equiv y^2 \pmod{p}$.

Problem 8.5.6.   Prove that $p - y \neq y$ and that $1 \leq p - y \leq p - 1$.   □

Problem 8.5.7.   Prove, on the other hand, that no quadratic residue has more than two distinct square roots. (*Hint:* assuming that $a^2 \equiv b^2 \pmod{p}$, consider $a^2 - b^2$.)   □

Problem 8.5.8.   Conclude from the preceding results that exactly half the integers between 1 and $p - 1$ are quadratic residues modulo p.   □

∗ Problem 8.5.9.   Prove that $x^{(p-1)/2} \equiv \pm 1 \pmod{p}$ for every integer $1 \leq x \leq p - 1$ and every odd prime p. Prove further that x is a quadratic residue modulo p if and only if $x^{(p-1)/2} \equiv +1 \pmod{p}$. (*Hint:* one direction follows immediately from Fermat's theorem: $x^{p-1} \equiv 1 \pmod{p}$; the other direction requires some knowledge of group theory.)   □

The preceding problem suggests an efficient algorithm for testing whether x is a quadratic residue modulo p: it suffices to use the fast exponentiation algorithm of Section 4.8 to calculate $x^{(p-1)/2} \bmod p$. Given an odd prime p and a quadratic residue x

modulo p, does there exist an efficient algorithm for calculating the two square roots of x modulo p? The problem is easy when $p \equiv 3 \pmod 4$, but no efficient deterministic algorithm is known to solve this problem when $p \equiv 1 \pmod 4$.

Problem 8.5.10.   Suppose that $p \equiv 3 \pmod 4$ and let x be a quadratic residue modulo p. Prove that $\pm x^{(p+1)/4} \bmod p$ are the two square roots of x modulo p. Calculate $55^{26} \bmod 103$ and verify that its square modulo 103 is indeed 55. □

There exists, however, an efficient Las Vegas algorithm to solve this problem when $p \equiv 1 \pmod 4$. Let us decide arbitrarily to denote by $\sqrt{x}$ the smaller of the two square roots of x. Even if the value of $\sqrt{x}$ is unknown, it is possible to carry out the *symbolic* multiplication of $a + b\sqrt{x}$ and $c + d\sqrt{x}$ modulo p, where a, b, c, and d are integers between 0 and $p-1$. This product is $((ac + bdx) \bmod p) + ((ad + bc) \bmod p)\sqrt{x}$. Note the similarity to a product of complex numbers. The symbolic exponentiation $(a + b\sqrt{x})^n$ can be calculated efficiently by adapting the algorithms of Section 4.8.

Example 8.5.1.   Let $p = 53 \equiv 1 \pmod 4$ and $x = 7$. A preliminary computation shows that x is a quadratic residue modulo p since $7^{26} \equiv 1 \pmod{53}$. Let us calculate symbolically $(1 + \sqrt{7})^{26} \bmod 53$. (All the following calculations are modulo 53.)

$$
\begin{aligned}
(1+\sqrt{7})^2 &= (1+\sqrt{7})(1+\sqrt{7}) & &= 8 + 2\sqrt{7} \\
(1+\sqrt{7})^3 &= (1+\sqrt{7})(8+2\sqrt{7}) & &= 22 + 10\sqrt{7} \\
(1+\sqrt{7})^6 &= (22+10\sqrt{7})(22+10\sqrt{7}) &&= 18 + 16\sqrt{7} \\
(1+\sqrt{7})^{12} &= (18+16\sqrt{7})(18+16\sqrt{7}) &&= 49 + 46\sqrt{7} \\
(1+\sqrt{7})^{13} &= (1+\sqrt{7})(49+46\sqrt{7}) & &= 0 + 42\sqrt{7} \\
(1+\sqrt{7})^{26} &= (0+42\sqrt{7})(0+42\sqrt{7}) & &= 52 + 0\sqrt{7}
\end{aligned}
$$
□

Thus we see that $(1+\sqrt{7})^{26} \equiv -1 \pmod{53}$. Since $26 = (p-1)/2$, we conclude that $(1+\sqrt{7}) \bmod 53$ is a quadratic nonresidue modulo 53 (Problem 8.5.9). But 7 has two square roots modulo p, and the symbolic calculation that we just carried out is valid regardless of which of them we choose to call $\sqrt{7}$. Consequently, $(1-\sqrt{7}) \bmod 53$ is also a quadratic nonresidue modulo 53. What happens if we calculate symbolically $(a+\sqrt{7})^{26} \equiv c + d\sqrt{7} \pmod{53}$ in a case when one of $(a+\sqrt{7}) \bmod 53$ and $(a-\sqrt{7}) \bmod 53$ is a quadratic residue modulo p and the other is not? Suppose, for instance, that $c + d\sqrt{7} \equiv 1 \pmod{53}$ and $c + d(-\sqrt{7}) = c - d\sqrt{7} \equiv -1 \pmod{53}$. Adding these two equations, we obtain $2c \equiv 0 \pmod{53}$ and hence $c = 0$ since $0 \le c \le 52$. Subtracting them, we find $2d\sqrt{7} \equiv 2 \pmod{53}$, and hence $d\sqrt{7} \equiv 1 \pmod{53}$.

Problem 8.5.11.   Using Example 8.5.1 as a guide, carry out the symbolic calculation $(2+\sqrt{7})^{26} \equiv 0 + 41\sqrt{7} \pmod{53}$ in detail. □

To obtain a square root of 7, the preceding problem shows that we need only find the unique integer y such that $1 \leq y \leq 52$ and $41y \equiv 1 \pmod{53}$. This can be done efficiently using a modification of Euclid's algorithm for calculating the greatest common divisor (Section 1.7.4).

* Problem 8.5.12.   Let u and v be two positive integers, and let d be their greatest common divisor.

i. Prove that there exist integers a and b such that $au + bv = d$. [*Hint:* Suppose without loss of generality that $u \geq v$. If $v = d$, the proof is trivial ($a = 0$ and $b = 1$). Otherwise, let $w = u \bmod v$. First show that d is also the greatest common divisor of v and w. (This is the heart of Euclid's algorithm.) By mathematical induction, now let a' and b' be such that $a'v + b'w = d$. Then we need only take $a = b'$ and $b = a' - b' \lfloor u/v \rfloor$.]

ii. Give an efficient iterative algorithm for calculating d, a, and b from u and v. Your algorithm should not calculate d before starting to work on a and b.

iii. If p is prime and $1 \leq a \leq p - 1$, prove that there exists a unique y such that $1 \leq y \leq p - 1$ and $ay \equiv 1 \pmod{p}$. Give an efficient algorithm for calculating y given p and a. $\square$

In our example (following Problem 8.5.11), we find $y = 22$ because $41 \times 22 \equiv 1 \pmod{53}$. This is indeed a square root of 7 modulo 53 since $22^2 \equiv 7 \pmod{53}$. The other square root is $53 - 22 = 31$. This suggests the following Las Vegas algorithm for calculating square roots.

```
procedure rootLV (x, p, var y, var success)
   { may find some y such that y² ≡ x (mod p)
     assuming p is a prime, p ≡ 1 (mod 4),
     and x is a quadratic residue modulo p }
   a ← uniform (1 .. p − 1)
   if a² ≡ x (mod p)    { very unlikely }
   then success ← true
        y ← a
   else compute c and d such that 0 ≤ c ≤ p − 1, 0 ≤ d ≤ p − 1
              and (a + √x )^(p − 1)/2 ≡ c + d √x (mod p )
        if d = 0 then success ← false
        else {c = 0}
             success ← true
             compute y such that 1 ≤ y ≤ p − 1 and dy ≡ 1 (mod p )
```

It remains to determine the probability of success of this algorithm.

* Problem 8.5.13.   Let $p \equiv 1 \pmod{4}$ be prime, and let x be a quadratic residue modulo p. An integer a, $1 \leq a \leq p - 1$, *gives the key* to $\sqrt{x}$ if $(a^2 - x) \bmod p$ is not a quadratic residue modulo p. Prove that

i. The Las Vegas algorithm finds a square root of x if and only if it randomly chooses an a that gives the key to $\sqrt{x}$; and

ii. Exactly $(p+3)/2$ of the $p-1$ possible random choices for a give the key to $\sqrt{x}$. [*Hint :* Consider the function

$$f:\{1,2,\ldots,p-1\}\setminus\{\sqrt{x},p-\sqrt{x}\}\rightarrow\{2,3,\ldots,p-2\}$$

defined by the equation $(a-\sqrt{x})f(a)\equiv a+\sqrt{x}\pmod p$. Prove that this function is one-to-one and that $f(a)$ is a quadratic residue modulo p if and only if a does not give the key to $\sqrt{x}$.] □

This shows that the Las Vegas algorithm succeeds with probability somewhat greater than one half, so that on the average it suffices to call it twice to obtain a square root of x. In view of the high proportion of integers that give a key to $\sqrt{x}$, it is curious that no known efficient deterministic algorithm is capable of finding even one of them with certainty.

Problem 8.5.14.   The previous problem suggests a modification to the algorithm *rootLV* : only carry out the symbolic calculation of $(a+\sqrt{x})^{(p-1)/2}$ if $(a^2-x)\bmod p$ is a quadratic nonresidue. This allows us to detect a failure more rapidly, but it takes longer in the case of a success. Give the modified algorithm explicitly. Is it to be preferred to the original algorithm? Justify your answer. □

*Problem 8.5.15.   The following algorithm increases the probability of success if $p\equiv 1\pmod 8$.

> **procedure** *rootLV* $2(x,p,$ **var** $y,$ **var** *success*)
> {assume that p is a prime and $p\equiv 1\pmod 4$ }
> $a\leftarrow uniform(1..p-1)$
> **if** $a^2\equiv -x\pmod p$ { very unlikely and unfortunate }
> **then** *success* $\leftarrow$ *false*
> **else let** odd t and $k\geq 2$ be such that $p=2^k t+1$
>   compute c and d such that $0\leq c\leq p-1,0\leq d\leq p-1$
>   and $(a+\sqrt{p-x})^t\equiv c+d\sqrt{p-x}\pmod p$
>   **if** $c=0$ **or** $d=0$
>   **then** *success* $\leftarrow$ *false*
>   **else while** $c^2\not\equiv d^2 x\pmod p$ **do**
> $b\leftarrow(c^2-d^2 x)\bmod p$
> $d\leftarrow 2cd\bmod p$
> $c\leftarrow b$
>   compute y such that $1\leq y\leq p-1$ and $yd\equiv 1\pmod p$
> $y\leftarrow cy\bmod p$
>   *success* $\leftarrow$ *true*

Prove that the probability of failure of this algorithm is exactly $(1/2)^{k-1}$ and that the **while** loop is executed at most $k-2$ times, where k is specified in the algorithm. □

Problem 8.5.16. An even more elementary problem for which no efficient deterministic algorithm is known is to find a quadratic nonresidue modulo p where $p \equiv 1 \pmod 4$ is a prime.

i. Give an efficient Las Vegas algorithm to solve this problem.

ii. Show that the problem is not more difficult than the problem of finding an efficient deterministic algorithm to calculate a square root. To do this, suppose that there exists a deterministic algorithm $root2(x,p)$ that is able to calculate efficiently $\sqrt{x} \bmod p$, where $p \equiv 1 \pmod 4$ is prime and x is a quadratic residue modulo p. Show that it suffices to call this algorithm less than $\lfloor \lg p \rfloor$ times to be certain of obtaining, in a way that is both deterministic and efficient, a quadratic nonresidue modulo p. (*Hint:* Let k be the largest integer such that 2^k divides $p-1$ exactly. Consider the sequence $x_1 = p-1$, $x_i = \sqrt{x_{i-1}} \bmod p$ for $2 \le i \le k$. Prove that x_i is a quadratic residue modulo p for $1 \le i < k$, but that x_k is not.) □

** Problem 8.5.17. The converse of Problem 8.5.16. Give an efficient deterministic algorithm $rootDET(x,p,z)$ to calculate a square root of x modulo p, provided p is an odd prime, x is a quadratic residue modulo p, and z is an arbitrary quadratic nonresidue modulo p. □

The two preceding problems show the computational equivalence between the efficient deterministic calculation of square roots modulo p and the efficient deterministic discovery of a quadratic nonresidue modulo p. This is an example of the technique called *reduction*, which we study in Chapter 10.

8.5.3 Factorizing Integers

Let n be an integer greater than 1. The problem of *factorizing* n consists of finding the unique decomposition $n = p_1^{m_1} p_2^{m_2} \cdots p_k^{m_k}$ such that $m_1, m_2, \ldots, m_k$ are positive integers and $p_1 < p_2 < \cdots < p_k$ are prime numbers. If n is composite, a *nontrivial factor* is an integer x, $1 < x < n$, that divides n exactly. Given composite n, the problem of *splitting* n consists of finding some nontrivial factor of n.

Problem 8.5.18. Suppose you have available an algorithm $prime(n)$, which tests whether or not n is prime, and an algorithm $split(n)$, which finds a nontrivial factor of n provided n is composite. Using these two algorithms as primitives, give an algorithm to factorize any integer. □

Section 8.6.2 concerns an efficient Monte Carlo algorithm for determining primality. Thus the preceding problem shows that the problem of factorization reduces to the problem of splitting. Here is the naive algorithm for the latter problem.

function *split* (n)
   { finds the smallest nontrivial factor of n if n is composite
   or returns 1 if n is prime }
   for $i \leftarrow 2$ **to** $\lfloor \sqrt{n} \rfloor$ **do**
      if $(n \bmod i) = 0$ **then return** i
   return 1

Problem 8.5.19.   Why is it sufficient to loop no further than $\sqrt{n}$?   □

The preceding algorithm takes a time in $\Omega(\sqrt{n})$ in the worst case to split n. It is therefore of no practical use even on medium-sized integers : it could take more than 3 million years in the worst case to split a number with forty or so decimal digits, counting just 1 microsecond for each trip round the loop. No known algorithm, whether deterministic or probabilistic, can split n in a time in $O(p(m))$ in the worst case, where p is a polynomial and $m = \lceil \log(1+n) \rceil$ is the size of n. Notice that $\sqrt{n} \approx 10^{m/2}$, which is not a polynomial in m. Dixon's probabilistic algorithm is nevertheless able to split n in a time in $O(2^{O(\sqrt{m \log m})})$.

Problem 8.5.20.   Prove that $O(m^k) \subset O(2^{O(\sqrt{m \log m})}) \subset O(10^{m/b})$ whatever the values of the positive constants k and b.   □

The notion of quadratic residue modulo a prime number (Section 8.5.2) generalizes to composite numbers. Let n be any positive integer. An integer x, $1 \le x \le n - 1$, is a *quadratic residue* modulo n if it is relatively prime to n (they have no nontrivial common factor) and if there exists an integer y, $1 \le y \le n - 1$, such that $x \equiv y^2 \pmod{n}$. Such a y is a *square root* of x modulo n. We saw that a quadratic residue modulo p has exactly two distinct square roots modulo p when p is prime. This is no longer true modulo n if n has at least two distinct odd prime factors. For instance, $8^2 \equiv 13^2 \equiv 22^2 \equiv 27^2 \equiv 29 \pmod{35}$.

*Problem 8.5.21.   Prove that if $n = pq$, where p and q are distinct odd primes, then each quadratic residue modulo n has exactly four square roots. Prove further that exactly one quarter of the integers x that are relatively prime to n and such that $1 \le x \le n - 1$ are quadratic residues modulo n.   □

Section 8.5.2 gave efficient algorithms for testing whether x is a quadratic residue modulo p, and if so for finding its square roots. These two problems can also be solved efficiently modulo a composite number n provided the factorization of n is given. If the factorization of n is not given, no efficient algorithm is known for either of these problems. The essential step in Dixon's factorization algorithm is to find two integers a and b relatively prime to n such that $a^2 \equiv b^2 \pmod{n}$ but $a \not\equiv \pm b \pmod{n}$. This implies that $a^2 - b^2 = (a - b)(a + b) \equiv 0 \pmod{n}$. Given that n is a divisor neither of $a + b$ nor of $a - b$, it follows that some nontrivial factor x of n must divide $a + b$ while n/x divides $a - b$. The greatest common divisor of n and $a + b$ is thus a nontrivial factor of n. In the previous example, $a = 8$, $b = 13$, and $n = 35$, and the greatest common divisor of $a + b = 21$ and $n = 35$ is $x = 7$, a nontrivial factor of 35. Here is an outline of Dixon's algorithm.

> **procedure** $Dixon(n, \textbf{var } x, \textbf{var } success)$
> { tries to find some nontrivial factor x of composite number n }
> **if** n is even **then** $x \leftarrow 2$, $success \leftarrow true$
> **else for** $i \leftarrow 2$ **to** $\lfloor \log_3 n \rfloor$ **do**
> > **if** $n^{1/i}$ is an integer **then** $x \leftarrow n^{1/i}$
> > > $success \leftarrow true$
> > > **return**
>
> { since n is assumed composite, we now know that it has
> at least two distinct odd prime factors }
> $a, b \leftarrow$ two integers such that $a^2 \equiv b^2 \pmod{n}$
> **if** $a \equiv \pm b \pmod{n}$ **then** $success \leftarrow false$
> **else** $x \leftarrow \gcd(a+b, n)$ { using Euclid's algorithm }
> > $success \leftarrow true$

So how we can find a and b such that $a^2 \equiv b^2 \pmod{n}$? Let k be an integer to be specified later. An integer is k-*smooth* if all its prime factors are among the first k prime numbers. For instance, $120 = 2^3 \times 3 \times 5$ is 3-smooth, but $35 = 5 \times 7$ is not. When k is small, k-smooth integers can be factorized efficiently by an adaptation of the naive algorithm $split(n)$ given earlier. In its first phase, Dixon's algorithm chooses integers x randomly between 1 and $n - 1$. A nontrivial factor of n is already found if by a lucky fluke x is not relatively prime to n. Otherwise, let $y = x^2 \bmod n$. If y is k-smooth, both x and the factorization of y are kept in a table. The process is repeated until we have $k + 1$ different integers for which we know the factorization of their squares modulo n.

Example 8.5.2. Let $n = 2,537$ and $k = 7$. We are thus concerned only with the primes 2, 3, 5, 7, 11, 13, and 17. A first integer $x = 1,769$ is chosen randomly. We calculate its square modulo n: $y = 1,240$. An attempt to factorize $1,240 = 2^3 \times 5 \times 31$ fails since 31 is not divisible by any of the admissible primes. A second attempt with $x = 2,455$ is more successful: its square modulo n is $y = 1,650 = 2 \times 3 \times 5^2 \times 11$. Continuing thus, we obtain

$$x_1 = 2,455 \quad y_1 = 1,650 = 2 \times 3 \times 5^2 \times 11$$
$$x_2 = 970 \quad y_2 = 2,210 = 2 \times 5 \times 13 \times 17$$
$$x_3 = 1,105 \quad y_3 = 728 = 2^3 \times 7 \times 13$$
$$x_4 = 1,458 \quad y_4 = 2,295 = 3^3 \times 5 \times 17$$
$$x_5 = 216 \quad y_5 = 990 = 2 \times 3^2 \times 5 \times 11$$
$$x_6 = 80 \quad y_6 = 1,326 = 2 \times 3 \times 13 \times 17$$
$$x_7 = 1,844 \quad y_7 = 756 = 2^2 \times 3^3 \times 7$$
$$x_8 = 433 \quad y_8 = 2,288 = 2^4 \times 11 \times 13 \ .$$

□

Problem 8.5.22.   Given that there are 512 integers x between 1 and 2,536 such that x^2 **mod** 2537 is 7-smooth, what is the average number of trials needed to obtain eight successes like those in Example 8.5.2 ?   □

The second phase of Dixon's algorithm finds a nonempty subset of the $k+1$ equations such that the product of the corresponding factorizations includes each of the k admissible prime numbers to an even power (including zero).

Example 8.5.3.   There are seven possible ways of doing this in Example 8.5.2, including

$$y_1 y_2 y_4 y_8 = 2^6 \times 3^4 \times 5^4 \times 7^0 \times 11^2 \times 13^2 \times 17^2$$

$$y_1 y_3 y_4 y_5 y_6 y_7 = 2^8 \times 3^{10} \times 5^4 \times 7^2 \times 11^2 \times 13^2 \times 17^2 \ .$$ □

Problem 8.5.23.   Find the other five possibilities.   □

Problem 8.5.24.   Why is there always at least one solution? Give an efficient algorithm for finding one. [*Hint*: Form a $(k+1) \times k$ binary matrix containing the parities of the exponents. The rows of this matrix cannot be independent (in arithmetic modulo 2) because there are more rows than columns. In Example 8.5.3, the first dependence corresponds to

$$(1,1,0,0,1,0,0) + (1,0,1,0,0,1,1) + (0,1,1,0,0,0,1) + (0,0,0,0,1,1,0)$$
$$\equiv (0,0,0,0,0,0,0) \text{ (mod 2) } .$$

Use Gauss-Jordan elimination to find a linear dependence between the rows.]   □

This gives us two integers a and b such that $a^2 \equiv b^2$ (mod n). The integer a is obtained by multiplying the appropriate x_i and the integer b by halving the powers of the primes in the product of the y_i . If $a \not\equiv \pm b$ (mod n), it only remains to calculate the greatest common divisor of $a+b$ and n to obtain a nontrivial factor. This occurs with probability at least one half.

Example 8.5.4.   The first solution of Example 8.5.3 gives

$$a = x_1 x_2 x_4 x_8 \textbf{ mod } n = 2,455 \times 970 \times 1,458 \times 433 \textbf{ mod } 2,537 = 1,127 \quad \text{and}$$

$$b = 2^3 \times 3^2 \times 5^2 \times 11 \times 13 \times 17 \textbf{ mod } 2,537 = 2,012 \not\equiv \pm a \text{ (mod } n \text{)}.$$

The greatest common divisor of $a+b = 3,139$ and $n = 2,537$ is 43, a nontrivial factor of n. On the other hand, the second solution gives

$$a = x_1 x_3 x_4 x_5 x_6 x_7 \textbf{ mod } n = 564 \quad \text{and}$$

$$b = 2^4 \times 3^5 \times 5^2 \times 7 \times 11 \times 13 \times 17 \textbf{ mod } 2,537 = 1,973 \equiv -a \text{ (mod } n \text{)} ,$$

which does not reveal a factor.   □

Problem 8.5.25.   Why is there at least one chance in two that $a \not\equiv \pm b$ (mod n)? In the case when $a \equiv \pm b$ (mod n), can we do better than simply starting all over again?   □

It remains to be seen how we choose the best value for the parameter k. The larger this parameter, the higher the probability that $x^2 \bmod n$ will be k-smooth when x is chosen randomly. On the other hand, the smaller this parameter, the faster we can carry out a test of k-smoothness and factorize the k-smooth values y_i, and the fewer such values we need to be sure of having a linear dependence. Set $L = e^{\sqrt{\ln n \ln \ln n}}$ and let $b \in \mathbb{R}^+$. It can be shown that if $k \approx L^b$, there are about $L^{1/2b}$ failures for every success when we try to factorize $x^2 \bmod n$. Since each unsuccessful attempt requires k divisions and since it takes $k+1$ successes to end the first phase, the latter takes an average time that is approximately in $O(L^{2b+1/2b})$, which is minimized by $b = \frac{1}{2}$. The second phase takes a time in $O(k^3) = O(L^{3b})$ by Problem 8.5.24 (it is possible to do better than this), which is negligible when compared to the first phase. The third phase can also be neglected. Thus, if we take $k \approx \sqrt{L}$, Dixon's algorithm splits n with probability at least one half in an approximate expected time in $O(L^2)$ $= O(e^{2\sqrt{\ln n \ln \ln n}})$ and in a space in $O(L)$.

Several improvements make the algorithm more practical. For example, the probability that y will be k-smooth is improved if x is chosen near $\lceil \sqrt{n} \rceil$, rather than being chosen randomly between 1 and $n-1$. A generalization of this approach, known as the *continued fraction algorithm*, has been used successfully. Unlike Dixon's algorithm, however, its rigorous analysis is unknown. It is therefore more properly called a heuristic. Another heuristic, the *quadratic sieve*, operates in a time in $O(L^{\sqrt{9/8}})$ and space in $O(L^{\sqrt{1/2}})$. In practice, we would *never* implement Dixon's algorithm because the heuristics perform so much better. More recently, H. W. Lenstra Jr. has proposed a factorization algorithm based on the theory of elliptic curves.

Problem 8.5.26.   Let $n = 10^{40}$ and $L = e^{\sqrt{\ln n \ln \ln n}}$. Compare $L^{\sqrt{9/8}}$, L^2, and $\sqrt{n}$ microseconds. Repeat the problem with $n = 10^{50}$.   □

8.5.4 Choosing a Leader

A number of identical synchronous processors are linked into a network in the shape of a ring, as in Figure 8.5.1. Each processor can communicate directly with its two immediate neighbours. Each processor starts in the same state with the same program and the same data in its memory. Such a network is of little use so long as all the processors do exactly the same thing at exactly the same time, for in this case a single processor would suffice (unless such duplication is intended to catch erratic behaviour from faulty processors in a sensitive real-time application). We seek a protocol that allows the network to *choose a leader* in such a way that all the processors are in agreement on its identity. The processor that is elected leader can thereafter break the symmetry of the network in whatever way it pleases by giving different tasks to different processors.

No deterministic algorithm can solve this problem, no matter how much time is available. Whatever happens, the processors continue to do the same thing at the same time. If one of them decides that it wants to be the leader, for example, then so do all

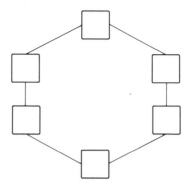

Figure 8.5.1.   A ring of identical processors.

the others simultaneously. We can compare the situation to the deadlock that arises when two people whose degree of courtesy is exactly equal try to pass simultaneously through the same narrow door. However, if each processor knows in advance how many others are connected in the network, there exists a Las Vegas algorithm that is able to solve this problem in linear expected time. The symmetry can be broken on the condition that the random generators used by the processors are independent. If the generators are in fact pseudorandom and not genuinely random, and if each processor starts from the same seed, then the technique will not work.

Suppose there are n processors in the ring. During phase zero, each processor initializes its local variable m to the known value n, and its Boolean indicator *active* to the value *true*. During phase k, $k > 0$, each active processor chooses an integer randomly, uniformly, and independently between 1 and m. Those processors that chose 1 inform the others by sending a one-bit message round the ring. (The inactive processors continue, nevertheless, to pass on messages.) After $n - 1$ clock pulses, each processor knows the number l of processors that chose 1. There are three possibilities. If $l = 0$, phase k produces no change in the situation. If $l > 1$, only those processors that chose 1 remain active, and they set their local variable m to the value l. In either case phase $k + 1$ is now begun. The protocol ends when $l = 1$ with the election of the single processor that just chose 1.

This protocol is classed as a Las Vegas algorithm, despite the fact that it never ends by admitting a failure, because there is no upper bound on the time required for it to succeed. However, it never gives an incorrect solution: it can neither end after electing more than one leader nor after electing none.

Let $l(n)$ be the expected number of phases needed to choose a leader among n processors using this algorithm (not counting phase zero, the initialization). Let $p(n,j) = \binom{n}{j} n^{-j} (1 - 1/n)^{n-j}$ be the probability that j processors out of n randomly choose the value 1 during the first stage. With probability $p(n,1)$ only a single phase is needed; with probability $p(n,0)$ we have to start all over again; and with probability $p(n,j)$, $2 \le j \le n$, $l(j)$ subsequent phases will be necessary on the average. Consequently,

$$l(n) = 1 + p(n,0)l(n) + \sum_{j=2}^{n} p(n,j)l(j), \quad n \ge 2.$$

A little manipulation gives

$$l(n) = (1 + \sum_{j=2}^{n-1} p(n,j)l(j)) / (1 - p(n,0) - p(n,n)).$$

* Problem 8.5.27.   Show that $l(n) < e \approx 2.718$ for every $n \ge 2$.   □

* Problem 8.5.28.   Show that $\lim_{n \to \infty} l(n) < 2.442$.   □

One phase of this protocol consists of single bit messages passed round the ring. Since each phase thus takes linear time, the preceding problems show that the choice of a leader also takes expected linear time.

** Problem 8.5.29.   Prove that no protocol (not even a Las Vegas protocol) can solve the problem of choosing a leader in a ring of n identical processors if they are not given the value n. Nevertheless, given an arbitrary parameter $p > 0$, give a Monte Carlo algorithm that is able to determine n exactly with a probability of error less than p whatever the number of processors.   □

8.6 MONTE CARLO ALGORITHMS

There exist problems for which no efficient algorithm is known that is able to obtain a correct solution every time, whether it be deterministic or probabilistic. A *Monte Carlo* algorithm occasionally makes a mistake, but it finds a correct solution with high probability whatever the instance considered. This is not the same as saying that it works correctly on a majority of instances, only failing now and again in some special cases. No warning is usually given when the algorithm makes a mistake.

Problem 8.6.1.   Prove that the following algorithm decides correctly whether or not an integer is prime in more than 80% of the cases. (It performs much better on small integers.) Show, on the other hand, that it is not a Monte Carlo algorithm by exhibiting an instance on which the algorithm systematically gives a wrong answer.

```
function wrong (n)
    if gcd(n, 30030) = 1    { using Euclid's algorithm }
    then return true
    else return false
```

What constant should be used instead of 30,030 to bring the proportion of successes above 85% even if very large integers are considered?   □

Let p be a real number such that $1/2 < p < 1$. A Monte Carlo algorithm is *p-correct* if it returns a correct solution with probability not less than p, whatever the instance considered. The *advantage* of such an algorithm is $p - 1/2$. The algorithm is *consistent* if it never gives two different correct solutions to the same instance. Some Monte Carlo algorithms take as a parameter not only the instance to be solved but also an upper bound on the probability of error that is acceptable. The time taken by such algorithms is then expressed as a function of both the size of the instance and the reciprocal of the acceptable error probability. To increase the probability of success of a consistent, *p*-correct algorithm, we need only call it several times and choose the most frequent answer. This increases our confidence in the result in a way similar to the world series calculation of Section 5.2.

Problem 8.6.2.   Let $MC(x)$ be a consistent, 75%-correct Monte Carlo algorithm. Consider the following algorithm.

> **function** $MC3(x)$
> $\quad t \leftarrow MC(x); u \leftarrow MC(x); v \leftarrow MC(x)$
> $\quad$**if** $t = u$ **or** $t = v$ **then return** t
> $\quad$**return** v

Prove that this algorithm is consistent and 27/32-correct, hence 84%-correct. Show on the other hand that it might not even be 71%-correct if MC, though 75%-correct, were not consistent.   □

More generally, let ε and δ be two positive real numbers such that $\varepsilon + \delta < \frac{1}{2}$. Let $MC(x)$ be a Monte Carlo algorithm that is consistent and $(\frac{1}{2} + \varepsilon)$-correct. Let $c_\varepsilon = -2 / \lg(1 - 4\varepsilon^2)$. Let x be some instance to be solved. It suffices to call $MC(x)$ at least $\lceil c_\varepsilon \lg 1/\delta \rceil$ times and to return the most frequent answer (ties can be broken arbitrarily) to obtain an algorithm that is consistent and $(1 - \delta)$-correct. This allows us to *amplify* the advantage of the algorithm, however small it may be, so as to obtain a new algorithm whose error probability is as small as we choose.

To prove the preceding claim, let $n \geq c_\varepsilon \lg 1/\delta$ be the number of times that the $(\frac{1}{2} + \varepsilon)$-correct algorithm is called. Let $m = \lfloor n/2 \rfloor + 1$, $p = \frac{1}{2} + \varepsilon$, and $q = 1 - p = \frac{1}{2} - \varepsilon$. The repetitive algorithm finds the correct answer if the latter is obtained at least m times. Its error probability is therefore at most

$$\sum_{i=0}^{m-1} \text{Prob}[\, i \text{ correct answers in } n \text{ tries}\,]$$

$$\leq \sum_{i=0}^{m-1} \binom{n}{i} p^i q^{n-i}$$

$$= (pq)^{n/2} \sum_{i=0}^{m-1} \binom{n}{i} (q/p)^{\frac{n}{2}-i}$$

$$\le (pq)^{n/2} \sum_{i=0}^{m-1} \binom{n}{i} \qquad \qquad \text{since } q/p < 1 \text{ and } \tfrac{n}{2} - i \ge 0$$

$$\le (pq)^{n/2} \sum_{i=0}^{n} \binom{n}{i} \; = \; (pq)^{n/2} 2^n$$

$$= (4pq)^{n/2} \; = \; (1-4\varepsilon^2)^{n/2}$$

$$\le (1-4\varepsilon^2)^{(c_\varepsilon/2)\lg 1/\delta} \qquad \qquad \text{since } 0 < 1-4\varepsilon^2 < 1$$

$$= 2^{-\lg 1/\delta} \; = \; \delta \qquad \qquad \text{since } \alpha^{1/\lg \alpha} = 2 \text{ for every } \alpha > 0.$$

The probability of success of the repetitive algorithm is therefore at least $1-\delta$.

Problem 8.6.3.   Prove that $c_\varepsilon < (\ln 2)/2\varepsilon^2$. $\square$

For example, suppose we have a consistent Monte Carlo algorithm whose advantage is 5% and we wish to obtain an algorithm whose probability of error is less than 5% (that is, we wish to go from a 55%-correct algorithm to a 95%-correct algorithm). The preceding theorem tells us that this can be achieved by calling the original algorithm about 600 times on the given instance. A more precise calculation shows that it is enough to repeat the original algorithm 269 times, and that repeating it 600 times yields an algorithm that is better that 99%-correct. [This is because some of the inequalities used in the proof are rather crude. A more complicated argument shows that if a consistent ($\tfrac{1}{2}+\varepsilon$)-correct Monte Carlo algorithm is repeated $2m-1$ times, the resulting algorithm is $(1-\delta)$-correct, where

$$\delta \; = \; \tfrac{1}{2} - \varepsilon \sum_{i=0}^{m-1} \binom{2i}{i} (\tfrac{1}{4} - \varepsilon^2)^i \; \le \; \frac{(1-4\varepsilon^2)^m}{4\varepsilon\sqrt{\pi m}} \; .$$

The first part of this formula can be used efficiently to find the exact number of repetitions required to reduce the probability of error below any desired threshold δ. Alternatively, a good upper bound on this number of repetitions is quickly obtained from the second part: find x such that $e^x \sqrt{x} \ge 1/(2\delta\sqrt{\pi})$ and then set $m = \lceil x/4\varepsilon^2 \rceil$.]

Repeating an algorithm several hundred times to obtain a reasonably small probability of error is not attractive. Fortunately, most Monte Carlo algorithms that occur in practice are such that we can increase our confidence in the result obtained much more rapidly. Assume for simplicity that we are dealing with a decision problem and that the original Monte Carlo algorithm is *biased* in the sense that it is always correct whenever it returns the answer *true*, errors being possible only when it returns the answer *false*. If we repeat such an algorithm several times to increase our confidence in the final result, it would be silly to return the most frequent answer: a single *true* outweighs any number of *falses*. As we shall see shortly, it suffices to repeat such an algorithm 4 times to improve it from 55%-correct to 95%-correct, or 6 times to obtain a 99%-correct algorithm. Moreover, the restriction that the original algorithm be p-correct for some $p > \frac{1}{2}$ no longer applies: arbitrarily high confidence can be obtained by repeating a biased p-correct algorithm enough times, even if $p < \frac{1}{2}$ (as long as $p > 0$).

More formally, let us return to an arbitrary problem (not necessarily a decision problem), and let y_0 be some distinguished answer. A Monte Carlo algorithm is y_0-*biased* if there exists a subset X of the instances such that

i. the solution returned by the algorithm is always correct whenever the instance to be solved is not in X, and

ii. the correct solution to all the instances that belong to X is y_0, but the algorithm may not always return the correct solution to these instances.

Although the distinguished answer y_0 is known explicitly, it is not required that an efficient test be known for membership in X. The following paragraph shows that this definition is tuned precisely to make sure that the algorithm is always correct when it answers y_0.

Let MC be a Monte Carlo algorithm that is consistent, y_0-biased and p-correct. Let x be an instance, and let y be the solution returned by $MC(x)$. What can we conclude if $y = y_0$?

- If $x \notin X$, the algorithm always returns the correct solution, so y_0 is indeed correct; and
- if $x \in X$, the correct solution is necessarily y_0.

In both cases, we may conclude that y_0 is a correct solution. What happens if on the other hand $y \neq y_0$?

- If $x \notin X$, y is indeed correct; and
- if $x \in X$, the algorithm has made a mistake since the correct solution is y_0; the probability of such an error occurring is not greater than $1 - p$ given that the algorithm is p-correct.

Now suppose that we call $MC(x)$ k times and that the answers obtained are y_1, $y_2, \ldots, y_k$.

- If there exists an i such that $y_i = y_0$, the preceding argument shows that this is indeed the correct solution;
- if there exist $i \neq j$ such that $y_i \neq y_j$, the only possible explanation is that $x \in X$ (because the algorithm is consistent by assumption), and therefore the correct solution is y_0; and
- if $y_i = y \neq y_0$ for all the i, it is still possible that the correct solution is y_0 and that the algorithm has made a mistake k times in succession on $x \in X$, but the probability of such an occurrence is at most $(1 - p)^k$.

Suppose, for example, that $p = \frac{1}{2}$ (once again, this is not allowed for general Monte Carlo algorithms, but it causes no problems with a biased algorithm). It suffices to repeat the algorithm at most 20 times to be either sure that the correct solution is y_0 (if either of the first two cases previously described is observed), or extremely confident that the solution obtained on every one of the trials is correct (since otherwise the probability of obtaining the results observed is less than one chance in a million). In general, k repetitions of a consistent, p-correct, y_0-biased algorithm yield an algorithm that is $(1 - (1-p)^k)$-correct and still consistent and y_0-biased.

Assume your consistent, p-correct, y_0-biased Monte Carlo algorithm has yielded k times in a row the same answer $y \neq y_0$ on some instance x. It is important to understand how to interpret such behaviour correctly. It may be tempting to conclude that "the probability that y is an incorrect answer is at most $(1-p)^k$". Such a conclusion of course makes no sense because either the correct answer is indeed y, or not. The probability in question is therefore either 0 or 1, despite the fact that we cannot tell for sure which it is. The correct interpretation is as follows: "I believe that y is the correct answer and if you quiz me enough times on different instances, my proportion of errors should not significantly exceed $(1-p)^k$".

The "proportion of errors" in question, however, is averaged over the *entire* sequence of answers given by the algorithm, *not* only over those occurrences in which the algorithm actually provides such probabilistic answers. Indeed, if you systematically quiz the algorithm with instances for which the correct solution is y_0, it will *always* be wrong whenever it "believes" otherwise. This last remark may appear trivial, but it is in fact crucial if the probabilistic algorithm is used to generate with high probability some random instance x on which the correct answer is a specific $y \neq y_0$.

To illustrate this situation, consider a nonempty finite set I of instances. We are interested in generating a random member of some subset $S \subseteq I$. (As a practical example, we may be interested in generating a random prime of a given length — see Section 8.6.2.) Let MC be a *false*-biased, p-correct Monte Carlo algorithm to decide, given any $x \in I$, whether $x \in S$. Let $q = 1-p$. By definition of a *false*-biased algorithm, $\text{Prob}[MC(x) = true] = 1$ for each instance $x \in S$ and $\text{Prob}[MC(x) = true] \leq q$ for each instance $x \notin S$. Consider the following algorithms.

```
function repeatMC (x, k)
    i ← 0
    ans ← true
    while ans and i < k do
        i ← i + 1
        ans ← MC (x)
    return ans
```

```
function genrand (k)
   repeat
      x ← uniform (I)
   until repeatMC (x, k)
   return x
```

It is tempting to think that $genrand(k)$ returns a random member of S with a probability of failure at most q^k. *This is wrong in general.* The problem is that we must not confuse the conditional probabilities $\text{Prob}[X \mid Y]$ and $\text{Prob}[Y \mid X]$, where X stands for "$repeatMC(x,k)$ returns *true*" and Y stands for "$x \notin S$". It is correct that $\text{Prob}[X \mid Y] \le q^k$, but we are in fact interested in $\text{Prob}[Y \mid X]$. To calculate this probability, we need an a priori probability that a call on $uniform(I)$ returns a member of S.

* Problem 8.6.4.   Let r denote the probability that $x \in S$ given that x is returned by a call on $uniform(I)$. Prove that the probability that a call on $genrand(k)$ erroneously returns some $x \notin S$ is at most

$$\frac{1}{1 + \dfrac{r}{1-r}\, q^{-k}} \; .$$

In particular, if the error probability of MC on instances not in S is exactly q (rather than at most q), then the probability that a call on $genrand(k)$ returns some $x \notin S$ is about q^k / r if $q^k \ll r \ll 1$, about $\frac{1}{2}$ if $q^k \approx r \ll 1$, and nearly 1 if $r \ll q^k$. This can be significant when the confidence one gets in the belief that x belongs to S from running MC several times must be weighed against the a priori likelihood that a randomly selected x does not belong to S. This situation illustrates dramatically the difference between the error probability of *genrand* and that of *repeatMC*.   □

We are not aware of any unbiased Monte Carlo algorithm sufficiently simple to feature in this introduction. Thus the section continues with some examples of biased Monte Carlo algorithms. They all involve the solution of a decision problem, that is, the only possible answers that they can return are *true* and *false*.

Problem 8.6.5.   Let A and B be two efficient Monte Carlo algorithms for solving the same decision problem. Algorithm A is p-correct and *true*-biased, whereas algorithm B is q-correct and *false*-biased. Give an efficient Las Vegas algorithm $LV(x, \textbf{var } y, \textbf{var } success)$ to solve the same problem. What is the best value of r you can deduce so that your Las Vegas algorithm succeeds with probability at least r on each instance?   □

8.6.1 Majority Element in an Array

The problem is to determine whether an array $T[1..n]$ has a majority element (see Problems 4.11.5, 4.11.6, and 4.11.7). Consider the following algorithm.

function $maj(T[1..n])$
$\quad i \leftarrow uniform(1..n)$
$\quad x \leftarrow T[i]$
$\quad k \leftarrow 0$
$\quad$ **for** $j \leftarrow 1$ **to** n **do if** $T[j] = x$ **then** $k \leftarrow k+1$
$\quad$ **return** $(k > n/2)$

We see that $maj(T)$ chooses an element of the array at random, and then checks whether this element forms a majority in T. If the answer returned is *true*, the element chosen is a majority element, and hence trivially there is a majority element in T. If, on the other hand, the answer returned is *false*, it is nonetheless possible that T contains a majority element, although in this case the element chosen randomly is in a minority. If the array does indeed contain a majority element, and if one of its elements is chosen at random, the probability of choosing an element that is in a minority is less than one-half, since majority elements occupy more than half the array. Therefore, if the answer returned by $maj(T)$ is *false*, we may reasonably suspect that T does indeed have no majority element. In sum, this algorithm is *true*-biased and $\frac{1}{2}$-correct, that is

$$T \text{ has a majority element} \quad \Rightarrow \quad maj(T) = true, \quad \text{with probability} > \tfrac{1}{2}$$
$$T \text{ has no majority element} \quad \Rightarrow \quad maj(T) = false, \text{ with certainty}.$$

An error probability of 50% is intolerable in practice. The general technique for biased Monte Carlo algorithms allows us to reduce this probability efficiently to any arbitrary value. First, consider

function $maj2(T)$
$\quad$ **if** $maj(T)$ **then return** *true*
$\quad\quad\quad\quad$ **else return** $maj(T)$.

If the array does not have a majority element, each call of $maj(T)$ is certain to return *false*, hence so does $maj2(T)$. If the array does have a majority element, the probability that the first call of $maj(T)$ will return *true* is $p > \frac{1}{2}$, and in this case $maj2(T)$ returns *true*, too. On the other hand, if the first call of $maj(T)$ returns *false*, which happens with probability $1-p$, the second call of $maj(T)$ may still with probability p return *true*, in which case $maj2(T)$ also returns *true*. Summing up, the probability that $maj2(T)$ will return *true* if the array T has a majority element is

$$p + (1-p)p = 1 - (1-p)^2 > 3/4.$$

The algorithm $maj2$ is therefore also *true*-biased, but $3/4$-correct. The probability of error decreases because the successive calls of $maj(T)$ are independent: the fact that

$maj(T)$ has returned *false* on an array with a majority element does not change the probability that it will return *true* on the following call on the same instance.

Problem 8.6.6.   Show that the probability that k successive calls of $maj(T)$ all return *false* is less than 2^{-k} if T contains a majority element. On the other hand, as soon as any call returns *true*, we can be certain that T contains a majority element.   □

The following Monte Carlo algorithm solves the problem of detecting the presence of a majority element with a probability of error less than ε for every $\varepsilon > 0$.

function $majMC(T, \varepsilon)$
    $k \leftarrow \lceil \lg(1/\varepsilon) \rceil$
    for $i \leftarrow 1$ **to** k **do**
        if $maj(T)$ **then return** *true*
    return *false*

The algorithm takes a time in $O(n \log(1/\varepsilon))$, where n is the number of elements in the array and ε is the acceptable probability of error. This is interesting only as an illustration of a Monte Carlo algorithm since a linear time deterministic algorithm is known (Problems 4.11.5 and 4.11.6).

8.6.2 Probabilistic Primality Testing

This classic Monte Carlo algorithm recalls the algorithm used to determine whether or not an array has a majority element. The problem is to decide whether a given integer is prime or composite. No deterministic or Las Vegas algorithm is known that can solve this problem in a reasonable time when the number to be tested has more than a few hundred decimal digits. (It is currently possible to establish with certainty the primality of numbers up to 213 decimal digits within approximately 10 minutes of computing time on a CDC CYBER 170/750.)

A first approach to finding a probabilistic algorithm might be

function $prime(n)$
    $d \leftarrow uniform(2 .. \lfloor \sqrt{n} \rfloor)$
    return $((n \bmod d) \neq 0)$.

If the answer returned is *false*, the algorithm has been lucky enough to find a nontrivial factor of n, and we can be certain that n is composite. Unfortunately, the answer *true* is returned with high probability even if n is in fact composite. Consider for example $n = 2,623 = 43 \times 61$. The algorithm chooses an integer randomly between 2 and 51. Thus there is only a meagre 2% probability that it will happen on $d = 43$ and hence return *false*. In 98% of calls the algorithm will inform us incorrectly that n is prime. For larger values of n the situation gets worse. The algorithm can be improved slightly by testing whether n and d are relatively prime, using Euclid's algorithm, but it is still unsatisfactory.

To obtain an efficient Monte Carlo algorithm for the primality problem, we need a theorem whose proof lies beyond the scope of this book. Let n be an odd integer greater than 4, and let s and t be positive integers such that $n-1 = 2^s t$, where t is odd. Let a be an integer such that $2 \leq a \leq n-2$. We say that n is a *strong pseudoprime* to the base a if $a^t \equiv 1 \pmod{n}$ or if there exists an integer i such that $0 \leq i < s$ and $a^{2^i t} \equiv -1 \pmod{n}$.

If n is prime, it is a strong pseudoprime to any base. There exist however composite numbers that are strong pseudoprimes to some bases. Such a base is then a *false witness of primality* for this composite number. For example, 158 is a false witness of primality for 289 because $288 = 9 \times 2^5$, $158^9 \equiv 131 \pmod{289}$, $158^{2\times9} \equiv 131^2 \equiv 110 \pmod{289}$, $158^{4\times9} \equiv 110^2 \equiv 251 \pmod{289}$, and finally, $158^{8\times9} \equiv 251^2 \equiv -1 \pmod{289}$.

The theorem assures us that if n is composite, it cannot be a strong pseudoprime to more than $(n-9)/4$ different bases. The situation is even better if n is composed of a large number r of distinct prime factors: in this case it cannot be a strong pseudoprime to more than $\phi(n)/2^{r-1}-2$ different bases, where $\phi(n) < n-1$ is Euler's *totient* function. The theorem is generally pessimistic. For instance, 289 has only 14 false witnesses of primality, whereas 737 does not even have one.

** Problem 8.6.7.   Prove this theorem.   □

Problem 8.6.8.   Give an efficient algorithm for testing whether n is a strong pseudoprime to the base a. Your algorithm should not take significantly longer (and sometimes even be faster) than simply calculating $a^{(n-1)/2}$ **mod** n with *dexpoiter* from Section 4.8.   □

This suggests the following algorithm.

function *Rabin* (n)
    { this algorithm is only called if $n > 4$ is odd }
    $a \leftarrow uniform\,(2..n-2)$
    if n is strongly pseudoprime to the base a
    then return *true*
    else return *false*

For an odd integer $n > 4$ the theorem ensures that n is composite if *Rabin* (n) returns *false*. This certainty is not accompanied however by any indication of what are the nontrivial factors of n. Contrariwise, we may begin to suspect that n is prime if *Rabin* (n) returns *true*. This Monte Carlo algorithm for deciding primality is *false*-biased and $^3/_4$-correct.

This test of primality has several points in common with the Las Vegas algorithm for finding square roots modulo p (Section 8.5.2). In both cases, an integer a is chosen at random. If n is composite, there is at least a 75% chance that n will not be a strong pseudoprime to the base a, in which case we obtain the correct answer with certainty. Similarly there is better than a 50% chance that a will provide a key for finding $\sqrt{x}$. Nevertheless the algorithm for testing primality is only a Monte Carlo algorithm whereas the one for finding square roots is Las Vegas. This difference is explained by the fact that the Las Vegas algorithm is able to detect when it has been unlucky: the fact that a does not provide a key for $\sqrt{x}$ is easy to test. On the other hand, if n is a strong pseudoprime to the base a, this can be due either to the fact that n is indeed prime or to the fact that a is a false witness of primality for the composite number n. The difference can also be explained using Problem 8.2.2.

As usual, the probability of error can be made arbitrarily small by repeating the algorithm. A philosophical remark is once again in order: the algorithm does not reply "this number is prime with probability $1-\varepsilon$", but rather, "I believe this number to be prime; otherwise I have observed a natural phenomenon whose probability of occurrence was not greater than ε". The first reply would be nonsense, since every integer larger than 1 is either prime or composite.

* Problem 8.6.9.   Consider the following nonterminating program.

program *print primes*
    print 2, 3
    $n \leftarrow 5$
    repeat
        if *repeatRabin*$(n, \lfloor \lg n \rfloor)$ **then print** n
        $n \leftarrow n + 2$
    ad nauseum

Clearly, every prime number will eventually be printed by this program. One might also expect composite numbers to be produced erroneously once in a while. Prove that this is unlikely to happen. More precisely, prove that the probability is better than 99% that not even a single composite number larger than 100 will *ever* be produced, regardless of how long the program is allowed to run. (*Note:* this figure of 99% is very conservative as it would still hold even if *Rabin*(n) had a flat 25% chance of failure on each composite integer.)   □

** Problem 8.6.10.   Find a *true*-biased Monte Carlo algorithm for primality testing whose running time is polynomial in the logarithm of the integer being tested. Notice that this combines with the *false*-biased algorithm described previously to yield a Las Vegas algorithm (by Problem 8.6.5).   □

8.6.3 A Probabilistic Test for Set Equality

We have a universe U of N elements, and a collection of n sets, not necessarily disjoint, all of which are empty at the outset. We suppose that N is quite large while n is small. The basic operation consists of adding x to the set S_i, where $x \in U \setminus S_i$ and $1 \le i \le n$. At any given moment the question "Does $S_i = S_j$?" may be asked.

The naive way to solve this problem is to keep the sets in arrays, lists, search trees, or hash tables. Whatever structure is chosen, each test of equality will take a time in $\Omega(k)$, if indeed it is not in $\Omega(k \log k)$, where k is the cardinality of the larger of the two sets concerned.

For any $\varepsilon > 0$ fixed in advance, there exists a Monte Carlo algorithm that is able to handle a sequence of m questions in an average total time in $O(m)$. The algorithm never makes an error when $S_i = S_j$; in the opposite case its probability of error does not exceed ε. This algorithm provides an interesting application of universal hashing (Section 8.4.4).

Let $\varepsilon > 0$ be the error probability that can be tolerated for each request to test the equality of two sets. Let $k = \lceil \lg(\max(m, 1/\varepsilon)) \rceil$. Let H be a universal$_2$ class of functions from U into $\{0,1\}^k$, the set of k-bit strings. The Monte Carlo algorithm first chooses a function at random in this class and then initializes a hash table that has U for its domain. The table is used to implement a random function $rand : U \rightarrow \{0,1\}^k$ as follows.

> **function** $rand(x)$
>   **if** x is in the table **then return** its associated value
> $y \leftarrow$ some random k-bit string
>   add x to the table and associate y to it
>   **return** y

Notice that this is a memory function in the sense of Section 5.7. Each call of $rand(x)$ returns a random string chosen with equal probability among all the strings of length k. Two different calls with the same argument return the same value, and two calls with different arguments are independent. Thanks to the use of universal hashing, each call of $rand(x)$ takes constant expected time.

To each set S_i we associate a variable $v[i]$ initialized to the binary string composed of k zeros. Here is the algorithm for adding an element x to the set S_i. We suppose that x is not already a member of S_i.

> **procedure** $add(i,x)$
> $v[i] \leftarrow v[i] \oplus rand(x)$

The notation $t \oplus u$ stands for the bit-by-bit exclusive-or of the binary strings t and u. The algorithm to test the equality of S_i and S_j is:

> **function** $test(i,j)$
>   **if** $v[i] = v[j]$
>   **then return** $true$
>   **else return** $false$.

It is obvious that $S_i \neq S_j$ if $v[i] \neq v[j]$. What is the probability that $v[i] = v[j]$ when $S_i \neq S_j$? Suppose without loss of generality that there exists an $x_0 \in S_i$ such that $x_0 \notin S_j$. Let $S'_i = S_i \setminus \{x_0\}$. For a set $S \subseteq U$, let

$XOR(S)$ be the exclusive-or of the $rand(x)$ for every $x \in S$. By definition, $v[i] = XOR(S_i) = rand(x_0) \oplus XOR(S_i')$ and $v[j] = XOR(S_j)$. Let $y_0 = XOR(S_i') \oplus XOR(S_j)$. The fact that $v[i] = v[j]$ implies that $rand(x_0) = y_0$; the probability of this happening is only 2^{-k} since the value of $rand(x_0)$ is chosen independently of those values that contribute to y_0. Notice the similarity to the use of signatures in Section 7.2.1.

This Monte Carlo algorithm differs from those in the two previous sections in that our confidence in an answer "$S_i = S_j$" cannot be increased by repeating the call of $test(i,j)$. It is only possible to increase our confidence in the set of answers obtained to a *sequence* of requests by repeating the application of the algorithm to the entire sequence. Moreover, the different tests of equality are not independent. For instance, if $S_i \neq S_j$, $x \notin S_i \cup S_j$, $S_k = S_i \cup \{x\}$, $S_l = S_j \cup \{x\}$, and if an application of the algorithm replies incorrectly that $S_i = S_j$, then it will also reply incorrectly that $S_k = S_l$.

Problem 8.6.11.   What happens with this algorithm if by mistake a call of $add(i,x)$ is made when x is already in S_i?   □

Problem 8.6.12.   Show how you could also implement a procedure $elim(i,x)$, which removes the element x from the set S_i. A call of $elim(i,x)$ is only permitted when x is already in S_i.   □

Problem 8.6.13.   Modify the algorithm so that it will work correctly (with probability of error ε) even if a call of $add(i,x)$ is made when $x \in S_i$. Also implement a request $member(i,x)$, which decides without ever making an error whether $x \in S_i$. A sequence of m requests must still be handled in an expected time in $O(m)$.   □

** Problem 8.6.14.   Universal hashing allows us to implement a random function $rand: U \rightarrow \{0,1\}^k$. The possibility that $rand(x_1) = rand(x_2)$ even though $x_1 \neq x_2$, which does not worry us when we want to test set equality, may be troublesome for other applications. Show how to implement a random *permutation*. More precisely, let N be an integer and let $U = \{1, 2, \ldots, N\}$. You must accept two kinds of request: *init* and $p(i)$ for $1 \leq i \leq N$. A call of *init* initializes a new permutation $\pi: U \rightarrow U$. A call of $p(i)$ returns the value $\pi(i)$ for the current permutation. Two calls $p(i)$ and $p(j)$ that are not separated by a call of *init* should therefore return two different answers if and only if $i \neq j$. Two such calls separated by a call of *init* should on the other hand be independent. Suppose that a call of $uniform(u..v)$ takes constant time for $1 \leq u \leq v \leq N$. Your implementation should satisfy each request in constant time in the worst case, whatever happens. You may use memory space in $O(N)$, but no request may consult or modify more than a constant number of memory locations: thus it is not possible to create the whole permutation when *init* is called. (Hint: reread Problem 5.7.2 and Example 8.4.1).   □

8.6.4 Matrix Multiplication Revisited

You have three $n \times n$ matrices A, B, and C and you would like to decide whether $AB = C$. Here is an intriguing *false*-biased, $\frac{1}{2}$-correct Monte Carlo algorithm that is capable of solving this problem in a time in $O(n^2)$. Compare this with the fastest known deterministic algorithm to compute the product AB (Section 4.9), which takes a time in $O(n^{2.376})$, and with the probabilistic algorithm mentioned in Section 8.3.5, which only computes the product approximately.

> **function** *goodproduct* (A, B, C, n)
>   **array** $X[1..n]$ { to be considered as a column vector }
>   **for** $i \leftarrow 1$ **to** n **do** $X[i] \leftarrow uniform(\{-1, 1\})$
>   **if** $ABX = CX$ **then return** *true*
>   **else return** *false*

In order to take a time in $O(n^2)$, we must compute ABX as A times BX, providing a dramatic example of the topic discussed in Section 5.3.

Problem 8.6.15.   It is obvious that *goodproduct* (A, B, C, n) returns *true* whenever $AB = C$. Prove that it returns *false* with probability at least $\frac{1}{2}$ whenever $AB \neq C$. (Hint: consider the columns of the matrix $AB - C$ and show that at least half the ways to add and subtract them yield a nonzero column vector, provided $AB \neq C$.)   □

Problem 8.6.16.   Given two $n \times n$ matrices A and B, adapt this algorithm to decide probabilistically whether B is the inverse of A.   □

Problem 8.6.17.   Given three polynomials $p(x)$, $q(x)$, and $r(x)$ of degrees n, n and $2n$, respectively, give a *false*-biased, $\frac{1}{2}$-correct Monte Carlo algorithm to decide whether $r(x)$ is the symbolic product of $p(x)$ and $q(x)$. Your algorithm should run in a time in $O(n)$. (In the next chapter we shall see a deterministic algorithm that is capable of computing the symbolic product of two polynomials of degree n in a time in $O(n \log n)$, but no such algorithm is known that only takes a time in $O(n)$.)   □

8.7 REFERENCES AND FURTHER READING

The experiment devised by Leclerc (1777) was carried out several times in the nineteenth century; see for instance Hall (1873). It is no doubt the earliest recorded probabilistic algorithm. The term "Monte Carlo" was introduced into the literature by Metropolis and Ulam (1949), but it was already in use in the secret world of atomic research during World War II, in particular in Los Alamos, New Mexico. Recall that it is often used to describe any probabilistic algorithm. The term "Las Vegas" was introduced by Babaï (1979) to distinguish probabilistic algorithms that occasionally make a mistake from those that reply correctly if they reply at all. The term "Sherwood" is our own. For the solution to Problem 8.2.3, see Anon. (1495).

Two encyclopaedic sources of techniques for generating pseudorandom numbers are Knuth (1969) and Devroye (1986). The former includes tests for trying to distinguish a pseudorandom sequence from one that is truly random. A more interesting generator from a cryptographic point of view is given by Blum and Micali (1984); this article and the one by Yao (1982) introduce the notion of an *unpredictable generator*, which can pass any statistical test that can be carried out in polynomial time. Under the assumption that it is infeasible to factorize large numbers, a more efficient unpredictable pseudorandom generator is proposed in Blum, Blum, and Shub (1986). More references on this subject can be found in Brassard (1988). General techniques are given in Vazirani (1986, 1987) to cope with generators that are only "semi-random".

For more information on numeric probabilistic algorithms, consult Sobol' (1974). A guide to simulation is provided by Bratley, Fox, and Schrage (1983). The point is made in Fox (1986) that pure Monte Carlo methods are not specially good for numerical integration with a *fixed* dimension: it is preferable to choose your points systematically so that they are well spaced, a technique known as quasi Monte Carlo. Problem 8.3.14 is solved in Klamkin and Newman (1967). The application of probabilistic counting to the Data Encryption Standard (Example 8.3.1) is described in Kaliski, Rivest, and Sherman (1988). Section 8.3.4 follows Flajolet and Martin (1985). Numeric probabilistic algorithms designed to solve problems from linear algebra are discussed in Curtiss (1956), Vickery (1956), Hammersley and Handscomb (1965), and Carasso (1971).

An early (1970) linear expected time probabilistic median finding algorithm is attributed to Floyd: see Exercise 5.3.3.13 in Knuth (1973). It predates the classic worst-case linear time deterministic algorithm described in Section 4.6. A probabilistic algorithm that is capable of finding the ith smallest among n elements in an expected number of comparisons in $n + i + O(\sqrt{n})$ is given in Rivest and Floyd (1973). Computation with encrypted instances (end of Section 8.4.2) is an idea originating in Feigenbaum (1986) and developed further in Abadi, Feigenbaum, and Kilian (1987). The technique for searching in an ordered list and its application to sorting (Problem 8.4.14) come from Janko (1976). An analysis of this technique (Problem 8.4.12) is given in Bentley, Stanat, and Steele (1981); the same reference gives the statement of Problem 8.4.4. Classic hash coding is described in Knuth (1968); many solutions to Problem 8.4.15 appear there. Universal hashing was invented by Carter and Wegman (1979); several universal$_2$ classes are described there, including the one from Problem 8.4.20. For solutions to Problem 8.4.21, see Wegman and Carter (1981), and Bennett, Brassard, and Robert (1988).

The probabilistic approach to the eight queens problem was suggested to the authors by Manuel Blum. The experiments on the twenty queens problem were carried out by Pierre Beauchemin. The algorithm of Section 8.5.2 for finding square roots modulo a prime number, including Problem 8.5.15, is due to Peralta (1986). Early algorithms to solve this problem are given in Lehmer (1969) and Berlekamp (1970). As a generalization, Rabin (1980a) gives an efficient probabilistic algorithm for computing roots of *arbitrary* polynomials over any finite field. For a solution to Problem 8.5.12, consult Aho, Hopcroft, and Ullman (1974). The solution to

Problem 8.5.17 is given by the algorithm of Shanks (1972) and Adleman, Manders, and Miller (1977). The integer factorization algorithm of Pollard (1975) has a probabilistic flavour. The probabilistic integer factorization algorithm discussed in Section 8.5.3 originated with Dixon (1981); for a comparison with other methods, refer to Pomerance (1982). The algorithm based on elliptic curves is discussed in Lenstra (1986). For efficiency considerations in factorization algorithms, consult Montgomery (1987). The algorithm for electing a leader in a network, including Problem 8.5.29, comes from Itai and Rodeh (1981).

Amplification of the advantage of an unbiased Monte Carlo algorithm is used to serve cryptographic ends in Goldwasser and Micali (1984). The probabilistic test of primality presented here is equivalent to the one in Rabin (1976, 1980b). The test of Solovay and Strassen (1977) was discovered independently. The expected number of false witnesses of primality for a random composite integer is investigated in Erdős and Pomerance (1986); see also Monier (1980). More information on number theory can be found in the classic Hardy and Wright (1938). The implication of Problem 8.6.4 for the generation of random numbers that are probably prime is explained in Beauchemin, Brassard, Crépeau, Goutier, and Pomerance (1988), which also gives a fast probabilistic splitting algorithm whose probability of success on any given composite integer is at least as large as the probability of failure of Rabin's test on the same integer. A theoretical solution to Problem 8.6.10 is given in Goldwasser and Kilian (1986) and Adleman and Huang (1987). For more information on tests of primality and their implementation, consult Williams (1978), Lenstra (1982), Adleman, Pomerance, and Rumely (1983), Kranakis (1986), and Cohen and Lenstra (1987). The probabilistic test for set equality comes from Wegman and Carter (1981); they also give a cryptographic application of universal hashing. The solution to Problem 8.6.14 is in Brassard and Kannan (1988). The Monte Carlo algorithm to verify matrix multiplication (Section 8.6.4) and the solution to Problem 8.6.17 are given in Freivalds (1979); also read Freivalds (1977).

Several interesting probabilistic algorithms have not been discussed in this chapter. We close by mentioning a few of them. Given the cartesian coordinates of points in the plane, Rabin (1976) gives an algorithm that is capable of finding the closest pair in expected linear time (contrast this with Problem 4.11.14). Rabin (1980a) gives an efficient probabilistic algorithm for factorizing polynomials over arbitrary finite fields, and for finding irreducible polynomials. A Monte Carlo algorithm is given in Schwartz (1978) to decide whether a multivariate polynomial over an infinite domain is identically zero and to test whether two such polynomials are identical. Consult Zippel (1979) for sparse polynomial interpolation probabilistic algorithms. An efficient probabilistic algorithm is given in Karp and Rabin (1987) to solve the string-searching problem discussed in Section 7.2. Our favourite unbiased Monte Carlo algorithm for a decision problem, which allows us to decide efficiently whether a given integer is a perfect number and whether a pair of integers is amicable, is described in Bach, Miller, and Shallit (1986). For an anthology of probabilistic algorithms, read Valois (1987).

9

Transformations
of the Domain

9.1 INTRODUCTION

It is sometimes useful to reformulate a problem before trying to solve it. If you were asked, for example, to multiply two large numbers given in Roman figures, you would probably begin by translating them into Arabic notation. (You would thus use an *algorism*, with this word's original meaning!) More generally, let D be the domain of objects to be manipulated in order to solve a given problem. Let $f : D^t \rightarrow D$ be a function to be calculated. An *algebraic transformation* consists of a *transformed domain R*, an invertible *transformation function* $\sigma : D \rightarrow R$ and a *transformed function* $g : R^t \rightarrow R$ such that

$$f(x_1, x_2, \ldots, x_t) = \sigma^{-1}(g(\sigma(x_1), \sigma(x_2), \ldots, \sigma(x_t)))$$

for all $x_1, x_2, \ldots, x_t$ in the domain D. Such a transformation is of interest if g can be calculated in the transformed domain more rapidly than f can be calculated in the original domain, and if the transformations σ and σ^{-1} can also be computed efficiently. Figure 9.1.1 illustrates this principle.

  Example 9.1.1. The most important transformation used before the advent of computers resulted from the invention of logarithms by Napier in 1614. Kepler found this discovery so useful that he dedicated his *Tabulae Rudolphinae* to Napier. In this case, $D = \mathbb{N}^+$ or $\mathbb{R}^+$, $f(u,v) = u \times v$, $R = \mathbb{R}$, $\sigma(u) = \ln u$ and $g(x,y) = x + y$. This allows a multiplication to be replaced by the calculation of a logarithm, an addition, and an exponentiation. Since the computation of σ and σ^{-1} would take more time than the original multiplication, this idea is only of interest when tables of logarithms are

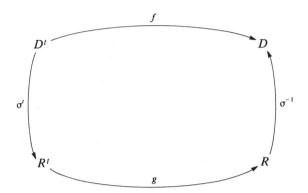

Figure 9.1.1. Transformation of the domain.

computed beforehand. Such tables, calculated once and for all, thus furnish a historical example of preconditioning (Chapter 7).   □

Example 9.1.2.   It is often useful to transform between Cartesian and polar coordinates.   □

Example 9.1.3.   Most computers that handle numerical data read these data and print the results in decimal but carry out their computations in binary.   □

Example 9.1.4.   You want to perform a symbolic multiplication of two polynomials, for example, $p(x) = 3x^3 - 5x^2 - x + 1$ and $q(x) = x^3 - 4x^2 + 6x - 2$. These polynomials are represented by their coefficients. The original domain of the problem is therefore $\mathbf{Z}^{d+1}$, where d is the degree of the polynomials involved. The naive algorithm for multiplying these polynomials resembles the classic algorithm for integer multiplication; it takes a time in $\Theta(d^2)$ if the scalar operations are taken to be elementary. An alternative way to represent the same polynomials is to give their values at $d+1$ distinct points, for instance, $0, 1, 2, \ldots, d$. The polynomials in our example would be represented by $p = (1, -2, 3, 34)$ and $q = (-2, 1, 2, 7)$. The transformed domain is still $\mathbf{Z}^{d+1}$, but its meaning has changed. The new representation suffices to define the original polynomials because there is exactly one polynomial of degree at most 3 that passes through any four given points.

Let $r(x) = 3x^6 - 17x^5 + 37x^4 - 31x^3 + 8x - 2$ be the product of $p(x)$ and $q(x)$. Using the transformed representation, we can carry out the multiplication in a time in $O(d)$ since $r(i)$ is calculated straightforwardly as $p(i)q(i)$ for $0 \le i \le d$. Thus we obtain $r = (-2, -2, 6, 238)$. However, this does not allow us to recover the coefficients of $r(x)$ because a polynomial of degree 6 is not uniquely defined by its value at only four points. We would have needed to use seven points from the outset, taking $p = (1, -2, 3, 34, 109, 246, 463)$ and $q = (-2, 1, 2, 7, 22, 53, 106)$, if the computation of r were to be carried out correctly using this representation.

In this example, calculating the transformed function (carrying out the pointwise multiplication) is much quicker than straightforwardly calculating the original function (performing a symbolic multiplication of two polynomials given by their coefficients). However, this is only useful if the transformation function (evaluation of the polynomials) and its inverse (interpolation) can be calculated rapidly. At first glance, it seems that the computation of the values $p(i)$ and $q(i)$ for $0 \le i \le 2d$ must take a time in $\Omega(d^2)$. Worse still, a naive implementation of Lagrange's algorithm to carry out the final interpolation would take a time in $\Omega(d^3)$, which appears to remove any interest this approach may have had. We shall see in Section 9.4 that this is not in fact the case, provided that the points at which we evaluate the polynomials are chosen judiciously.   □

9.2 THE DISCRETE FOURIER TRANSFORM

For the rest of this chapter all the arithmetic operations involved are carried out either modulo some integer m to be determined or in the field of complex numbers — or more generally in any commutative ring. They are assumed to be executed at unit cost unless it is explicitly stated otherwise. Let $n > 1$ be a power of 2. We denote by ω some constant such that $\omega^{n/2} = -1$. If $n = 8$, for example, then $\omega = 4$ is a possible value if we are using arithmetic modulo 257, and $\omega = (1+i)/\sqrt{2}$ is a possible value in the field of complex numbers.

Consider an n-tuple $a = (a_0, a_1, \ldots, a_{n-1})$. This defines in the natural way a polynomial $p_a(x) = a_{n-1}x^{n-1} + a_{n-2}x^{n-2} + \cdots + a_1 x + a_0$ of degree less than n. The *discrete Fourier transform* of a with respect to ω is the n-tuple $F_\omega(a) = (p_a(1), p_a(\omega), p_a(\omega^2), \ldots, p_a(\omega^{n-1}))$. As in Example 9.1.4, it appears at first glance that the number of scalar operations needed to calculate this transform is in $\Omega(n^2)$. However, this is not in fact the case, thanks to an algorithm known as the Fast Fourier Transform (FFT). This algorithm is vitally important for a variety of applications, particularly in the area of signal processing (Section 1.7.6).

Suppose that $n > 2$ and set $t = n/2$. The t-tuples b and c defined by $b = (a_0, a_2, \ldots, a_{n-4}, a_{n-2})$ and $c = (a_1, a_3, \ldots, a_{n-3}, a_{n-1})$ are such that $p_a(x) = p_b(x^2) + x p_c(x^2)$. In particular, $p_a(\omega^i) = p_b(\alpha^i) + \omega^i p_c(\alpha^i)$, where $\alpha = \omega^2$. Clearly, $\alpha^{t/2} = (\omega^2)^{t/2} = \omega^t = \omega^{n/2} = -1$, so it is legitimate to talk about $F_\alpha(b)$ and $F_\alpha(c)$. Furthermore, $\alpha^t = 1$ and $\omega^t = -1$, hence $\alpha^{t+i} = \alpha^i$ and $\omega^{t+i} = -\omega^i$, so that $p_a(\omega^{t+i}) = p_b(\alpha^i) - \omega^i p_c(\alpha^i)$. The Fourier transform $F_\omega(a)$ is calculated using the divide-and-conquer technique.

```
function FFT (a [0..n − 1], ω): array [0..n − 1]
    { n is a power of 2 and ω^{n/2} = −1 }
    array A [0..n − 1]    { the answer is computed in this array }
    if n = 1 then A [0] ← a [0]
    else t ← n /2
        arrays b, c, B, C [0..t−1]    { intermediate arrays }
        { creation of the sub-instances }
        for i ← 0 to t−1 do b [i] ← a [2i ]
                            c [i] ← a [2i + 1]
```

{ recursive Fourier transform computation of the sub-instances }
$B \leftarrow FFT(b, \omega^2)$
$C \leftarrow FFT(c, \omega^2)$
{ Fourier transform computation of the original instance }
$\alpha \leftarrow 1$
for $i \leftarrow 0$ **to** $t-1$ **do**
$\quad \{\alpha = \omega^i\}$
$\quad A[i] \leftarrow B[i] + \alpha C[i]$
$\quad A[t+i] \leftarrow B[i] - \alpha C[i]$
$\quad \alpha \leftarrow \alpha \omega$
return A

Problem 9.2.1.   Show that the execution time of this algorithm is in $\Theta(n \log n)$. $\qquad\qquad\qquad\qquad\qquad\qquad\qquad\qquad\qquad\qquad\qquad\qquad\qquad$ □

Example 9.2.1.   Let $n = 8$ and $a = (255, 8, 0, 226, 37, 240, 3, 0)$. Let us calculate $F_\omega(a)$ in arithmetic modulo $m = 257$, where $\omega = 4$. This can be done because $4^4 \equiv -1 \pmod{257}$. First a is split into $b = (255, 0, 37, 3)$ and $c = (8, 226, 240, 0)$. The recursive calls, which use $t = 4$ and $\omega^2 = 16$, yield $B = F_{16}(b) = (38, 170, 32, 9)$ and $C = F_{16}(c) = (217, 43, 22, 7)$. We combine these results to obtain A.

$$A[0] \leftarrow 38+217 \;\;= 255 \qquad A[4] \leftarrow 38-217 \;\;= 78$$
$$A[1] \leftarrow 170+43\omega = \;\;85 \qquad A[5] \leftarrow 170-43\omega = 255$$
$$A[2] \leftarrow 32+22\omega^2 = 127 \qquad A[6] \leftarrow 32-22\omega^2 = 194$$
$$A[3] \leftarrow 9+7\omega^3 \;\;= 200 \qquad A[7] \leftarrow 9-7\omega^3 \;\;= 75$$

The final result is therefore $A = (255, 85, 127, 200, 78, 255, 194, 75)$. $\qquad\qquad$ □

**Problem 9.2.2.   Despite the conceptual simplicity of this recursive algorithm, it is preferable in practice to use an iterative version. Give such an iterative algorithm. [*Hint*: Let $p(x)$ be an arbitrary polynomial, and let $q(x) = (x-x_1)(x-x_2) \cdots (x-x_t)$ be a polynomial of degree t such that $q(x_1) = q(x_2) = \cdots = q(x_t) = 0$, where $x_1, x_2, \ldots, x_t$ are arbitrary distinct values. Let $r(x)$ be the remainder of the symbolic division of $p(x)$ by $q(x)$. Show that $p(x_i) = r(x_i)$ for $1 \le i \le t$. In particular, the remainder of the symbolic division of $p(x)$ by the monomial $(x-x_i)$ is the constant polynomial whose value is $p(x_i)$.] $\qquad$ □

9.3 THE INVERSE TRANSFORM

Despite its importance in signal processing, our concern here is to use the discrete Fourier transform as a tool for transforming the domain of a problem. Our principal aim is to save the idea proposed in Example 9.1.4. Using the fast Fourier transform allows us to evaluate the polynomials $p(x)$ and $q(x)$ at the points $1, \omega, \omega^2, \ldots, \omega^{n-1}$ in a time in $O(n \log n)$, where n is a power of 2 greater than the degree of the product polynomial. The pointwise multiplication can be carried out in a time in $O(n)$. To obtain the final result, we still have to interpolate the unique polynomial $r(x)$ of degree less than n that passes through these points. Thus we have in fact to invert the Fourier transformation. With this in view we say that ω is a *principal n th root of unity* if it fulfills three conditions:

1. $\omega \neq 1$
2. $\omega^n = 1$, and
3. $\displaystyle\sum_{j=0}^{n-1} \omega^{jp} = 0$ for every $1 \leq p < n$.

When n is a power of 2, it turns out that these conditions are automatically satisfied whenever $\omega^{n/2} = -1$, as we have assumed already.

Theorem 9.3.1.   Consider any commutative ring. Let $n > 1$ be a power of 2, and let ω be an element of the ring such that $\omega^{n/2} = -1$. Then

i. ω is a principal n th root of unity.

ii. ω^{n-1} is the multiplicative inverse of ω in the ring in question; we shall denote it by ω^{-1}. More generally we use ω^{-i} to denote $(\omega^{-1})^i$ for any integer i.

iii. ω^{-1} is also a principal n th root of unity.

iv. As a slight abuse of notation, let us use "n" also to denote the element in our ring obtained by adding n ones together (the "1" from our ring). Assuming this n is not zero, $1 = \omega^0, \omega^1, \omega^2, \ldots, \omega^{n-1}$ are all distinct — they are called the *n th roots of unity.*

v. Assuming the existence of a multiplicative inverse n^{-1} for n in our ring, $\omega^{n/2} = -1$ is a consequence of ω being a principal n th root of unity.   □

*Problem 9.3.1.   Prove Theorem 9.3.1.   *Hints*:

i. Conditions (1) and (2) are obviously fulfilled. To show condition (3), let $n = 2^k$ and decompose $p = 2^u v$ where u and v are integers, and v is odd. Let $s = 2^{k-u-1}$. Show that $\omega^{jp} = -\omega^{(j+s)p}$ for every integer j. Conclude by splitting $\sum_{j=0}^{n-1} \omega^{jp}$ into 2^u sub-sums of $2s$ elements, each summing to zero.

ii. Obvious.

iii. Notice that $(\omega^{-1})^{n/2} = -1$.

iv. Assume $\omega^i = \omega^j$ for $0 \leq i < j < n$ and let $p = j - i$. Then $\omega^p = 1$ and $1 \leq p < n - 1$. Use condition (3) and $n \neq 0$ to obtain a contradiction.

v. Use $p = n/2$ in condition (3), and use the existence of n^{-1}.   □

Problem 9.3.2.   Prove that $e^{2i\pi/n}$ is a principal n th root of unity in the field of complex numbers.   □

Problem 9.3.3. Let n and ω be positive powers of 2, and let $m = \omega^{n/2} + 1$. Prove that ω is a principal n th root of unity in arithmetic modulo m. Prove further that n^{-1} exists modulo m, by showing that $n^{-1} = m - (m-1)/n$.   □

Problem 9.3.4. When m is of the form $2^u + 1$, as in Problem 9.3.3, multiplications modulo m can be carried out on a binary computer without using division operations. Let a and b be two integers such that $0 \le a < m$ and $0 \le b < m$; we wish to obtain their product modulo m. Let $c = ab$ be the ordinary product obtained from a multiplication. Decompose c into two blocks of u bits: $c = 2^u j + i$, where $0 \le i < 2^u$ and $0 \le j \le 2^u$. (The only possibility for $j = 2^u$ is when $a = b = m - 1$.) If $i \ge j$, set $d = i - j$, and otherwise set $d = m + i - j$. Prove that $0 \le d < m$ and $ab \equiv d \pmod{m}$. More generally, show how $x \bmod m$ can be calculated in a time in $O(u)$ when $m = 2^u + 1$, provided that $-(2^{2u} + 2^u) < x < 2^{2u} + 2^{u+1}$.   □

From now on, assume that ω is a principal n th root of unity and that n^{-1} exists in the algebraic structure considered. Let A be the $n \times n$ matrix defined by $A_{ij} = \omega^{ij}$ for $0 \le i < n$ and $0 \le j < n$. The main theorem asserts that A has an inverse, namely the matrix B defined by $B_{ij} = n^{-1} \omega^{-ij}$.

Theorem 9.3.2. Let A and B be the matrices just defined. Then $AB = I_n$, the $n \times n$ identity matrix.

Proof. Let $C = AB$. By definition, $C_{ij} = \sum_{k=0}^{n-1} A_{ik} B_{kj} = n^{-1} \sum_{k=0}^{n-1} \omega^{(i-j)k}$. There are three cases to consider.

i. If $i = j$, then $\omega^{(i-j)k} = \omega^0 = 1$, and so $C_{ij} = n^{-1} \sum_{k=0}^{n-1} 1 = n \times n^{-1} = 1$.

ii. If $i > j$, let $p = i - j$. Now $C_{ij} = n^{-1} \sum_{k=0}^{n-1} \omega^{kp} = 0$ by property (3) of a principal n th root of unity, since $1 \le p < n$.

iii. If $i < j$, let $p = j - i$. Now $C_{ij} = n^{-1} \sum_{k=0}^{n-1} (\omega^{-1})^{kp} = 0$ because ω^{-1} is also a principal n th root of unity by Theorem 9.3.1(iii).   □

This matrix A provides us with another equivalent definition of the discrete Fourier transform: $F_\omega(a) = aA$. Theorem 9.3.2 justifies the following definition. Let $a = (a_0, a_1, \ldots, a_{n-1})$ be an n-tuple. The *inverse Fourier transform* of a with respect to ω is the n-tuple

$$F_\omega^{-1}(a) = aB = (n^{-1} p_a(1), n^{-1} p_a(\omega^{-1}), n^{-1} p_a(\omega^{-2}), \ldots, n^{-1} p_a(\omega^{-(n-1)})) .$$

Problem 9.3.5. Prove that $F_\omega^{-1}(F_\omega(a)) = F_\omega(F_\omega^{-1}(a)) = a$ for every a.   □

The inverse Fourier transform can be calculated efficiently by the following algorithm, provided that n^{-1} is either known or easily calculable.

function $FFTinv\,(a\,[0\,..\,n-1],\omega):$ **array** $[0\,..\,n-1]$
    $\{\,n$ is a power of 2 and $\omega^n = 1\,\}$
    array $F\,[0\,..\,n-1]$
    $F \leftarrow FFT\,(a\,,\omega^{n-1})$
    for $i \leftarrow 0$ **to** $n-1$ **do** $F\,[i] \leftarrow n^{-1}F\,[i]$
    return F

Example 9.3.1.   Let $n = 8$ and $a = (255,85,127,200,78,255,194,75)$. Let us calculate $F_\omega^{-1}(a)$ in arithmetic modulo $m = 257$, where $\omega = 4$. By Problem 9.3.3, ω is indeed a principal nth root of unity. First we calculate $FFT\,(a\,,\omega^{-1})$, where $\omega^{-1} = \omega^7 = 193$. To do this, a is decomposed into $b = (255,127,78,194)$ and $c = (85,200,255,75)$. The recursive calls with $\omega^{-2} = 241$ yield $B = FFT\,(b\,,\omega^{-2}) = (140,221,12,133)$ and $C = FFT\,(c\,,\omega^{-2}) = (101,143,65,31)$. Combined, these results give $A = (241,64,0,9,39,121,24,0)$. There remains the multiplication by $n^{-1} = m - (m-1)/n = 225$ (Problem 9.3.3). The final result is thus $F = (255,8,0,226,37,240,3,0)$, which is consistent with Example 9.2.1.   □

If the Fourier transform is calculated in the field of complex numbers (Problem 9.3.2), rounding errors may occur on the computer. On the other hand, if the transform is calculated modulo m (Problem 9.3.3), it may be necessary to handle large integers. For the rest of this section we no longer suppose that arithmetic operations can be performed at unit cost: the addition of two numbers of size l takes a time in $O(l)$. We already know (Problem 9.3.4) that reductions modulo m can be carried out in a time in $O(\log m)$, thanks to the particular form chosen for m. Furthermore, the fact that ω is a power of 2 means that multiplications in the FFT algorithm can be replaced by shifts. For this it is convenient to modify the algorithm slightly. First, instead of giving ω as the second argument, we supply the base 2 logarithm of ω, denoted by γ. Secondly, the recursive calls are made with 2γ rather than ω^2 as the second argument. The final loop becomes

$\beta \leftarrow 0$
for $i \leftarrow 0$ **to** $t-1$ **do**
    $\{\,\beta = i\gamma\,\}$
    $A\,[i] \leftarrow B\,[i] + C\,[i]\!\uparrow\!\beta$
    $A\,[t+i\,] \leftarrow B\,[i] - C\,[i]\!\uparrow\!\beta$
    $\beta \leftarrow \beta + \gamma$,

where $x\!\uparrow\!y$ denotes the value of x shifted left y binary places, that is, $x \times 2^y$. All the arithmetic is carried out modulo $m = \omega^{n/2} + 1$ using Problem 9.3.4.

The heart of the algorithm consists of executing instructions of the form $A \leftarrow (B \pm C \uparrow \beta)\,\mathbf{mod}\,m$, where $0 \leq B < m$ and $0 \leq C < m$. The value of the shift β never exceeds $(\frac{n}{2} - 1)\lg\omega$, even when the recursive calls are taken into account. Consequently $-\omega^{n-1} \leq B \pm C\!\uparrow\!\beta \leq \omega^{n-1} + \omega^{n/2}$, which means that it can be reduced modulo m in a time in $O(\log m) = O(n \log \omega)$ by Problem 9.3.4. Since the number of operations of this type is in $O(n \log n)$, the complete computation of the Fourier transform modulo m can be carried out in a time in $O(n^2 \log n \log \omega)$. (From a practical point of view, if m is sufficiently small that arithmetic modulo m can be considered to be elementary, the algorithm takes a time in $O(n \log n)$.)

Problem 9.3.6.   Show that the inverse transform modulo $m = \omega^{n/2} + 1$ can also be computed in a time in $O(n^2 \log n \log \omega)$. (The algorithm *FFTinv* has to be modified. Otherwise a direct call on the new *FFT* with $\gamma = (n-1) \lg \omega$, corresponding to the use of $\omega^{-1} = \omega^{n-1}$ as a principal root of unity, causes shifts that can go up to $\beta = (\frac{n}{2} - 1)(n-1) \lg \omega$, which means that Problem 9.3.4 can no longer be applied. Similarly, the final multiplication by n^{-1} can profitably be replaced by a multiplication by $-n^{-1}$ followed by a change of sign, since $-n^{-1} = \omega^{n/2}/n$ is a power of 2.)   □

9.4 SYMBOLIC OPERATIONS ON POLYNOMIALS

We now have available the tools that are necessary to finish Example 9.1.4. Let $p(x) = a_s x^s + a_{s-1} x^{s-1} + \cdots + a_1 x + a_0$ and $q(x) = b_t x^t + b_{t-1} x^{t-1} + \cdots + b_1 x + b_0$ be two polynomials of degrees s and t, respectively. We want to calculate symbolically the product polynomial $r(x) = c_d x^d + c_{d-1} x^{d-1} + \cdots + c_1 x + c_0 = p(x)q(x)$ of degree $d = s + t$. Let n be the smallest power of 2 greater than d, and let ω be a principal nth root of unity. Let a, b, and c be the n-tuples defined by $a = (a_0, a_1, \ldots, a_s, 0, 0, \ldots, 0)$, $b = (b_0, b_1, \ldots, b_t, 0, 0, \ldots, 0)$, and $c = (c_0, c_1, \ldots, c_d, 0, 0, \ldots, 0)$, respectively. (Padding c with zeros is unnecessary if $d-1$ is a power of 2.) Let $A = F_\omega(a)$, $B = F_\omega(b)$, and $C = F_\omega(c)$. By definition of the Fourier transform, $C_i = r(\omega^i) = p(\omega^i)q(\omega^i) = A_i B_i$. Therefore C is the pointwise product of A and B. By Problem 9.3.5, $c = F_\omega^{-1}(C)$.

Putting all this together, the coefficients of the product polynomial $r(x)$ are given by the first $d+1$ entries in $c = F_\omega^{-1}(F_\omega(a) \times F_\omega(b))$. Notice that this reasoning made no use of the classic unique interpolation theorem, and this is fortunate because unique interpolation does not always hold when the arithmetic is performed in a ring rather than in a field. (Consider, for instance, $p_1(x) = 2x + 1$ and $p_2(x) = 5x + 1$ in the ring of integers modulo 9. Both of these degree 1 polynomials evaluate to 1 and 7 at the points 0 and 3, respectively.)   □

Problem 9.4.1.   Give explicitly the algorithm we have just sketched. Show that it can be used to multiply two polynomials whose product is of degree d with a number of scalar operations in $O(d \log d)$, provided that a principal nth root of unity and the multiplicative inverse of n are both easily obtainable, where n is the smallest power of 2 greater than d.   □

In order to implement this algorithm for the symbolic multiplication of polynomials, we need to be able to calculate efficiently a principal nth root of unity. The easiest approach is to use the complex number field and Problem 9.3.2. It is somewhat surprising that efficient symbolic multiplication of polynomials with integer coefficients should require operations on complex numbers. If an exact answer is required, it becomes necessary to take precautions against the possibility of rounding errors on the computer. In this case a more thorough analysis of the possible build-up of rounding errors is needed. For this reason it may be more attractive to carry out the arithmetic modulo a sufficiently large number (Problem 9.4.2), and to use Problem 9.3.3 to obtain a principal n th root of unity. This may require the use of multiple-precision arithmetic.

Problem 9.4.2.   Let $p(x)$ and $q(x)$ be two polynomials with integer coefficients. Let a and b be the maxima of the absolute values of the coefficients of $p(x)$ and $q(x)$, respectively. Let u be the maximum of the degrees of the two polynomials. Prove that no coefficient of the product polynomial $p(x)q(x)$ exceeds $ab(u+1)$ in absolute value. (In Example 9.1.4, $a = 5$, $b = 6$, and $u = 3$, so no coefficient of $r(x)$ can exceed 120 in absolute value.)   □

Example 9.4.1. (Continuation of Example 9.1.4)   We wish to multiply symbolically the polynomials $p(x) = 3x^3 - 5x^2 - x + 1$ and $q(x) = x^3 - 4x^2 + 6x - 2$. Since the product is of degree 6, it suffices to take $n = 8$. By Problem 9.4.2, all the coefficients of the product polynomial $r(x) = p(x)q(x)$ lie between -120 and 120; thus it suffices to calculate them modulo $m = 257$. By Problem 9.3.3, $\omega = 4$ is a principal n th root of unity in arithmetic modulo 257, and $n^{-1} = 225$.

Let $a = (1, -1, -5, 3, 0, 0, 0, 0)$ and $b = (-2, 6, -4, 1, 0, 0, 0, 0)$. Two applications of the algorithm *FFT* yield

$$A = F_\omega(a) = (255, 109, 199, 29, 251, 247, 70, 133)$$

and

$$B = F_\omega(b) = (1, 22, 82, 193, 244, 103, 179, 188) \ .$$

The pointwise product of these two transforms, still working modulo 257, is $C = (255, 85, 127, 200, 78, 255, 194, 75)$. By Example 9.3.1, the vector c such that $F_\omega(c) = C$ is $c = (255, 8, 0, 226, 37, 240, 3, 0)$. Since all the coefficients of $r(x)$ lie between -120 and 120, the integers 255, 226, and 240 correspond to -2, -31, and -17, respectively. The final answer is therefore

$$r(x) = 3x^6 - 17x^5 + 37x^4 - 31x^3 + 8x - 2 \ . \qquad □$$

Problem 9.4.3.   Generalize this idea: give explicitly an algorithm $mul(a[0..s], b[0..t]): \textbf{array}[0..s+t]$ that carries out symbolic multiplication of polynomials with integer coefficients. Among other things, your algorithm should determine suitable values for n, ω, and m. (Use Problems 9.3.3 and 9.4.2 for this.)   □

The analysis of the algorithm of Problem 9.4.3 depends on the degrees s and t of the polynomials to be multiplied and on the size of their coefficients. If the latter are sufficiently small that it is reasonable to consider operations modulo m to be elementary, the algorithm multiplies $p(x)$ and $q(x)$ symbolically in a time in $O(d \log d)$, where $d = s + t$. The naive algorithm would have taken a time in $O(st)$. On the other hand, if we are obliged to use multiple-precision arithmetic, the initial computation of the Fourier transforms and the final calculation of the inverse take a time in $O(d^2 \log d \log \omega)$, and the intermediate pointwise multiplication of the transforms takes a time in $O(d M(d \log \omega))$, where $M(l)$ is the time required to multiply two integers of size l. Since $M(l) \in \Theta(l \log l \log \log l)$ with the best-known algorithm for integer multiplication (Section 9.5), the first term in this analysis can be neglected. The total time is therefore in $O(d M(d \log \omega))$, where $\omega = 2$ suffices if none of the coefficients of the polynomials to be multiplied exceeds $2^{n/4}/\sqrt{2(1 + \max(s, t))}$ in absolute value. (Remember that n is the smallest power of 2 greater than d.) By comparison, the naive algorithm takes a time in $O(s t M(l))$, where l is the size of the largest coefficient in the polynomials to be multiplied. It is possible for this time to be in $O(st)$ in practice, if arithmetic can be carried out on integers of size l at unit cost. The naive algorithm is therefore preferable to the "fast" algorithm if d is very large and l is reasonably small. In every case, the algorithm that uses $\omega = e^{2i\pi/n}$ can multiply approximately the two polynomials in a time in $O(d \log d)$.

Problem 9.4.4.   Let $x_1, x_2, \ldots, x_n$ be n distinct points. Give an efficient algorithm to calculate the coefficients of the unique monic polynomial $p(x)$ of degree n such that $p(x_i) = 0$ for every $1 \le i \le n$. (The polynomial is *monic* if the coefficient of x^n is 1.) Your algorithm should take a time in $O(n \log^2 n)$ provided all the necessary operations are taken to be elementary. (*Hint*: see Problem 4.11.2.)   □

*Problem 9.4.5.   Let $p(x)$ be a polynomial of degree n, and let $x_1, x_2, \ldots, x_n$ be n distinct points. Give an efficient algorithm to calculate each $y_i = p(x_i)$ for $1 \le i \le n$. Your algorithm should take a time in $O(n \log^2 n)$. (*Hint*: the hint to Problem 9.2.2 is relevant here too.)   □

**Problem 9.4.6.   Let $x_1, x_2, \ldots, x_n$ be n distinct points, and let $y_1, y_2, \ldots, y_n$ be n values, not necessarily distinct. Give an efficient algorithm to calculate the coefficients of the unique polynomial $p(x)$ of degree less than n such that $p(x_i) = y_i$ for every $1 \le i \le n$. Your algorithm should take a time in $O(n \log^2 n)$.   □

9.5 MULTIPLICATION OF LARGE INTEGERS

We return once more to the problem of multiplying large integers (Sections 1.1, 1.7.2, and 4.7). Let a and b be two n-bit integers whose product we wish to calculate. Suppose for simplicity that n is a power of 2 (nonsignificant leading zeros are added at the left of the operands if necessary). The classic algorithm takes a time in $\Omega(n^2)$, whereas the algorithm using divide-and-conquer requires only a time in $O(n^{1.59})$, or even in $O(n^\alpha)$ for any $\alpha > 1$ (Problem 4.7.8). We can do better than this thanks to a *double* transformation of the domain. The original integer domain is first transformed into the domain of polynomials represented by their coefficients; then the symbolic product of these polynomials is obtained using the discrete Fourier transform.

We denote by $p_a(x)$ the polynomial of degree less than n whose coefficients are given by the successive bits of the integer a. For instance, $p_{53}(x) = x^5 + x^4 + x^2 + 1$ because 53 in binary is 00110101. Clearly, $p_a(2) = a$ for every integer a. To obtain the product of the integers a and b, we need only calculate symbolically the polynomial $r(x) = p(x)q(x)$ using the fast Fourier transform (Section 9.4), and then evaluate $r(2)$. The algorithm is recursive because one of the stages in the symbolic multiplication of polynomials consists of a pointwise multiplication of Fourier transforms.

Example 9.5.1.   To make the illustration simpler, we perform the computation here in decimal rather than in binary. For the purpose of this example only, let $p_a(x)$ therefore denote the polynomial whose coefficients are given by the successive *digits* of a, so that $p_a(10) = a$. Let $a = 2301$ and $b = 1095$. Thus $p_a(x) = 2x^3 + 3x^2 + 1$ and $p_b(x) = x^3 + 9x + 5$. The symbolic product is

$$r(x) = p_a(x)p_b(x) = 2x^6 + 3x^5 + 18x^4 + 38x^3 + 15x^2 + 9x + 5$$

and we obtain the desired product ab as $r(10) = 2{,}519{,}595$.   □

The recursive nature of this algorithm obliges us to refine the analysis of symbolic multiplication of polynomials given following Problem 9.4.3. Let $M(n)$ be the time required to multiply two n-bit integers, where n is a power of 2. The central step in the symbolic multiplication of two polynomials of degree less than n consists of d multiplications of integers less than $\omega^{d/2} + 1$, where $d = 2n$ is a power of 2 greater than the degree of the product polynomial. Unfortunately, even if we take $\omega = 2$, these integers are of size $1 + \frac{1}{2}d \lg \omega = n + 1$. The original multiplication of two integers of size n therefore requires $2n$ multiplications of slightly larger integers!

To correct this, we must reduce the degree of the polynomials used to represent the integers to be multiplied, even if this means increasing the size of their coefficients. As an extreme case, the algorithm of Section 4.7 consists of representing each integer by a polynomial of degree 1 whose two coefficients lie between 0 and $2^{n/2} - 1$. However, in order that using the discrete Fourier transform should be attractive, the polynomials considered must have a sufficiently high degree. For instance, suppose we redefine $p_a(x)$ to be the polynomial whose coefficients are given by the successive figures of the representation of a in base 4. This polynomial is thus of degree less than $n/2$, its coefficients lie between 0 and 3, and $p_a(4) = a$. As before, the polynomial $r(x) = p_a(x)p_b(x)$ is calculated symbolically using the Fourier transform, and the final answer is obtained by evaluating $r(4)$.

This time the degree of the product polynomial $r(x)$ is less than n. The central step in the symbolic multiplication of $p_a(x)$ and $p_b(x)$ therefore only requires n multiplications of integers less than $m = \omega^{n/2} + 1$. This can be carried out in a time in $n M(\frac{1}{2}n \lg \omega) + O(n^2)$. (The second term is added in case certain operands are exactly $\omega^{n/2}$, since this integer of length $1 + \frac{1}{2}n \lg \omega$ has to be treated specially.) Taking into account also the time spent in computing the two initial Fourier transforms and the final inverse transform, the symbolic multiplication takes a time in $n M(\frac{1}{2}n \lg \omega) + O(n^2 \log n \log \omega)$.

Before analysing the new multiplication algorithm, we must choose a principal nth root of unity ω. Since the coefficients of the polynomials $p_a(x)$ and $p_b(x)$ lie between 0 and 3, and since these polynomials are of degree less than $n/2$, the largest coefficient possible in $r(x)$ is $9n/2$ (Problem 9.4.2). As the computations are carried out modulo $m = \omega^{n/2} + 1$, it is thus sufficient that $9n/2 \leq \omega^{n/2}$. The choice $\omega = 2$ is adequate provided $n \geq 16$. The symbolic computation of $r(x)$ therefore takes a time in $n M(n/2) + O(n^2 \log n)$. The last stage of the multiplication of a and b, namely the evaluation of $r(4)$, consists of n shifts and n additions of integers whose size is not more than $\lg(9n/2)$, which takes a negligible time in $O(n \log n)$. We thus obtain the asymptotic recurrence $M(n) \in n M(n/2) + O(n^2 \log n)$.

Problem 9.5.1.   Consider the recurrence

$$\begin{cases} t(n) = nt(n/2), & n = 2^k, k \in \mathbb{N}^+ \\ t(1) = 1 \ . \end{cases}$$

When n is a power of 2, prove that $t(n) = n^{(1+\lg n)/2}$. Show that $t(n) \notin O(n^k)$, whatever the value of the constant k.   □

The preceding problem shows that the modified algorithm is still bad news, even if we do not take into account the time required to compute the Fourier transforms! This is explained by the fact that we used the "fast" algorithm for multiplying two polynomials in exactly the circumstances when it should be avoided: the polynomials are of high degree and their coefficients are small. To correct this, we must lower the degree of the polynomials still further.

Let $l = 2^{\lceil \frac{1}{2} \lg n \rceil}$; that is, $l = \sqrt{n}$ or $l = \sqrt{2n}$, depending on whether $\lg n$ is even or odd. Let $k = n/l$. Note that l and k are powers of 2. This time, denote by $p_a(x)$ the polynomial of degree less than k whose coefficients correspond to the k blocks of l successive bits in the binary representation of a. Thus we have that $p_a(2^l) = a$. To calculate the product of the integers a and b, we need only calculate symbolically the polynomial $r(x) = p_a(x)p_b(x)$, using Fourier transforms, and then evaluate $r(2^l)$.

Example 9.5.2.   Let $a = 9885$ and $b = 21{,}260$, so that $n = 16$, $l = 4$, and $k = 4$. We form the polynomials $p_a(x) = 2x^3 + 6x^2 + 9x + 13$ and $p_b(x) = 5x^3 + 3x^2 + 12$. The first of these polynomials, for instance, comes from the decomposition into four blocks of the binary representation of a: 0010 0110 1001 1101. The symbolic product is

$$r(x) = p_a(x)p_b(x) = 10x^6 + 36x^5 + 63x^4 + 116x^3 + 111x^2 + 108x + 156$$

and the final evaluation yields $r(16) = 210{,}155{,}100 = 9885 \times 21{,}260$.   □

Let $d = 2k$, a power of 2 greater than the degree of the product polynomial $r(x)$. This time we need to choose a principal dth root of unity ω. Since the coefficients of the polynomials $p_a(x)$ and $p_b(x)$ lie between 0 and $2^l - 1$, and the degree of these polynomials is less than k, the largest coefficient possible in $r(x)$ is $k(2^l - 1)^2$. It suffices therefore that $k2^{2l} < m = \omega^{d/2} + 1$, that is $\lg \omega \geq (2l + \lg k)/(d/2)$. In the case when $\lg n$ is even $l = k = d/2 = \sqrt{n}$ and $\lg \omega \geq (2\sqrt{n} + \lg \sqrt{n})/\sqrt{n} = 2 + (\lg \sqrt{n})/\sqrt{n}$. Similarly, when n is odd, we obtain $\lg \omega \geq 4 + (\lg \sqrt{n/2})/\sqrt{n/2}$. Consequently, $\omega = 8$ suffices to guarantee that the computation of the coefficients of $r(x)$ will be correct when $\lg n$ is even, and $\omega = 32$ is sufficient when $\lg n$ is odd.

The multiplication of two n-bit integers is thus carried out using a symbolic multiplication of two polynomials, which takes a time in $dM(\frac{1}{2}d \lg \omega) + O(d^2 \log d \log \omega)$. As far as the final evaluation of $r(2^l)$ is concerned, this can easily be carried out in a time in $O(d^2 \log \omega)$, which is negligible. When n is even, $d = 2\sqrt{n}$ and $\omega = 8$, which gives $M(n) \in 2\sqrt{n} M(3\sqrt{n}) + O(n \log n)$. When n is odd, $d = \sqrt{2n}$ and $\omega = 32$; hence, $M(n) \in \sqrt{2n} M(\frac{5}{2}\sqrt{2n}) + O(n \log n)$.

*Problem 9.5.2.   Let $\gamma > 0$ be a real constant, and let $t(n)$ be a function satisfying the asymptotic recurrence $t(n) \in \gamma t(O(\sqrt{n})) + O(\log n)$. Prove that

$$t(n) \in \begin{cases} O(\log n) & \text{if } \gamma < 2 \\ O(\log n \log\log n) & \text{if } \gamma = 2 \\ O((\log n)^{\lg \gamma}) & \text{if } \gamma > 2 \ . \end{cases}$$

[*Hints*: For the second case use the fact that $\lg\lg(\beta \sqrt{n}) \leq (\lg\lg n) - \lg(5/3)$ provided that $n \geq \beta^{10}$ for every real constant $\beta > 1$. For the third case prove by constructive induction that $t(n) \leq \delta[(\lg n)^{\lg \gamma} - \psi(\lg n)^{(\lg \gamma) - 1}] - \rho \lg n$, for some constants δ, ψ, and ρ that you must determine and for n sufficiently large. Also use the fact that

$$(\lg \beta \sqrt{n})^{\lg \gamma} \leq \frac{1}{\gamma}(\lg n)^{\lg \gamma} + 2\lg \gamma \lg \beta (\lg \beta \sqrt{n})^{(\lg \gamma) - 1}$$

provided $n \geq \gamma^{2\lg \beta}$, for all real constants $\beta \geq 1$ and $\gamma \geq 2$.] $\square$

Let $t(n) = M(n)/n$. The equations obtained earlier for $M(n)$ lead to $t(n) \in 6t(3\sqrt{n}) + O(\log n)$ when $\lg n$ is even, and $t(n) \in 5t(\frac{5\sqrt{2}}{2}\sqrt{n}) + O(\log n)$ when n is odd. By Problem 9.5.2, $t(n) \in O((\log n)^{\lg 6}) \subset O((\log n)^{2.59})$. Consequently, this algorithm can multiply two n-bit integers in a time in $M(n) = nt(n) \in O(n(\log n)^{2.59})$.

Problem 9.5.3.   Prove that $O(n(\log n)^{2.59}) \subset n^\alpha$ whatever the value of the real constant $\alpha > 1$. This algorithm therefore outperforms all those we have seen previously, provided n is sufficiently large. $\square$

Is this algorithm optimal? To go still faster using a similar approach, Problem 9.5.2 suggests that we should reduce the constant $\gamma = 6$, which arises here as the maximum of 2×3 and $\sqrt{2} \times \frac{5}{2}\sqrt{2}$. This is possible provided we increase slightly the size of the coefficients of the polynomials used in order to decrease their degree. More precisely, we split the n-bit integers to be multiplied into k blocks of l bits, where

$l = 2^{i + \lceil \frac{1}{2} \lg n \rceil}$ and $k = n/l$ for an arbitrary constant $i \geq 0$. Detailed analysis shows that this gives rise to $2^{1-i} \sqrt{n}$ recursive calls on integers of size $(2^{i+1} + 2^{-i}) \sqrt{n}$ if $\lg n$ is even, and $2^{-i} \sqrt{2n}$ recursive calls on integers of size $(2^{i+1} + 2^{-i-1}) \sqrt{2n}$ if $\lg n$ is odd provided n is sufficiently large. The corresponding γ is thus

$$\gamma = \max(2^{1-i}(2^{i+1} + 2^{-i}), \ 2^{-i} \sqrt{2}(2^{i+1} + 2^{-i-1}) \sqrt{2}) = 4 + 2^{1-2i} .$$

The algorithm obtained takes a time in $O(n (\log n)^{\alpha})$, where

$$\alpha = 2 + \lg(1 + 2^{-1-2i}) < 2 + 2^{-1-2i} / \ln 2$$

can be reduced arbitrarily close to 2. Needless to say, increasing the parameter i reduces the exponent α at the expense of increasing the hidden constant in the asymptotic notation.

     This is still not optimal, but the algorithms that are even faster are too complicated to be described in detail here. We mention simply that it is possible to obtain $\gamma = 4$ by calculating the coefficients of the polynomial $r(x)$ modulo $2^{2l} + 1$ (using the Fourier transform and proceeding recursively) and modulo k (using the algorithm for integer multiplication of Section 4.7). Because $2^{2l} + 1$ and k are coprime, it is then possible to obtain the coefficients of $r(x)$ by the Chinese remainder theorem, and finally to evaluate $r(2^{2l})$. The outcome is an algorithm that is capable of multiplying two n-bit integers in a time in $O(n \log^2 n)$. To go faster still, at least asymptotically, we have to redefine the notion of the "product" of two polynomials so as to avoid doubling the degree of the result obtained. This approach allowed Schönhage and Strassen to obtain $\gamma = 2$, that is, an algorithm that takes a time in $O(n \log n \log \log n)$. The complexity of this algorithm is such that it is of theoretical interest only.

     An integer multiplication algorithm based on the fast Fourier transform has been used by the Japanese to calculate π to 10 million decimal places. Rather than resorting to modulo computations in a finite ring, they used a variant involving operations in the complex number field. Their approach requires care to avoid problems due to rounding errors but gives rise to a simpler algorithm. It also allows the computation to be carried out directly in decimal, thus avoiding a costly conversion when printing out the result. Even more decimals of π have been calculated since.

9.6 REFERENCES AND FURTHER READING

The first published algorithm for calculating discrete Fourier transforms in a time in $O(n \log n)$ is by Danielson and Lanczos (1942). These authors mention that the source of their method goes back to Runge and König (1924). In view of the great practical importance of Fourier transforms, it is astonishing that the existence of a fast algorithm remained almost entirely unknown until its rediscovery nearly a quarter of a century later by Cooley and Tukey (1965). For a more complete account of the history of the fast Fourier transform, read Cooley, Lewis, and Welch (1967). An efficient implementation and numerous applications are suggested in Gentleman and Sande (1966), and Rabiner and Gold (1974). The book by Brigham (1974) is also worth mentioning. The nonrecursive algorithm suggested in Problem 9.2.2 is described in several references, for example Aho, Hopcroft, and Ullman (1974).

Pollard (1971) studies the computation of Fourier transforms in a finite field. The solution to Problems 9.4.5 and 9.4.6 is given in Aho, Hopcroft, and Ullman (1974). Further ideas concerning the symbolic manipulation of polynomials, evaluation, and interpolation can be found in Borodin and Munro (1971, 1975), Horowitz and Sahni (1978), and Turk (1982).

The second edition of Knuth (1969) includes a survey of algorithms for integer multiplication. A practical algorithm for the rapid multiplication of integers with up to 10 thousand decimal digits is given in Pollard (1971). The algorithm that is able to multiply two integers of size n in a time in $O(n \log^2 n)$ is attributed to Karp and described in Borodin and Munro (1975). The details of the algorithm by Schönhage and Strassen (1971) are spelled out in Brassard, Monet, and Zuffellato (1986), although the solution given there to Problem 9.3.3 is unnecessarily complicated in the light of Theorem 9.3.1. Also read Turk (1982). The algorithm used by the Japanese to compute π to 10 million decimal places is described in Kanada, Tamura, Yoshino and Ushiro (1986); Cray Research (1986) mentions an even more precise computation of the decimals of π, but does not explain the algorithm used. The empire struck back shortly thereafter when the Japanese computed 134 million decimals, which is the world record at the time of this writing; read Gleick (1987). And the saga goes ever on.

10

Introduction

to Complexity

Up to now, we have been interested in the systematic development and analysis of specific algorithms, each more efficient than its predecessors, to solve some given problem. Computational complexity, a field of study that runs in parallel with algorithmics, considers globally the class of all algorithms that are able to solve a given problem. Using algorithmics, we can prove, by giving an explicit algorithm, that a certain problem can be solved in a time in $O(f(n))$ for some function $f(n)$ that we aim to reduce as much as possible. Using complexity, we try to find a function $g(n)$ as large as possible and to prove that *any* algorithm that is capable of solving our problem correctly on all of its instances must necessarily take a time in $\Omega(g(n))$. Our satisfaction is complete when $f(n) \in \Theta(g(n))$, since then we know that we have found the most efficient algorithm possible (except perhaps for changes in the hidden multiplicative constant). In this case we say that the complexity of the problem is known exactly; unfortunately, this does not happen often. In this chapter we introduce only a few of the principal techniques and concepts used in the study of computational complexity.

10.1 DECISION TREES

This technique applies to a variety of problems that use the concept of comparisons between elements. We illustrate it with the sorting problem. Thus we ask the following question: what is the minimum number of comparisons that are necessary to sort n elements? For simplicity we count only comparisons between the elements to be sorted, ignoring those that may be made to control the loops in our program. Consider first the following algorithm.

```
procedure countsort (T [1 .. n ])
    i ← min(T ), j ← max(T )
    array C [i .. j ] ← 0
    for k ← 1 to n do C [T [k]] ← C [T [k]]+1
    k ← 1
    for p ← i to j do
        for q ← 1 to C [p ] do
            T [k] ← p
            k ← k +1
```

Problem 10.1.1.   Simulate the operation of this algorithm on an array $T [1 .. 10]$ containing the values $3, 1, 4, 1, 5, 9, 2, 6, 5, 3$. $\Box$

This algorithm is very efficient if the difference between the largest and the smallest values in the array to be sorted is not too large. For example, if $\max(T) - \min(T) \approx \#T$, the algorithm provides an efficient and practical way of sorting an array in linear time. However, it becomes impractical, on account of both the memory and the time it requires, when the difference between the elements to be sorted is large. In this case, variants such as radix sort or lexicographic sort (not discussed here) can sometimes be used to advantage. However, only in rare applications will these algorithms prove preferable to quicksort or heapsort. The most important characteristic of *countsort* and its variations is that they work using *transformations*: arithmetic operations are carried out on the elements to be sorted. On the other hand, all the sorting algorithms considered in the preceding chapters work using *comparisons*: the only operation allowed on the elements to be sorted consists of comparing them pairwise to determine whether they are equal and, if not, which is the greater. This difference resembles that between binary search and hash coding. In this book we pay no further attention to algorithms for sorting by transformation.

Problem 10.1.2.   Show exactly how *countsort* can be said to carry out arithmetic operations on the elements to be sorted. As a function of n, the number of elements to be sorted, how many comparisons between elements are made? $\Box$

Coming back to the question we asked at the beginning of this section: what is the minimum number of comparisons that are necessary in any algorithm for sorting n elements *by comparison*? Although the theorems set out in this section still hold even if we consider probabilistic sorting algorithms (Section 8.4.1), we shall for simplicity confine our discussion to deterministic algorithms. A *decision tree* is a labelled, directed binary tree. Each internal node contains a comparison between two of the elements to be sorted. Each leaf contains an ordering of the elements. Given a total order relation between the elements, a *trip* through the tree consists of starting from the root and asking oneself the question that is found there. If the answer is "yes", the trip continues recursively in the left-hand subtree; otherwise it continues recursively in the right-hand subtree. The trip ends when it reaches a leaf; this leaf contains the *verdict* associated with the order relation used. A decision tree for sorting n elements is *valid* if to each possible order relation between the elements it associates a verdict that is compatible with this relation. Finally, a decision tree is *pruned* if all its leaves are accessible from the root by making some consistent sequence of decisions. The following problem will help you grasp these notions.

Problem 10.1.3.   Verify that the decision tree given in Figure 10.1.1 is valid for sorting three elements *A*, *B*, and *C*.   ☐

Every valid decision tree for sorting *n* elements gives rise to an *ad hoc* sorting algorithm for the same number of elements. For example, to the decision tree of Figure 10.1.1 there corresponds the following algorithm.

```
procedure adhocsort3(T [1 .. 3])
   A ← T [1], B ← T [2], C ← T [3]
   if A < B  then if B < C  then {already sorted}
                            else if A < C
                                    then T ← A ,C ,B
                                    else T ← C ,A ,B
             else if B < C  then if A < C
                                    then T ← B ,A ,C
                                    else T ← B ,C ,A
                            else T ← C ,B ,A
```

Similarly, to every deterministic algorithm for sorting by comparison there corresponds, for each value of *n*, a decision tree that is valid for sorting *n* elements. Figures 10.1.2 and 10.1.3 give the trees corresponding to the insertion sorting algorithm (Section 1.4) and to heapsort (Section 1.9.4 and Problem 2.2.3), respectively, when three elements are to be sorted. (The annotations on the trees are intended to help follow the progress of the corresponding algorithms.) Notice that heapsort

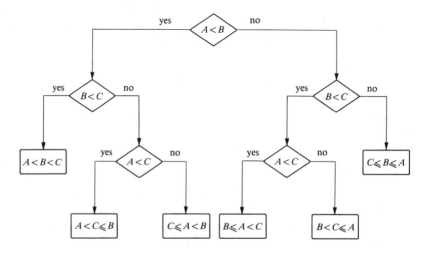

Figure 10.1.1.   A valid decision tree for sorting three elements.

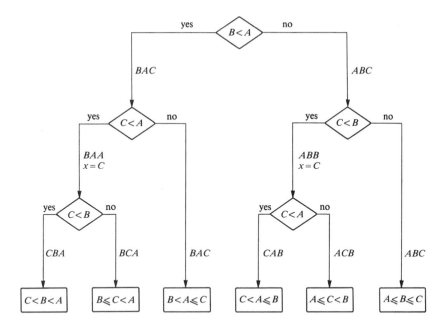

Figure 10.1.2.   The three element insertion sort decision tree.

sometimes makes unnecessary comparisons. For instance, if $B \leq A < C$, the decision tree of Figure 10.1.3 first tests whether $B > A$ (answer: no), and then whether $C > A$ (answer: yes). It would now be possible to establish the correct verdict, but it, nonetheless, asks again whether $B > A$ before reaching its conclusion. (Despite this, the tree is pruned: the leaf that would correspond to a contradictory answer "yes" to the third question has been removed, so that every leaf can be reached by some consistent sequence of decisions.) Thus heapsort is not optimal insofar as the number of comparisons is concerned. This situation does not occur with the decision tree of Figure 10.1.2, but beware of appearances: it occurs even more frequently with the insertion sorting algorithm than with heapsort when the number of elements to be sorted increases.

Problem 10.1.4.   Give the pruned decision trees corresponding to the algorithms for sorting by selection (Section 1.4) and by merging (Section 4.4), and to quicksort (Section 4.5) for the case of three elements. In the two latter cases do not stop the recursive calls until there remains only a single element to be "sorted".   ☐

Problem 10.1.5.   Give the pruned decision trees corresponding to the insertion sorting algorithm and to heapsort for the case of four elements. (You will need a big piece of paper!)   ☐

The following observation is crucial: the height of the pruned decision tree corresponding to any algorithm for sorting n elements by comparison, that is, the distance from the root to the most distant leaf, gives the number of comparisons carried

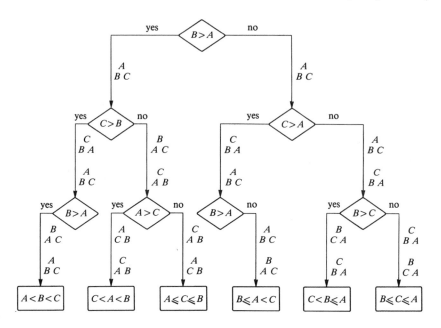

Figure 10.1.3.   The three element heapsort decision tree.

out by this algorithm in the worst case. For example, a possible worst case for sorting three elements by insertion is encountered if the array is already sorted into descending order ($C < B < A$); in this case the three comparisons $B < A$?, $C < A$?, and $C < B$? situated on the path from the root to the appropriate verdict in the decision tree all have to be made.

   The decision trees we have seen for sorting three elements are all of height 3. Can we find a valid decision tree for sorting three elements whose height is less? If so, we shall have an ad hoc algorithm for sorting three elements that is more efficient in the worst case. Try it: you will soon see that this cannot be done. We now prove more generally that such a tree is impossible.

Lemma 10.1.1.   Any binary tree with k leaves has a height of at least $\lceil \lg k \rceil$.

   Proof. It is easy to show (by mathematical induction on the total number of nodes in the tree) that any binary tree with k leaves must have at least $k - 1$ internal nodes. To say the same thing differently, a binary tree with t nodes in all cannot have more than $\lceil t/2 \rceil$ leaves. Now a binary tree of height h can have at most $2^{h+1} - 1$ nodes in all (by another simple argument using mathematical induction, this time on the height of the tree), and hence it has at most 2^h leaves. The lemma follows immediately.   □

Lemma 10.1.2.   Any valid decision tree for sorting n elements contains at least $n!$ leaves. (It may have more than $n!$ leaves if it is not pruned or if some of the leaves can only be reached when some keys are equal. The upper limit on the number of leaves of any pruned decision tree can be computed with Problem 5.8.6.)

Proof. A valid tree must be able to produce at least one verdict corresponding to each of the $n!$ possible orderings of the n elements to be sorted.   □

Theorem 10.1.1.   Any deterministic algorithm for sorting by comparison takes a time in $\Omega(n \log n)$ to sort n elements in the worst case.

Proof. To every deterministic algorithm for sorting by comparison there corresponds a pruned valid decision tree for sorting n elements. This tree contains at least $n!$ leaves by Lemma 10.1.2. Its height is therefore at least $\lceil \lg(n!) \rceil$ by Lemma 10.1.1. By the crucial observation that precedes Lemma 10.1.1, the algorithm thus needs at least $\lceil \lg(n!) \rceil$ comparisons in the worst case to sort n elements. Since each comparison takes a time in $\Omega(1)$, and $\lg(n!) \in \Omega(n \log n)$ (Problem 2.1.17), the algorithm takes a time in $\Omega(n \log n)$ in the worst case.   □

This proof shows that any deterministic algorithm for sorting by comparison must make at least $\lceil \lg(n!) \rceil$ comparisons in the worst case when sorting n elements. This certainly does not mean that it is always possible to sort n elements with as few as $\lceil \lg(n!) \rceil$ comparisons in the worst case. In fact, it has been proved that 30 comparisons are necessary and sufficient in the worst case for sorting 12 elements, and yet $\lceil \lg(12!) \rceil = 29$. In the worst case, the insertion sorting algorithm makes 66 comparisons when sorting 12 elements, whereas heapsort makes 59 (of which the first 18 are made during construction of the heap).

Problem 10.1.6.   Give exact formulas for the number of comparisons carried out in the worst case by the insertion sorting algorithm and by the selection sorting algorithm when sorting n elements. How well do these algorithms do when compared to the lower bound $\lceil \lg(n!) \rceil$ for $n = 50$?   □

** **Problem 10.1.7.**   Prove that the number of comparisons carried out by heapsort on n elements, $n \geq 2$, is never greater than $2n \lg n$. Prove further that if n is a power of 2, then mergesort makes $n \lg n - n + 1$ comparisons in the worst case when sorting n elements. What can you say about sorting by merging in the general case?   □

More precise analysis shows that $\lceil \lg(n!) \rceil \in n \lg n - \Theta(n)$. The previous problem therefore shows that heapsort is optimal to within a factor of 2 as far as the number of comparisons needed in the worst case is concerned, and that sorting by merging almost attains the lower bound. (Some modifications of heapsort come very close to being optimal for the worst-case number of comparisons.)

Problem 10.1.8.   Suppose we ask our sorting algorithm not merely to determine the order of the elements but also to determine which ones, if any, are equal. For example, a verdict such as $A < B \leq C$ is not acceptable: the algorithm must answer either $A < B < C$ or $A < B = C$. Give a lower bound on the number of comparisons required in the worst case to handle n elements. Rework this problem assuming that there are three possible outcomes of a comparison between A and B: $A < B$, $A = B$, or $A > B$.   □

Problem 10.1.9. Let $T[1..n]$ be an array sorted into ascending order, and let x be some element. How many comparisons between elements are needed in the worst case to locate x in the array? As in Section 4.3, the problem is to find an index i such that $0 \le i \le n$ and $T[i] \le x < T[i+1]$, with the logical convention that $T[0] = -\infty$ and $T[n+1] = +\infty$. How does binary search compare to this lower bound? What lower bound on the number of comparisons do you obtain using the decision tree technique if the problem is simply to determine whether x is in the array, rather than to determine its position?   □

Decision trees can also be used to analyse the complexity of a problem on the average rather than in the worst case. Let T be a binary tree. Define the *average height* of T as the sum of the depths of all the leaves divided by the number of leaves. For example, the decision tree of Figure 10.1.1 has an average height $(2+3+3+3+3+2)/6 = 8/3$. If each verdict is equally likely, then 8/3 is the average number of comparisons made by the sorting algorithm associated with this tree. Suppose for simplicity that the n elements are all distinct.

Lemma 10.1.3. Any binary tree with k leaves has an average height of at least $\lg k$. (By comparison with Lemma 10.1.1, we see that there is little difference between the worst case and the average.)

Proof. Let T be a binary tree with k leaves. Define $H(T)$ as the sum of the depths of the leaves. For example, $H(T) = 16$ for the tree in Figure 10.1.1. By definition, the average height of T is $H(T)/k$. The root of T can have 0, 1, or 2 children. In the first case the root is the only leaf in the tree and $H(T) = 0$. In the second case, the single child is the root of a subtree A, which also has k leaves. Since the distance from each leaf to the root of A is one less than the distance from the same leaf to the root of T, we have $H(T) = H(A)+k$. In the third case the tree T is composed of a root and of two subtrees B and C with i and $k-i$ leaves, respectively, for some $1 \le i < k$. By a similar argument we obtain this time $H(T) = H(B)+H(C)+k$.

For $k \ge 1$, define $h(k)$ as the smallest value possible for $H(X)$ for all the binary trees X with k leaves. In particular, $h(1)=0$. If we define $h(0) = 0$, the preceding discussion and the principle of optimality used in dynamic programming lead to

$$h(k) = \min \{ h(i)+h(k-i)+k \mid 0 \le i \le k \}$$

for every $k > 1$. At first sight this recurrence is not well founded since it defines $h(k)$ in terms of itself (when we take $i=0$ or $i=k$ in the minimum, which corresponds to the root having only one child). This difficulty disappears because it is impossible that $h(k) = h(k)+k$. We can thus reformulate the recurrence that defines $h(k)$.

$$h(k) = \begin{cases} 0 & \text{if } k \le 1 \\ k + \min \{ h(i)+h(k-i) \mid 1 \le i \le k-1 \} & \text{if } k > 1 \end{cases}$$

Now, consider the function $g(x) = x \lg x + (k-x)\lg(k-x)$, where $x \in \mathbb{R}$ is such that $1 \le x \le k-1$. Calculating the derivative gives $g'(x) = \lg x - \lg(k-x)$, which is zero if and only if $x = k - x$; that is, if $x = k/2$. Since the second derivative is positive, $g(x)$ attains its minimum at $x = k/2$. This minimum is $g(k/2) = (k \lg k) - k$.

The proof that $h(k) \geq k \lg k$ for every *integer* $k \geq 1$ now follows by mathematical induction. The base $k = 1$ is immediate. Let $k > 1$. Suppose by the induction hypothesis that $h(j) \geq j \lg j$ for every strictly positive integer $j \leq k - 1$. By definition,

$$h(k) = k + \min \{ h(i) + h(k-i) \mid 1 \leq i \leq k-1 \}.$$

By the induction hypothesis, $h(k) \geq k + \min \{ g(i) \mid 1 \leq i \leq k-1, i \in \mathbb{N} \}$.

Because $\min X \geq \min Y$ for any two nonempty sets X and Y such that $X \subseteq Y$, it follows that $h(k) \geq k + \min \{ g(x) \mid 1 \leq x \leq k-1, x \in \mathbb{R} \}$. Using the result obtained previously, we have $h(k) \geq k + g(k/2) = k \lg k$.

This shows that $H(T) \geq k \lg k$ for every tree T with k leaves. The average height of T being $H(T)/k$, it is therefore at least $\lg k$.   □

* Problem 10.1.10.   Let $t = \lfloor \lg k \rfloor$ and $l = k - 2^t$. Prove that $h(k) = kt + 2l$, where $h(k)$ is the function used in the proof of Lemma 10.1.3. Prove that this also implies that $h(k) \geq k \lg k$. (Optionally: give an intuitive interpretation of this formula in the context of the average height of a tree with k leaves.)   □

Theorem 10.1.2.   Any deterministic algorithm for sorting by comparison makes at least $\lg(n!)$ comparisons on the average to sort n elements. It therefore takes an average time in $\Omega(n \log n)$.

Proof.   Follows immediately from Lemmas 10.1.2 and 10.1.3.   □

Problem 10.1.11.   Determine the number of comparisons performed on the average by the insertion sorting algorithm and by the selection sorting algorithm when sorting n elements. How do these values compare to the number of comparisons performed by these algorithms in the worst case?   □

* Problem 10.1.12.   Let $T[1..n]$ be an array and $k \leq n$ an integer. The problem consists of returning in descending order the k largest elements of T. Prove that any deterministic algorithm that solves this problem using comparisons between the elements must make at least $(k/2) \lg (n/2)$ comparisons, both in the worst case and on the average. Conclude that it must take a time in $\Omega(k \log n)$. On the other hand, give an algorithm able to solve this problem in a time in $O(n \log k)$ and a space in $O(k)$ in the worst case. Your algorithm should make no more than one sequential pass through the array T; it is therefore efficient even if n is large and if the array is supplied on a magnetic tape. Justify your analysis of the time and space used by your algorithm.   □

10.2 REDUCTION

We have just shown that any algorithm for sorting by comparison takes a minimum time in $\Omega(n \log n)$ to sort n elements, both on the average and in the worst case. On the other hand, we know that heapsort and mergesort both solve the problem in a time in $O(n \log n)$. Except for the value of the multiplicative constant, the question of the complexity of sorting by comparison is therefore settled: a time in $\Theta(n \log n)$ is both necessary and sufficient for sorting n elements. Unfortunately, it does not often happen in the present state of our knowledge that the bounds derived from algorithmics and complexity meet so satisfactorily.

Because it is so difficult to determine the exact complexity of most of the problems we meet in practice, we often have to be content to compare the relative difficulty of different problems. There are two reasons for doing this. Suppose we are able to prove that a certain number of problems are equivalent in the sense that they have about the same complexity. Any algorithmic improvement in the method of solution of one of these problems now automatically yields, at least in theory, a more efficient algorithm for all the others. From a negative point of view, if these problems have all been studied independently in the past, and if all the efforts to find an efficient algorithm for any one of them have failed, then the fact that the problems are equivalent makes it even more unlikely that such an algorithm exists. Section 10.3 goes into this second motivation in more detail.

Definition 10.2.1. Let A and B be two solvable problems. A is *linearly reducible* to B, denoted $A \leq^l B$, if the existence of an algorithm for B that works in a time in $O(t(n))$, for any function $t(n)$, implies that there exists an algorithm for A that also works in a time in $O(t(n))$. When $A \leq^l B$ and $B \leq^l A$ both hold, A and B are linearly equivalent, denoted $A \equiv^l B$. □

Even if we are not able to determine the complexities of A and B exactly, when $A \equiv^l B$ we can be sure that they are the same. In the remainder of this section we shall see a number of examples of reduction from a variety of application areas.

*Problem 10.2.1. A less restrictive definition of linear reduction is obtained if we content ourselves with comparing the efficiency of algorithms for A on instances of size n with the efficiency of algorithms for B on instances of size in $O(n)$. For the purposes of this problem only, write $A \leq^g B$ if the existence of an algorithm for B that works in a time in $O(t(n))$, for any function $t(n)$, implies that there exists an algorithm for A that works in a time in $O(t(O(n)))$. Show with the help of an explicit example that the notions $A \leq^l B$ and $A \leq^g B$ are not equivalent even if there exists an algorithm for B that works in a time in $O(p(n))$, where $p(n)$ is a polynomial. □

Problem 10.2.2. Prove that the relations $\leq^l$ and $\equiv^l$ are transitive. □

Let us extend the notion of smooth functions (introduced in Section 2.1.5) to algorithms and problems. An *algorithm* is *smooth* if it takes a time in $\Theta(t(n))$ for some smooth function t. Even though a smooth function must be eventually nondecreasing by definition, this does not imply that the actual time taken by a specific implementation of the algorithm must also be given by an eventually nondecreasing function. Consider for instance the modular exponentiation algorithm *dexpo* of Section 4.8. We have seen (Problem 4.8.5) that the time it takes to compute $a^n \bmod m$ is a linear function both of $\lg n$ and the number of 1s in the binary representation of n. In particular, it takes longer to compute a 31st power than a 32nd. The actual time taken by any reasonable implementation of this algorithm is *not* an eventually nondecreasing function of the exponent. Nonetheless, this algorithm *is* smooth because it takes a time in $\Theta(\log n)$, counting the multiplications at unit cost, and $\log n$ is smooth.

A *problem* is *smooth* if any reasonable algorithm that solves it is smooth. By "reasonable", we mean an algorithm that does not purposely waste time. No problem could be smooth without this restriction to reasonable algorithms because any problem that can be solved at all can be solved by an algorithm that takes a time in $\Omega(2^n)$, which cannot be smooth.

A function $t : \mathbb{N} \rightarrow \mathbb{R}^*$ is *at least quadratic* if $t(n) \in \Omega(n^2)$. It is *strongly* at least quadratic (*strongly quadratic* for short) if it is eventually nondecreasing and if $t(an) \geq a^2 t(n)$ for every positive integer a and every sufficiently large integer n. Finally, it is *supra quadratic* if it is eventually nondecreasing and if there exists an $\varepsilon \in \mathbb{R}^+$ such that $t(an) \geq a^{2+\varepsilon} t(n)$ for every positive integer a and every sufficiently large integer n. At least, strongly and supra *linear* functions are defined similarly. These notions extend to algorithms and problems as in the case of smooth functions.

Problem 10.2.3.

i. Prove that any strongly quadratic function is at least quadratic. (*Hint:* apply Problem 2.1.20.)

ii. Give an explicit example of an eventually nondecreasing function that is at least quadratic but not strongly quadratic.

iii. Show that $n^2 \log n$ is strongly quadratic but not supra quadratic.   □

Most theorems in this section are stated conditionally on a "reasonable" assumption, such as "$A \leq^l B$, assuming B is smooth". This can be interpreted literally as meaning that $A \leq^l B$ under the assumption that B is smooth. From a more practical point of view it also means that the existence of an algorithm for B that works in a time in $O(t(n))$, for any smooth function $t(n)$, implies that there exists an algorithm for A that also works in a time in $O(t(n))$. Moreover, all these theorems are constructive: the algorithm for B follows from the algorithm for A and the proof of the corresponding theorem.

10.2.1 Reductions Among Matrix Problems

An *upper triangular* matrix is a square matrix M whose entries below the diagonal are all zero, that is, $M_{ij} = 0$ when $i > j$. We saw in Section 4.9 that a time in $O(n^{2.81})$ (or even $O(n^{2.376})$) is sufficient to multiply two arbitrary $n \times n$ matrices, contrary to the intuition that may suggest that this problem will inevitably require a time in $\Omega(n^3)$. Is it possible that multiplication of upper triangular matrices could be carried out significantly faster than the multiplication of two arbitrary square matrices? From another point of view, experience might well lead us to believe that inverting non-singular upper triangular matrices should be an operation inherently more difficult than multiplying them.

We denote these three problems, that is, multiplication of arbitrary square matrices, multiplication of upper triangular matrices, and inversion of nonsingular upper triangular matrices, by MQ, MT, and IT, respectively. We shall show under rea-sonable assumptions that $\text{MQ} \equiv^l \text{MT} \equiv^l \text{IT}$. (The problem of inverting an arbitrary nonsingular matrix is also linearly equivalent to the three preceding problems (Problem 10.2.9), but the proof of this is much more difficult, it requires a slightly stronger assumption, and the resulting algorithm is numerically unstable.) Once again this means that any new algorithm that allows us to multiply upper triangular matrices more efficiently will also provide us with a new, more efficient algorithm for inverting arbitrary nonsingular matrices (at least in theory). In particular, it implies that we can invert any nonsingular $n \times n$ matrix in a time in $O(n^{2.376})$.

In what follows we measure the complexity of algorithms that manipulate $n \times n$ matrices in terms of n, referring to an algorithm that runs in a time in $\Theta(n^2)$ as qua-dratic. Formally speaking, this is incorrect because the running time should be given as a function of the *size* of the instance, so that a time in $\Theta(n^2)$ is really linear. No con-fusion should arise from this. Notice that the problems considered are at least qua-dratic in the worst case because any algorithm that solves them must look at each entry of the matrix or matrices concerned.

Theorem 10.2.1. $\text{MT} \leq^l \text{MQ}$.

Proof. Any algorithm that can multiply two arbitrary square matrices can be used directly for multiplying upper triangular matrices.   □

Theorem 10.2.2. $\text{MQ} \leq^l \text{MT}$, assuming MT is smooth.

Proof. Suppose there exists an algorithm that is able to multiply two $n \times n$ upper triangular matrices in a time in $O(t(n))$, where $t(n)$ is a smooth function. Let A and B be two arbitrary $n \times n$ matrices to be multiplied. Consider the following matrix product:

$$\begin{bmatrix} 0 & A & 0 \\ 0 & 0 & 0 \\ 0 & 0 & 0 \end{bmatrix} \times \begin{bmatrix} 0 & 0 & 0 \\ 0 & 0 & B \\ 0 & 0 & 0 \end{bmatrix} = \begin{bmatrix} 0 & 0 & AB \\ 0 & 0 & 0 \\ 0 & 0 & 0 \end{bmatrix}$$

where the "0" are $n \times n$ matrices all of whose entries are zero. This product shows us how to obtain the desired result AB by multiplying two upper triangular $3n \times 3n$ matrices. The time required for this operation is in $O(n^2)$ for the preparation of the two big matrices and the extraction of AB from their product, plus $O(t(3n))$ for the multiplication of the two upper triangular matrices. By the smoothness of $t(n)$, $t(3n) \in O(t(n))$. Because $t(n)$ is at least quadratic, $n^2 \in O(t(n))$. Consequently, the total time required to obtain the product AB is in $O(t(n))$.   □

Theorem 10.2.3.   MQ $\leq^l$ IT, assuming IT is smooth.

Proof. Suppose there exists an algorithm that is able to invert a nonsingular $n \times n$ upper triangular matrix in a time in $O(t(n))$, where $t(n)$ is a smooth function. Let A and B be two arbitrary $n \times n$ matrices to be multiplied. Consider the following matrix product:

$$\begin{bmatrix} I & A & 0 \\ 0 & I & B \\ 0 & 0 & I \end{bmatrix} \times \begin{bmatrix} I & -A & AB \\ 0 & I & -B \\ 0 & 0 & I \end{bmatrix} = \begin{bmatrix} I & 0 & 0 \\ 0 & I & 0 \\ 0 & 0 & I \end{bmatrix}$$

where I is the $n \times n$ identity matrix. This product shows us how to obtain the desired result AB by inverting the first of the $3n \times 3n$ upper triangular matrices. As in the proof of the previous theorem, this operation takes a time in $O(t(n))$.   □

Theorem 10.2.4.   IT $\leq^l$ MQ, assuming MQ is strongly quadratic. (In fact, a weaker but less natural hypothesis suffices: it is enough that MQ be supra linear.)

Proof. Suppose there exists an algorithm that is able to multiply two arbitrary $n \times n$ matrices in a time in $O(t(n))$, where $t(n)$ is strongly quadratic. Let A be a nonsingular $n \times n$ upper triangular matrix to be inverted. Suppose for simplicity that n is a power of 2. (See Problem 10.2.4 for the general case.) If $n = 1$, inversion is trivial. Otherwise decompose A into three submatrices B, C, and D each of size $n/2 \times n/2$ such that

$$A = \begin{bmatrix} B & C \\ 0 & D \end{bmatrix}$$

where B and D are upper triangular and C is arbitrary. By Problem 10.2.5, the matrices B and D are nonsingular. Now consider the following product:

$$\begin{bmatrix} B & C \\ 0 & D \end{bmatrix} \times \begin{bmatrix} B^{-1} & -B^{-1}CD^{-1} \\ 0 & D^{-1} \end{bmatrix} = \begin{bmatrix} I & 0 \\ 0 & I \end{bmatrix} .$$

This product shows us how to obtain A^{-1} by first calculating B^{-1}, and D^{-1}, and then multiplying the matrices B^{-1}, C, and D^{-1}. The upper triangular matrices B and D, which we now have to invert, are smaller than the original matrix A. Using the divide-and-conquer technique suggests a recursive algorithm for inverting A in a time in $O(g(n))$ where $g(n) \in 2g(n/2) + 2t(n/2) + O(n^2)$. The fact that $t(n) \in \Omega(n^2)$ and the assumption that $t(n)$ is eventually nondecreasing (since it is strongly quadratic) yield $g(n) \in 2g(n/2) + O(t(n))$ when n is a power of 2. By Problem 10.2.6, using the assumption that $t(n)$ is strongly quadratic (or at least supra linear), this implies that $g(n) \in O(t(n) \mid n$ is a power of 2). $\qquad\qquad\qquad\square$

Problem 10.2.4.   Let IT2 be the problem of inverting nonsingular upper triangular matrices whose size is a power of 2. All that the proof of theorem 10.2.4 really shows is that IT2 $\leq^l$ MQ. Complete the proof that IT $\leq^l$ MQ. $\qquad\square$

Problem 10.2.5.   Prove that if A is a nonsingular upper triangular matrix whose size is even, and if B and D are defined as in the proof of theorem 10.2.4, then B and D are nonsingular. $\qquad\qquad\qquad\qquad\qquad\qquad\qquad\square$

Problem 10.2.6.   Prove that if $g(n) \in 2g(n/2) + O(t(n))$ when n is a power of 2, and if $t(n)$ is strongly quadratic, then $g(n) \in O(t(n) \mid n$ is a power of 2). (*Hint:* apply Problem 2.3.13(iv); note that it is enough to assume that $t(n)$ is supra linear.) $\quad\square$

Problem 10.2.7.   An upper triangular matrix is *unitary* if all the entries on its diagonal are 1. Denote by SU the problem of squaring a unitary upper triangular matrix. Prove that SU $\equiv^l$ MQ under suitable assumptions. What assumptions do you need? $\qquad\qquad\qquad\qquad\qquad\qquad\qquad\qquad\qquad\qquad\qquad\square$

Problem 10.2.8.   A matrix A is *symmetric* if $A_{ij} = A_{ji}$ for all i and j. Denote by MS the problem of multiplying symmetric matrices. Prove that MS $\equiv^l$ MQ under suitable assumptions. What assumptions do you need? $\qquad\qquad\qquad\square$

**Problem 10.2.9.   Denote by IQ the problem of inverting an arbitrary nonsingular matrix. Assume that both IQ and MQ are smooth and supra quadratic. Prove that IQ $\equiv^l$ MQ. (*Note:* this reduction would *not* go through should an algorithm that is capable of multiplying $n \times n$ matrices in a time in $O(n^2 \log n)$ exist — see Problem 10.2.3(iii).) $\qquad\qquad\qquad\qquad\qquad\qquad\qquad\qquad\qquad\qquad\qquad\square$

10.2.2 Reductions Among Graph Problems

In this section $\mathbb{R}^\infty$ denotes $\mathbb{R}^* \cup \{+\infty\}$, with the natural conventions that $x + (+\infty) = +\infty$ and $\min(x, +\infty) = x$ for all $x \in \mathbb{R}^\infty$.

Let X, Y, and Z be three sets of nodes. Let $f : X \times Y \to \mathbb{R}^\infty$ and $g : Y \times Z \to \mathbb{R}^\infty$ be two functions representing the cost of going directly from one node to another. An infinite cost represents the absence of a direct route. Denote by *fg* the function $h : X \times Z \to \mathbb{R}^\infty$ defined for every $x \in X$ and $z \in Z$ by $h(x,z) = \min \{ f(x,y) + g(y,z) \mid y \in Y \}$. This represents the minimum cost of going from x to z passing through exactly one node in Y. Notice the analogy between this definition and ordinary matrix multiplication (where addition and multiplication are replaced by the minimum operation and addition, respectively), but do not confuse this operation with the composition of functions.

The preceding notation becomes particularly interesting when the sets X, Y, and Z, and also the functions f and g, coincide. In this case *ff*, which we shall write f^2, gives the minimum cost of going from one node of X to another (possibly the same) while passing through exactly one intermediate node (possibly the same, too). Similarly, $\min(f, f^2)$ gives the minimum cost of going from one node of X to another either directly or by passing through exactly one intermediate node. The meaning of f^i is similar for any $i > 0$. By analogy, f^0 represents the cost of going from one node to another while staying in the same place, so that

$$ f^0(x,y) = \begin{cases} 0 & \text{if } x = y \\ +\infty & \text{otherwise.} \end{cases} $$

The minimum cost of going from one node to another without restrictions on the number of nodes on the path, which we write f^*, is therefore $f^* = \min \{ f^i \mid i \geq 0 \}$. This definition is not practical because it apparently implies an infinite computation; it is not even immediately clear that f^* is well defined. However, f never takes negative values. Any path that passes twice through the same node can therefore be shortened by taking out the loop thus formed, without increasing the cost of the resulting path. Consequently, it suffices to consider only those paths whose length is less than the number of nodes in X. Let this number be n. We thus have that $f^* = \min \{ f^i \mid 0 \leq i < n \}$. At first sight, computing f^* for a given function f seems to need more time than calculating a simple product *fg*.

The straightforward algorithm for calculating *fg* takes a time in $\Theta(n^3)$ if the three sets of nodes concerned are of cardinality n. Unfortunately, there is no obvious way of adapting to this problem Strassen's algorithm for ordinary matrix multiplication (Section 4.9). (The intuitive reason is that Strassen's algorithm does subtractions. There is no equivalent to this operation in the present context since taking the minimum is not a reversible operation. Nevertheless, there exist more efficient algorithms for this problem. Because they are quite complicated and have only theoretical advantages, we do not discuss them here.) So the definition of f^* can be taken to give us a direct algorithm for calculating its value in a time in $\Theta(n^4)$.

However, we saw in Section 5.4 a dynamic programming algorithm for calculating shortest paths in a graph, namely Floyd's algorithm. This calculation is nothing other than the calculation of f^*. Thus it *is* possible to get away with a time in $\Theta(n^3)$ after all. Could it be that the problems of calculating *fg* and f^* are of the same complexity? The following two theorems show that this is indeed the case: these two problems are linearly equivalent. The existence of algorithms asymptotically more efficient than $\Theta(n^3)$ for solving the problem of calculating fg therefore implies that Floyd's algorithm for calculating shortest routes is not optimal, at least in theory.

Denote by MUL and TRC the problems consisting of calculating fg and f^*, respectively. As in the previous section, time complexities will be measured as a function of the number of nodes in the graphs concerned. An algorithm such as Dijkstra's, for instance, would be considered quadratic even though it is linear in the number of edges (for dense graphs). Again, the problems considered are at least quadratic in the worst case because any algorithm that solves them must look at each edge concerned.

Theorem 10.2.5. MUL $\leq^l$ TRC, assuming TRC is smooth.

Proof. Suppose there exists an algorithm that is able to calculate h^* in a time in $O(t(n))$, for a smooth function $t(n)$, where n is the cardinality of the set W such that $h : W \times W \to \mathbb{R}^\infty$. Let X, Y, and Z be three sets of nodes of cardinality n_1, n_2 and n_3, respectively, and let $f : X \times Y \to \mathbb{R}^\infty$ and $g : Y \times Z \to \mathbb{R}^\infty$ be two functions for which we wish to calculate fg.

Suppose without loss of generality that X, Y, and Z are disjoint. Let $W = X \cup Y \cup Z$. Define the function $h : W \times W \to \mathbb{R}^\infty$ as follows:

$$h(u,v) = \begin{cases} f(u,v) & \text{if } u \in X \text{ and } v \in Y \\ g(u,v) & \text{if } u \in Y \text{ and } v \in Z \\ +\infty & \text{otherwise} . \end{cases}$$

Notice in particular that $h(x,z) = +\infty$ when $x \in X$ and $z \in Z$.

Now, let us find the value of $h^2(u,v)$. By definition, $h^2(u,v) = \min\{h(u,w) + h(w,v) \mid w \in W\}$. By the definition of h, it is impossible that $h(u,w) \neq +\infty$ and $h(w,v) \neq +\infty$ simultaneously unless $w \in Y$. Consequently, $h^2(u,v) = \min\{h(u,y) + h(y,v) \mid y \in Y\}$. But the only way to have $h(u,y) \neq +\infty$ when $y \in Y$ is to have $u \in X$, and the only way to have $h(y,v) \neq +\infty$ when $y \in Y$ is to have $v \in Z$. If $u \notin X$ or if $v \notin Z$, it therefore follows that $h^2(u,v) = +\infty$. In the case when $u \in X$, $y \in Y$, and $v \in Z$ it suffices to note that $h(u,y) = f(u,y)$ and $h(y,v) = g(y,v)$ to conclude that $h^2(u,v) = \min\{f(u,y) + g(y,v) \mid y \in Y\}$, which is precisely the definition of $fg(u,v)$. Summing up, we have

$$h^2(u,v) = \begin{cases} fg(u,v) & \text{if } u \in X \text{ and } v \in Z \\ +\infty & \text{otherwise} . \end{cases}$$

The calculation of $h^3(u,v)$ is easier. By definition, $h^3(u,v) = hh^2(u,v) = \min\{h(u,w) + h^2(w,v) \mid w \in W\}$. But $h(u,w) = +\infty$ when $w \in X$ whereas $h^2(w,v) = +\infty$ when $w \notin X$. Therefore $h(u,w) + h^2(w,v) = +\infty$ for every $w \in W$, which implies that $h^3(u,v) = +\infty$ for all $u,v \in W$. The same holds for $h^i(u,v)$ for all $i > 3$.

The conclusion from all this is that $h^* = \min(h^0, h, h^2)$ is given by the following equation.

$$h^*(u,v) = \begin{cases} 0 & \text{if } u = v \\ f(u,v) & \text{if } u \in X \text{ and } v \in Y \\ g(u,v) & \text{if } u \in Y \text{ and } v \in Z \\ fg(u,v) & \text{if } u \in X \text{ and } v \in Z \\ +\infty & \text{otherwise} \end{cases}$$

Therefore the restriction of h^* to $X \times Z$ is precisely the product fg we wished to calculate. Let $n = n_1 + n_2 + n_3$ be the cardinality of W. Using the algorithm for calculating h^* thus allows us to compute fg in a time in

$$t(n) + O(n^2) \subseteq O(t(3\max(n_1,n_2,n_3))) + O(n^2) \subseteq O(t(\max(n_1,n_2,n_3)))$$

because $t(n)$ is smooth and at least quadratic.   □

Theorem 10.2.6.   TRC $\leq^l$ MUL, assuming MUL is strongly quadratic. (In fact, a weaker but less natural hypothesis suffices again: it is enough that MUL be supra linear.)

Proof. Suppose there exists an algorithm that is able to calculate fg in a time in $O(t(\max(n_1,n_2,n_3)))$, where n_1, n_2, and n_3 are the cardinalities of the sets X, Y, and Z such that $f : X \times Y \to \mathbb{R}^\infty$ and $g : Y \times Z \to \mathbb{R}^\infty$. Assume $t(n)$ is strongly quadratic. Let H be a set of cardinality n and let $h : H \times H \to \mathbb{R}^\infty$ be a cost function for which we wish to calculate h^*. Suppose for simplicity that n is a power of 2 (see Problem 10.2.11 for the general case). If $n = 1$, it is obvious that $h^*(u,u) = 0$ for the unique $u \in H$. Otherwise split H into two disjoint subsets J and K, each containing half the nodes. Define $a : J \times J \to \mathbb{R}^\infty$, $b : J \times K \to \mathbb{R}^\infty$, $c : K \times K \to \mathbb{R}^\infty$, and $d : K \times J \to \mathbb{R}^\infty$ as the restrictions of h to the corresponding subdomains.

Let $e : J \times J \to \mathbb{R}^\infty$ be the function given by $e = \min(a, bc^*d)$. Notice that $e(u,v)$, for u and v in J, represents the minimum cost for going from u to v without passing through any node in J (not counting the endpoints); however, the nodes in K can be used as often as necessary, or not used at all if this is preferable. The minimum cost for going from u to v with no restrictions on the path is therefore $e^*(u,v)$ when u and v are both in J. In other words, e^* is the restriction to $J \times J$ of the h^* that we wish to calculate. The other restrictions are obtained similarly.

$$h^*(u,v) = \begin{cases} e^*(u,v) & \text{if } u \in J \text{ and } v \in J \\ e^*bc^*(u,v) & \text{if } u \in J \text{ and } v \in K \\ c^*de^*(u,v) & \text{if } u \in K \text{ and } v \in J \\ (\min(c^*, c^*de^*bc^*))(u,v) & \text{if } u \in K \text{ and } v \in K \end{cases}$$

To calculate h^* using divide-and-conquer, we therefore solve recursively two instances of the TRC problem, each of size $n/2$, in order to obtain c^* and e^*. The desired result is then obtained after a number of instances of the MUL problem. As in theorem 10.2.4, the time $g(n)$ required by this approach is characterized by the equation $g(n) \in 2g(n/2) + O(t(n))$, which implies that $g(n) \in O(t(n) \mid n$ is a power of 2).   □

*Problem 10.2.10.   Prove formally that the preceding formula for $h*$ is correct. □

Problem 10.2.11.   Let TRC2 be the problem of calculating $h*$ for $h : X \times X \rightarrow \mathbb{R}^\infty$ when the cardinality of X is a power of 2. All the proof of theorem 10.2.6 really shows is that TRC2 $\leq^l$ MUL. Complete the proof that TRC $\leq^l$ MUL.   □

When the range of the cost functions is restricted to $\{0, +\infty\}$, calculating $f*$ comes down to determining for each pair of nodes whether or not there is a path joining them, regardless of the cost of the path. We saw that Warshall's algorithm (Problem 5.4.2) solves this problem in a time in $\Theta(n^3)$. Let MULB and TRCB be the problems consisting of calculating fg and $h*$, respectively, when the cost functions are restricted in this way. It is clear that MULB $\leq^l$ MUL and TRCB $\leq^l$ TRC since the general algorithms can also be used to solve instances of the restricted problems. Furthermore, the proof that MUL $\equiv^l$ TRC can easily be adapted to show that MULB $\equiv^l$ TRCB. This is interesting because MULB $\leq^l$ MQ, where MQ is the problem of multiplying arbitrary arithmetic square matrices (Problem 10.2.12). Unlike the case of arbitrary cost functions, Strassen's algorithm can therefore be used to solve the problems MULB and TRCB in a time in $O(n^{2.81})$, thus showing that Warshall's algorithm is not optimal. Note, however, that using Strassen's algorithm requires a number of *arithmetic* operations in $O(n^{2.81})$; the time in $O(n^3)$ taken by Warshall's algorithm counts only *Boolean* operations as elementary. No algorithm is known that can solve MULB faster than MQ.

Problem 10.2.12.   Let $f : X \times Y \rightarrow \{0, +\infty\}$ and $g : Y \times Z \rightarrow \{0, +\infty\}$ be two restricted cost functions. Assuming we count arithmetic operations at unit cost, show how to transform the problem of calculating fg into the computation of an ordinary arithmetic matrix multiplication. Conclude that MULB $\leq^l$ MQ. Show that the arithmetic can be done modulo p, where p is any prime number larger than n.   □

*Problem 10.2.13.   A cost function $f : X \times X \rightarrow \mathbb{R}^\infty$ is *symmetric* if $f(u, v) = f(v, u)$ for every $u, v \in X$. Each of the four problems discussed earlier has a symmetric version that arises when the cost functions involved are symmetric. Call these four problems MULS, TRCS, MULBS, and TRCBS. Prove that MULBS $\equiv^l$ MULB. Do you believe that MULBS $\equiv^l$ TRCBS? If not, does one of these two problems appear strictly more difficult than the other? Which one? Justify your answer.   □

10.2.3 Reductions Among Arithmetic and Polynomial Problems

We return to the problems posed by the arithmetic of large integers (sections 1.7.2, 4.7, and 9.5). We saw that it is possible to multiply two integers of size n in a time in $O(n^{1.59})$ and even in $O(n \log n \log\log n)$. What can we say about integer division and taking square roots? Our everyday experience leads us to believe that the second of these problems, and probably the first one, too, is genuinely more difficult than multiplication. Once again this turns out not to be true. Let SQR, MLT, and DIV be the problems consisting of squaring an integer of size n, of multiplying two integers of size n, and of determining the quotient when an integer of size $2n$ is divided by an integer of size n, respectively. Clearly, these problems are at least linear because any algorithm that solves them must take into account every bit of the operands involved. (For simplicity we measure the size of integers in bits. As mentioned in Section 1.7.2, however, this choice is not critical: the time taken by the various algorithms would be in the same order if given as a function of the size of their operands in decimal digits or computer words. This is the case *precisely* because we assume all these algorithms to be smooth.)

Theorem 10.2.7.   SQR $\equiv^l$ MLT $\equiv^l$ DIV, assuming these three problems are smooth and MLT is strongly linear (weaker but more complicated assumptions would suffice).

Proof outline.   The full proof of this theorem is long and technical. Its conceptual beauty is also defaced in places by the necessity of using an inordinate number of ad hoc tricks to circumvent the problems caused by integer truncation (see Problem 10.2.22). For this reason we content ourselves in the rest of this section with showing the equivalence of these operations in the "cleaner" domain of polynomial arithmetic. Nonetheless, we take a moment to prove that SQR $\equiv^l$ MLT, assuming SQR is smooth (a weaker assumption would do).

Clearly, SQR $\leq^l$ MLT, since squaring is only a special case of multiplication. To show that MLT $\leq^l$ SQR, suppose there exists an algorithm that is able to square an integer of size n in a time in $O(t(n))$, where $t(n)$ is smooth (it is enough to assume that $t(n+1) \in O(t(n))$). Let x and y be two integers of size n to be multiplied. Assume without loss of generality that $x \geq y$. The following formula enables us to obtain their product by carrying out two squaring operations of integers of size at most $n+1$, a few additions, and a division by 4:

$$xy = ((x+y)^2 - (x-y)^2)/4.$$

Since the additions and the division by 4 can be carried out in a time in $O(n)$, we can solve MLT in a time in $2t(n+1) + O(n) \subseteq O(t(n))$ because $t(n)$ is smooth and $t(n) \in \Omega(n)$.   □

We have seen in Section 9.4 how to multiply two polynomials of degree n in a time in $O(n \log n)$, provided that the necessary arithmetic operations on the coefficients can be counted at unit cost. We now show that the problem of polynomial division is linearly equivalent to that of multiplication. Notice that a direct approach using discrete Fourier transforms, which works so well for multiplying two polynomials, is inapplicable in the case of division unless the two polynomials concerned divide one another exactly with no remainder. For example, let $p(x) = x^3 + 3x^2 + x + 2$ and $d(x) = x^2 + x + 2$. The quotient of the division of $p(x)$ by $d(x)$ is $q(x) = x + 2$. In this case $p(2) = 24$ and $d(2) = 8$, but $q(2) = 4 \neq 24/8$. We even have that $p(1) = 7$ is not divisible by $d(1) = 4$. This is all due to the remainder of the division $r(x) = -3x - 2$. Despite this difficulty, it is possible to determine the quotient and the remainder produced when a polynomial of degree $2n$ is divided by a polynomial of degree n in a time in $O(n \log n)$ by reducing these problems to a certain number of polynomial multiplications calculated using the Fourier transform.

Recall that $p(x) = \sum_{i=0}^{n} a_i x^i$ is a polynomial of *degree* n provided that $a_n \neq 0$. By convention the polynomial $p(x) = 0$ is of degree -1. Let $p(x)$ be a polynomial of degree n, and let $d(x)$ be a nonzero polynomial of degree m. Then there exists a unique polynomial $r(x)$ of degree strictly less than m and a unique polynomial $q(x)$ such that $p(x) = q(x)d(x) + r(x)$. The polynomial $q(x)$ is of degree $n - m$ if $n \geq m - 1$; otherwise $q(x) = 0$. We call $q(x)$ and $r(x)$, respectively, the *quotient* and the *remainder* of the division of $p(x)$ by $d(x)$. By analogy with the integers, the quotient is denoted by $q(x) = \lfloor p(x)/d(x) \rfloor$.

Problem 10.2.14.   Prove the existence and the uniqueness of the quotient and the remainder. Show that if both $p(x)$ and $d(x)$ are monic polynomials (the coefficient of highest degree is 1) with integer coefficients then both $q(x)$ and $r(x)$ have integer coefficients and $q(x)$ is monic (unless $q(x) = 0$).   □

We also need the notion of an inverse. Let $p(x)$ be a nonzero polynomial of degree n. The *inverse* of $p(x)$, which we denote $p^*(x)$, is defined by $p^*(x) = \lfloor x^{2n}/p(x) \rfloor$. For example, if $p(x) = x^3 + 3x^2 + x + 2$, then $p^*(x) = x^3 - 3x^2 + 8x - 23$. Notice that $p(x)$ and $p^*(x)$ are always of the same degree.

Problem 10.2.15.   Let $p(x) = x^3 + x^2 + 5x + 1$ and $d(x) = x - 2$. Calculate $\lfloor p(x)/d(x) \rfloor$, $p^*(x)$, and $d^*(x)$.   □

Problem 10.2.16.   Prove that if $p(x)$ is a nonzero polynomial and if $q(x) = p^*(x)$ then $q^*(x) = p(x)$. (There is a very simple proof.)   □

*Problem 10.2.17.   Prove that if $p(x)$, $p_1(x)$ and $p_2(x)$ are three arbitrary polynomials, and if $d(x)$, $d_1(x)$ and $d_2(x)$ are three nonzero polynomials, then

i. $\left\lfloor \dfrac{p_1(x)}{d_1(x)} \right\rfloor \pm \left\lfloor \dfrac{p_2(x)}{d_2(x)} \right\rfloor = \left\lfloor \dfrac{p_1(x)d_2(x) \pm p_2(x)d_1(x)}{d_1(x)d_2(x)} \right\rfloor$;

in particular, $\lfloor p_1(x)/d(x) \rfloor \pm \lfloor p_2(x)/d(x) \rfloor = \lfloor (p_1(x) \pm p_2(x))/d(x) \rfloor$;

ii. $\lfloor \lfloor p(x)/d_1(x) \rfloor /d_2(x) \rfloor = \lfloor p(x)/(d_1(x)d_2(x)) \rfloor$; and

iii. $\lfloor p_1(x)/\lfloor p_2(x)/d(x) \rfloor \rfloor = \lfloor (p_1(x)d(x))/p_2(x) \rfloor$, provided that the degree of $p_1(x)$ is not more than twice the degree of $\lfloor p_2(x)/d(x) \rfloor$.   □

Consider the four following problems: SQRP consists of squaring a polynomial of degree n, MLTP of multiplying two polynomials of degree at most n, INVP of determining the inverse of a polynomial of degree n and DIVP of calculating the quotient of the division of a polynomial of degree at most $2n$ by a polynomial of degree n. We now prove under suitable assumptions that these four problems are linearly equivalent using the following chain of reductions: MLTP $\leq^l$ SQRP $\leq^l$ INVP $\leq^l$ MLTP and INVP $\leq^l$ DIVP $\leq^l$ INVP. Again, all these problems are at least linear. We assume arithmetic operations can be carried out at unit cost on the coefficients.

Theorem 10.2.8.   MLTP $\leq^l$ SQRP, assuming SQRP is eventually nondecreasing.

Proof.   Essentially the same as the proof that MLT $\leq^l$ SQR given in theorem 10.2.7. There is no need for a smoothness assumption this time because the sum or difference of two polynomials of degree n cannot exceed degree n.   □

Theorem 10.2.9.   SQRP $\leq^l$ INVP, assuming INVP is smooth.

Proof.   The intuitive idea is given by the following formula, where x is a nonzero real number:

$$x^2 = (x^{-1} - (x+1)^{-1})^{-1} - x \ .$$

A direct attempt to calculate the square of a polynomial $p(x)$ using the analogous formula $(p^*(x) - (p(x)+1)^*)^* - p(x)$ has no chance of working: the degree of this expression cannot be greater than the degree of $p(x)$. This failure is caused by truncation errors, which we can, nevertheless, eliminate using an appropriate scaling factor.

Suppose there exists an algorithm that is able to calculate the inverse of a polynomial of degree n in a time in $O(t(n))$, where $t(n)$ is a smooth function. Let $p(x)$ be a polynomial of degree $n \geq 1$ whose square we wish to calculate. The polynomial $x^{2n} p(x)$ is of degree $3n$, so

$$[x^{2n} p(x)]^* = \lfloor x^{6n} / x^{2n} p(x) \rfloor = \lfloor x^{4n} / p(x) \rfloor \ .$$

Similarly

$$[x^{2n} (p(x)+1)]^* = \lfloor x^{4n} / (p(x)+1) \rfloor \ .$$

By Problem 10.2.17

$$[x^{2n} p(x)]^* - [x^{2n} (p(x)+1)]^* = \lfloor x^{4n} / p(x) \rfloor - \lfloor x^{4n} / (p(x)+1) \rfloor$$

$$= \lfloor (x^{4n} (p(x)+1) - x^{4n} p(x)) / (p(x)(p(x)+1)) \rfloor$$

$$= \lfloor x^{4n} / (p^2(x) + p(x)) \rfloor$$

$$= [p^2(x) + p(x)]^* \ .$$

The last equality follows from the fact that $p^2(x) + p(x)$ is of degree $2n$. By Problem 10.2.16, we conclude finally that

$$p^2(x) = [[x^{2n} p(x)]^* - [x^{2n} (p(x)+1)]^*]^* - p(x) \ .$$

This gives us an algorithm for calculating $p^2(x)$ by performing two inversions of polynomials of degree $3n$, one inversion of a polynomial of degree $2n$, and a few operations (additions, subtractions, multiplications by powers of x) that take a time in $O(n)$. This algorithm can therefore solve SQRP in a time in

$$2t(3n)+t(2n)+O(n) \subseteq O(t(n))$$

because $t(n)$ is smooth and at least linear.   □

Theorem 10.2.10.   INVP $\le^l$ DIVP.

Proof.   To calculate $p^*(x)$, where $p(x)$ is a polynomial of degree n, we evaluate $\lfloor x^{2n}/p(x) \rfloor$, an instance of size n of the problem of polynomial division.   □

Theorem 10.2.11.   DIVP $\le^l$ INVP, assuming INVP is smooth.

Proof.   The intuitive idea is given by the following formula, where x and y are real numbers and $y \neq 0$:

$$x/y = xy^{-1} \ .$$

If we try to calculate the quotient of a polynomial $p(x)$ divided by a nonzero polynomial $d(x)$ using directly the analogous formula $p(x)d^*(x)$, the degree of the result is too high. To solve this problem we divide the result by an appropriate scaling factor.

Suppose there exists an algorithm that is able to calculate the inverse of a polynomial of degree n in a time in $O(t(n))$, where $t(n)$ is a smooth function. Let $p(x)$ be a polynomial of degree less than or equal to $2n$, and let $d(x)$ be a polynomial of degree n. We wish to calculate $\lfloor p(x)/d(x) \rfloor$. Let $r(x)$ be the remainder of the division of x^{2n} by $d(x)$, which is to say that $d^*(x) = \lfloor x^{2n}/d(x) \rfloor = (x^{2n}-r(x))/d(x)$ and that the degree of $r(x)$ is strictly less than n. Now consider

$$\left\lfloor \frac{d^*(x)p(x)}{x^{2n}} \right\rfloor = \left\lfloor \frac{x^{2n}p(x)-r(x)p(x)}{x^{2n}d(x)} \right\rfloor$$

$$= \left\lfloor \frac{x^{2n}p(x)}{x^{2n}d(x)} \right\rfloor - \left\lfloor \frac{r(x)p(x)}{x^{2n}d(x)} \right\rfloor$$

by Problem 10.2.17(i). But the degree of $r(x)p(x)$ is strictly less than $3n$, whereas the degree of $x^{2n}d(x)$ is equal to $3n$, and so $\lfloor (r(x)p(x))/(x^{2n}d(x)) \rfloor = 0$. Consequently, $\lfloor (d^*(x)p(x))/x^{2n} \rfloor = \lfloor p(x)/d(x) \rfloor$, which allows us to obtain the desired quotient by performing an inversion of a polynomial of degree n, the multiplication of two polynomials of degree at most $2n$, and the calculation of the quotient from a division by a power of x. This last operation corresponds to a simple shift and can be carried out in a time in $O(n)$. The multiplication can be performed in a time in $O(t(2n))$ thanks to theorems 10.2.8 and 10.2.9 (using the assumption that $t(n)$ is smooth).

The calculation of $\lfloor p(x)/d(x) \rfloor$ can therefore be carried out in a time in $t(n) + O(t(2n)) + O(n) \subseteq O(t(n))$ because $t(n)$ is smooth and at least linear. $\square$

Theorem 10.2.12.   INVP $\leq^l$ MLTP, assuming MLTP is strongly linear (a weaker assumption will do — see the hint of Problem 10.2.19).

Proof. This reduction is more difficult than the previous ones. Once again, we appeal to an analogy with the domain of real numbers, namely, Newton's method for finding the zero of $f(w) = 1 - xw$. Let x be a positive real number for which we want to calculate x^{-1}. Let y be an approximation to x^{-1} in the sense that $xy = 1 - \delta$, for $-1 < \delta < 1$. We can improve the approximation y by calculating $z = 2y - y^2 x$. Indeed, $xz = x(2y - y^2 x) = xy(2 - xy) = (1 - \delta)(1 + \delta) = 1 - \delta^2$. From our assumption on δ, δ^2 is smaller than δ in absolute value, so that z is closer than y to x^{-1}. To calculate the inverse of a polynomial, we proceed similarly, first finding a good approximation to this inverse and then correcting the error.

Suppose there exists an algorithm able to multiply two polynomials of degrees less than or equal to n in a time in $O(t(n))$ where $t(n)$ is strongly linear. Let $p(x)$ be a nonzero polynomial of degree n whose inverse we wish to calculate. Suppose for simplicity that $n + 1$ is a power of 2 (see Problem 10.2.18 for the general case). If $n = 0$, the inverse $p^*(x)$ is easy to calculate. Otherwise let $k = (n + 1)/2$.

During the first stage of the polynomial inversion algorithm we find an approximation $h(x)$ to $p^*(x)$ such that the degree of $x^{2n} - p(x)h(x)$ is less than $3k - 1$. (Note that the degree of $x^{2n} - p(x)p^*(x)$ can be as high as $n - 1 = 2k - 2$.) The idea is to rid ourselves provisionally of the k coefficients of lowest degree in the polynomial $p(x)$ by dividing the latter by x^k. Let $h(x) = x^k \lfloor p(x)/x^k \rfloor^*$. Note first that the degree of $\lfloor p(x)/x^k \rfloor$ is $n - k = k - 1$, so $\lfloor p(x)/x^k \rfloor^* = \lfloor x^{2k-2}/\lfloor p(x)/x^k \rfloor \rfloor = \lfloor x^{3k-2}/p(x) \rfloor$ by Problem 10.2.17(iii). Let $r(x)$ be the polynomial of degree less than n such that $\lfloor x^{3k-2}/p(x) \rfloor = (x^{3k-2} - r(x))/p(x)$. Then we have

$$x^{2n} - p(x)h(x) = x^{4k-2} - p(x)x^k (x^{3k-2} - r(x))/p(x) = x^k r(x) ,$$

which is indeed a polynomial of degree less than $3k - 1$.

During the second stage, we improve the approximation $h(x)$ in order to obtain $p^*(x)$ exactly. Taking into account the appropriate scaling factor, the analogy introduced at the beginning of this proof suggests that we should calculate $q(x) = 2h(x) - \lfloor h^2(x)p(x)/x^{2n} \rfloor$. Let $s(x)$ be the polynomial of degree less that $2n$ such that $\lfloor h^2(x)p(x)/x^{2n} \rfloor = (h^2(x)p(x) - s(x))/x^{2n}$. Now calculate

$$p(x)q(x) = 2p(x)h(x) - (p^2(x)h^2(x) - p(x)s(x))/x^{2n}$$

$$= [(p(x)h(x))(2x^{2n} - p(x)h(x)) + p(x)s(x)]/x^{2n}$$

$$= [(x^{2n} - x^k r(x))(x^{2n} + x^k r(x)) + p(x)s(x)]/x^{2n}$$

$$= [x^{4n} - x^{2k} r^2(x) + p(x)s(x)]/x^{2n}$$

$$= x^{2n} + (p(x)s(x) - x^{2k} r^2(x))/x^{2n} .$$

It remains to remark that the polynomials $p(x)s(x)$ and $x^{2k}r^2(x)$ are of degree at most $3n-1$ to conclude that the degree of $x^{2n}-p(x)q(x)$ is less than n, hence $q(x)=p^*(x)$, which is what we set out to calculate.

Combining these two stages, we obtain the following recursive formula:

$$p^*(x) = 2x^k \lfloor p(x)/x^k \rfloor^* - \lfloor p(x)[\lfloor p(x)/x^k \rfloor^*]^2/x^{n-1} \rfloor \ .$$

Let $g(n)$ be the time taken to calculate the inverse of a polynomial of degree n by the divide-and-conquer algorithm suggested by this formula. Taking into account the recursive evaluation of the inverse of $\lfloor p(x)/x^k \rfloor$, the two polynomial multiplications that allow us to improve our approximation, the subtractions, and the multiplications and divisions by powers of x, we see that

$$g(n) \in g((n-1)/2)+t((n-1)/2)+t(n)+O(n) \subseteq g((n-1)/2)+O(t(n))$$

because $t(n)$ is strongly linear. Using Problem 10.2.19, we conclude that $g(n) \in O(t(n))$.   □

Problem 10.2.18.   Let INVP2 be the problem of calculating $p^*(x)$ when $p(x)$ is a polynomial of degree n such that $n+1$ is a power of 2. All that the proof of theorem 10.2.12 really shows is that INVP2 $\leq^l$ MLTP. Complete the proof that INVP $\leq^l$ MLTP.   □

Problem 10.2.19.   Prove that if $g(n) \in g((n-1)/2)+O(t(n))$ when $n+1$ is a power of 2, and if $t(n)$ is strongly linear, then $g(n) \in O(t(n) \mid n+1$ is a power of 2). (*Hint*: apply Problem 2.3.13(iv) with the change of variable $T(n)=g(n-1)$; note that it is enough to assume the existence of a real constant $\alpha>1$ such that $t(2n) \geq \alpha t(n)$ for all sufficiently large n — strong linearity of $t(n)$ would unnecessarily impose $\alpha=2$.)   □

Problem 10.2.20.   Let $p(x)=x^3+x^2+5x+1$. Calculate $p^*(x)$ using the approach described in the proof of theorem 10.2.12. You may carry out directly the intermediate calculation of the inverse of $\lfloor p(x)/x^2 \rfloor$ rather than doing so recursively. Compare your answer to the one obtained as a solution to Problem 10.2.15.   □

*Problem 10.2.21.   We saw in Section 9.4 how Fourier transforms can be used to perform the multiplication of two polynomials of degrees not greater than n in a time in $O(n \log n)$. Theorems 10.2.11 and 10.2.12 allow us to conclude that this time is also sufficient to determine the quotient obtained when a polynomial of degree at most $2n$ is divided by a polynomial of degree n. However, the proof of theorem 10.2.11 depends crucially on the fact that the degree of the dividend is not more than double the degree of the divisor. Generalize this result by showing how we can divide a polynomial of degree m by a polynomial of degree n in a time in $O(m \log n)$.   □

**Problem 10.2.22.   Following the general style of theorems 10.2.9 to 10.2.12, complete the proof of theorem 10.2.7. You will have to define the notion of inverse for an integer: if i is an n-bit integer (that is, $2^{n-1} \le i \le 2^n - 1$), define $i^* = \lfloor 2^{2n-1}/i \rfloor$. Notice that i^* is also an n-bit integer, unless i is a power of 2. The problem INV is defined on the integers in the same way as INVP on the polynomials.

The difficulties start with the fact that $(i^*)^*$ is not always equal to i, contrary to Problem 10.2.16. (For example, $13^* = 9$ but $9^* = 14$.) This hinders all the proofs. For example, consider how we prove that DIV $\le^l$ INV. Let i be an integer of size $2n$ and let j be an integer of size n; we want to calculate $\lfloor i/j \rfloor$. If we define $z = \lfloor i j^*/2^{2n-1} \rfloor$ by analogy with the calculation of $\lfloor (p(x)d^*(x))/x^{2n} \rfloor$ in the proof of theorem 10.2.11, we no longer obtain automatically the desired result $z = \lfloor i/j \rfloor$. Detailed analysis shows, however, that $z \le \lfloor i/j \rfloor \le z + 2$. The exact value of $\lfloor i/j \rfloor$ can therefore be obtained by a correction loop that goes around at most three times.

$$z \leftarrow \lfloor i j^*/2^{2n-1} \rfloor$$
$$t \leftarrow (z+1) \times j$$
while $t \le i$ **do**
$\qquad t \leftarrow t + j$
$\qquad z \leftarrow z + 1$
return z

The other proofs have to be adapted similarly.   □

** Problem 10.2.23.   Let SQRT be the problem of computing the largest integer less than or equal to the square root of a given integer of size n. Prove under suitable assumptions that SQRT $\equiv^l$ MLT. What assumptions do you need? (*Hint*: for the reduction SQRT $\le^l$ MLT, follow the general lines of theorem 10.2.12 but use Newton's method to find the positive zero of $f(w) = w^2 - x$; for the inverse reduction, use the fact that

$$\sqrt{x + \sqrt{x+1} - \sqrt{x-1}} - \sqrt{x - \sqrt{x+1} + \sqrt{x-1}} \approx 1/x.)$$ □

Problem 10.2.24.   Let MOD be the problem of computing the remainder when an integer of size $2n$ is divided by an integer of size n. Prove that MOD $\le^l$ MLT.   □

** Problem 10.2.25.   Let GCD be the problem of computing the greatest common divisor of two integers of size at most n. Prove or disprove GCD $\equiv^l$ MLT. (*Warning*: at the time of writing, this is an open problem.)   □

10.3 INTRODUCTION TO NP-COMPLETENESS

There exist many real-life, practical problems for which no efficient algorithm is known, but whose intrinsic difficulty no one has yet managed to prove. Among these are such different problems as the travelling salesperson (Sections 3.4.2, 5.6, and 6.6.3), optimal graph colouring (Section 3.4.1), the knapsack problem, Hamiltonian circuits (Example 10.3.2), integer programming, finding the longest simple path in a graph (Problem 5.1.3), and the problem of satisfying a Boolean expression. (Some of these problems are described later.) Should we blame algorithmics or complexity? Maybe there do in fact exist efficient algorithms for these problems. After all, computer science is a relative newcomer: it is certain that new algorithmic techniques remain to be discovered.

This section presents a remarkable result: an efficient algorithm to solve any one of the problems we have listed in the previous paragraph would automatically provide us with efficient algorithms for all of them. We do not know whether these problems are easy or hard to solve, but we do know that they are all of similar complexity. The practical importance of these problems ensured that each of them separately has been the object of sustained efforts to find an efficient method of solution. For this reason it is widely conjectured that such algorithms do not exist. If you have a problem to solve and you are able to show that it is equivalent (see Definition 10.3.1) to one of those mentioned previously, you may take this result as convincing evidence that your problem is hard (but evidence is not a proof). At the very least you will be certain that nobody else claims to be able to solve your problem efficiently at the moment.

10.3.1 The Classes P and NP

Before going further it will help to define what we mean by an efficient algorithm. Does this mean it takes a time in $O(n \log n)$? $O(n^2)$? $O(n^{2.81})$? It all depends on the problem to be solved. A sorting algorithm taking a time in $\Theta(n^2)$ is inefficient, whereas an algorithm for matrix multiplication taking a time in $O(n^2 \log n)$ would be an astonishing breakthrough. So we might be tempted to say that an algorithm is efficient if it is better than the obvious straightforward algorithm, or maybe if it is the best possible algorithm to solve our problem. But then what should we say about the dynamic programming algorithm for the travelling salesperson problem (Section 5.6) or the branch-and-bound algorithm (Section 6.6.3)? Although more efficient than an exhaustive search, in practice these algorithms are only good enough to solve instances of moderate size. If there exists no significantly more efficient algorithm to solve this problem, might it not be reasonable to decide that the problem is inherently intractable?

For our present purposes we answer this question by stipulating that an algorithm is *efficient* (or *polynomial-time*) if there exists a polynomial $p(n)$ such that the algorithm can solve any instance of size n in a time in $O(p(n))$. This definition is motivated by the comparison in Section 1.6 between an algorithm that takes a time in $\Theta(2^n)$ and one that only requires a time in $O(n^3)$, and also by sections 1.7.3, 1.7.4, and 1.7.5. An exponential-time algorithm becomes rapidly useless in practice, whereas generally speaking a polynomial-time algorithm allows us to solve much larger instances. The definition should, nevertheless, be taken with a grain of salt. Given two algorithms requiring a time in $\Theta(n^{\lg \lg n})$ and in $\Theta(n^{10})$, respectively, the first, not being polynomial, is "inefficient". However, it will beat the polynomial algorithm on all instances of size less than 10^{300}, assuming that the hidden constants are similar. In fact, it is not reasonable to assert that an algorithm requiring a time in $\Theta(n^{10})$ is efficient in practice. Nonetheless, to decree that $O(n^3)$ is efficient whereas $\Omega(n^4)$ is not, for example, seems rather too arbitrary.

In this section, it is crucial to avoid pathological algorithms and analyses such as those suggested in Problem 1.5.1. Hence no algorithm is allowed to perform arithmetic operations at unit cost on operands whose size exceeds some fixed polynomial in the size of the instance being solved. (The polynomial may depend on the algorithm but not of course on the instance.) If the algorithm needs larger operands (as would be the case in the solution of Problem 1.5.1), it must break them into sections, keep them in an array, and spend the required time to carry out multiprecision arithmetic. Without loss of generality, we also restrict all arrays to contain a number of elements at most polynomial in the size of the instance considered.

The notion of linear reduction and of linear equivalence considered in Section 10.2 is interesting for problems that can be solved in quadratic or cubic time. It is, however, too restrictive when we consider problems for which the best-known algorithms take exponential time. For this reason we introduce a different kind of reduction.

Definition 10.3.1.   Let X and Y be two problems. Problem X is *polynomially reducible* to problem Y *in the sense of Turing*, denoted $X \leq_T^p Y$, if there exists an algorithm for solving X in a time that would be polynomial if we took no account of the time needed to solve arbitrary instances of problem Y. In other words, the algorithm for solving problem X may make whatever use it chooses of an imaginary procedure that can somehow magically solve problem Y at no cost. When $X \leq_T^p Y$ and $Y \leq_T^p X$ simultaneously, then X and Y are *polynomially equivalent in the sense of Turing*, denoted $X \equiv_T^p Y$. (This notion applies in a natural way to unsolvable problems — see Problem 10.3.32.) $\square$

Example 10.3.1.   Let SMALLFACT(n) be the problem of finding the smallest integer $x \geq 2$ such that x divides n (for $n \geq 2$), let PRIME(n) be the problem of determining whether $n \geq 2$ is a prime number, and let NBFACT(n) be the problem of counting the number of distinct primes that divide n. Then both PRIME $\leq_T^p$ SMALLFACT and NBFACT $\leq_T^p$ SMALLFACT. Indeed, imagine solutions to the problem SMALLFACT can be obtained at no cost by a call on *SolveSF*; then the following procedures solve the other two problems in a time polynomial in the size of their operand.

function *DecidePRIME* (n)
    { we assume $n \geq 2$ }
    if $n = SolveSF(n)$ **then return** *true*
                         else return *false*

function *SolveNBF* (n)
    $nb \leftarrow 0$
    while $n > 1$ **do** $nb \leftarrow nb + 1$
                    $x \leftarrow SolveSF(n)$
                    while x divides n **do** $n \leftarrow n/x$
    return nb

Notice that *SolveNBF* works in polynomial time (counting calls of *SolveSF* at no cost) because no integer n can have more than $\lfloor \lg n \rfloor$ prime factors, even taking repetitions into account.   □

The usefulness of this definition is brought out by the two following exercises.

Problem 10.3.1.   Let X and Y be two problems such that $X \leq_T^p Y$. Suppose there exists an algorithm that is able to solve problem Y in a time in $O(t(n))$, where $t(n)$ is a nonzero, nondecreasing function. Prove that there exist a polynomial $p(n)$ and an algorithm that is able to solve problem X in a time in $O(p(n)t(p(n)))$.   □

Problem 10.3.2.   Let X and Y be two problems such that $X \leq_T^p Y$. Prove that the existence of an algorithm to solve problem Y in polynomial time implies that there also exists a polynomial-time algorithm to solve problem X.   □

In particular, the equivalence mentioned in the introduction to this section implies that either all the problems listed there can be solved in polynomial time, or none of them can.

For technical reasons we confine ourselves from now on to the study of decision problems. For example, "Is n a prime number?" is a decision problem, whereas "find the smallest prime factor of n" is not. A decision problem can be thought of as defining a subset X of the set I of all its instances. Then the problem consists of deciding, given some $x \in I$, whether or not $x \in X$. We generally assume that the set of all instances is easy to recognize, such as $\mathbb{N}$, or "the set of all possible graphs". We also assume that the instances can be coded efficiently in the form of strings of bits. When no confusion can arise, we may sometimes omit to state explicitly the set of instances for the decision problem under consideration.

Definition 10.3.2.   **P** is the class of decision problems that can be solved by a polynomial-time algorithm.   □

Problem 10.3.3.   Let X and Y be two decision problems. Prove that if $X \leq_T^p Y$ and $Y \in \mathbf{P}$, then $X \in \mathbf{P}$.   □

The restriction to decision problems allows us to introduce a simplified notion of polynomial reduction.

Definition 10.3.3.   Let $X \subseteq I$ and $Y \subseteq J$ be two decision problems. Problem X is *many-one polynomially reducible* to problem Y, denoted by $X \leq_m^p Y$, if there exists a function $f : I \rightarrow J$ computable in polynomial time, known as the *reduction function* between X and Y, such that

$$(\forall x \in I)[x \in X \Leftrightarrow f(x) \in Y] .$$

When $X \leq_m^p Y$ and $Y \leq_m^p X$ both hold, then X and Y are *many-one polynomially equivalent*, denoted $X \equiv_m^p Y$.   □

Example 10.3.2.   Let TSPD and HAM be the travelling salesperson decision problem and the Hamiltonian circuit problem, respectively. An instance of TSPD consists of a directed graph with costs on the edges, together with some bound L used to turn the travelling salesperson optimization problem (as in Sections 5.6 and 6.6.3) into a decision problem: the question is to decide whether there exists a tour in the graph that begins and ends at some node, after having visited each of the other nodes exactly once, and whose cost does not exceed L. An instance of HAM is a directed graph, and the question is to decide whether there exists a circuit in the graph passing exactly once through each node n (with no optimality constraint).

To prove that HAM $\leq_m^p$ TSPD, let $G = <N, A>$ be a directed graph for which you would like to decide if it has a Hamiltonian circuit. Define $f(G)$ as the instance for TSPD consisting of the complete graph $H = <N, N \times N>$, the cost function

$$c(u,v) = \begin{cases} 1 & \text{if } (u,v) \in A \\ 2 & \text{otherwise} \end{cases}$$

and the bound $L = \#N$, the number of nodes in G. Clearly, $G \in$ HAM if and only if $<H, c, L> \in$ TSPD. It is also the case that TSPD $\leq_m^p$ HAM, but this is significantly harder to prove.   □

Lemma 10.3.1.   If X and Y are two decision problems such that $X \leq_m^p Y$, then $X \leq_T^p Y$.

Proof. Imagine solutions to problem Y can be obtained at no cost by a call on *DecideY* and let f be the polynomial-time computable reduction function between X and Y. Then the following procedure solves X in polynomial time.

```
function DecideX (x)
    y ← f (x)
    if DecideY (y) then return true
                   else return false
```
                                                                                  □

* Problem 10.3.4.   Prove that the converse of Lemma 10.3.1 does not necessarily hold by giving explicitly two decision problems X and Y for which you can prove that $X \leq_T^p Y$ whereas it is not the case that $X \leq_m^p Y$.   □

Problem 10.3.5.   Prove that the relations $\leq_T^p$, $\leq_m^p$, $\equiv_T^p$, and $\equiv_m^p$ are transitive.   □

The introduction of TSPD in Example 10.3.2 shows that the restriction to decision problems is not a severe constraint. In fact, most optimization problems are polynomially equivalent in the sense of Turing to an analogous decision problem, as the following exercise illustrates.

*Problem 10.3.6.   Let $G = <N, A>$ be an undirected graph, let k be an integer, and $c: N \to \{1, 2, \ldots, k\}$ a function. This function is a *valid colouring* of G if there do not exist nodes $u, v \in N$ such that $\{u, v\} \in A$ and $c(u) = c(v)$ (Section 3.4.1). The graph G can be *coloured with k colours* if there exists such a valid colouring. The smallest integer k such that G can be coloured with k colours is called the *chromatic number* of G, and in this case a colouring with k colours is called an *optimal colouring*. Consider the three following problems.

COLD: Given a graph G and an integer k, can G be coloured with k colours?

COLO: Given a graph G, find the chromatic number of G.

COLC: Given a graph G, find an optimal colouring of G.

Prove that COLD $\equiv_T^p$ COLO $\equiv_T^p$ COLC. Conclude that there exists a polynomial-time algorithm to determine the chromatic number of a graph, and even to find an optimal colouring, if and only if COLD $\in \mathbf{P}$.   □

These graph colouring problems have the characteristic that although it is perhaps difficult to *decide* whether or not a graph can be coloured with a given number of colours, it is easy to *check* whether a suggested colouring is valid.

Definition 10.3.4.   Let X be a decision problem. Let Q be a set, arbitrary for the time being, which we call the *proof space* for X. A *proof system* for X is a subset $F \subseteq X \times Q$ such that $(\forall x \in X)(\exists q \in Q)[<x, q> \in F]$. Any q such that $<x, q> \in F$ is known as a *proof* or a *certificate* that $x \in X$. Intuitively, each true statement of the type $x \in X$ has a proof in F, whereas no false statement of this type has one (because if $x \notin X$, there does not exist a $q \in Q$ such that $<x, q> \in F$).   □

Example 10.3.3.   Let $I = \mathbb{N}$ and COMP $= \{n \mid n$ is a composite integer$\}$. We can take $Q = \mathbb{N}$ as the proof space and $F = \{<n, q> \mid 1 < q < n$ and q divides n exactly$\}$ as the proof system. Notice that some problems may have more than one natural proof system. In this example we could also use the ideas of Section 8.6.2 to define

$F' = \{<n, q> \mid (n$ is even **and** $n > 2)$ **or**

$\qquad\qquad\qquad (1 < q < n$ **and** n is not a strong pseudoprime to the base $q)\}$,

which offers a large number of proofs for all odd composite numbers.   □

Example 10.3.4.   Consider the set of instances $I = \{<G, k> \mid G$ is an undirected graph and k is an integer$\}$ and the problem COLD $= \{<G, k> \in I \mid G$ can be coloured with k colours$\}$. As proof space we may take $Q = \{c: N \to \{1, 2, \ldots, k\} \mid N$ is a set of nodes and k is an integer$\}$. Then a proof system is given by

$F = \{ \ <<G,k>,c> \mid G = <N,A> \text{ is an undirected graph,}$

k is an integer,

$c : N \rightarrow \{1, 2, \ldots, k\}$ is a function and

$(\forall u, v \in N)[\{u,v\} \in A \Rightarrow c(u) \neq c(v)] \}$.   □

Problem 10.3.7.   Let $G = <N,A>$ be an undirected graph. A *clique* in G is a set of nodes $K \subseteq N$ such that $\{u,v\} \in A$ for every pair of nodes $u,v \in K$. Given a graph G and an integer k, the CLIQUE problem consists of determining whether there exists a clique of k nodes in G. Give a proof space and a proof system for this decision problem.   □

Definition 10.3.5.   **NP** is the class of decision problems for which there exists a proof system such that the proofs are succinct and easy to check. More precisely, a decision problem X is in **NP** if and only if there exist a proof space Q, a proof system $F \subseteq X \times Q$, and a polynomial $p(n)$ such that

i. $(\forall x \in X)(\exists q \in Q)[<x,q> \in F \text{ and } |q| \leq p(|x|)],$

where $|q|$ and $|x|$ denote the sizes of q and x, respectively; and

ii. $F \in \mathbf{P}.$

We do not require that there should exist an efficient way to *find* a proof of x when $x \in X$, only that there should exist an efficient way to *check* the validity of a proposed short proof.   □

Example 10.3.5.   The conceptual distinction between **P** and **NP** is best grasped with an example. Let COMP be the problem in Example 10.3.3. In order to have COMP $\in \mathbf{P}$, we would need an algorithm

function *DecideCOMP* (n)
    {decides whether n is a composite number or not}
    .
    .
    . $\cdots$ **return** *true*
    .
    .
    . $\cdots$ **return** *false*
    .
    .
    .

whose running time is polynomial in the size of n. No such algorithm is currently known. However, to show that COMP $\in \mathbf{NP}$, we need only exhibit the following (obvious) polynomial-time algorithm.

function *VerifyCOMP* (n, q)
   if $1 < q < n$ **and** q divides n **then return** *true*
                     else return *false*

By definition of **NP**, any run of *VerifyCOMP* (n, q) that returns *true* is a proof that n is composite, and every composite number has at least one such proof (but prime numbers have none). However, the situation is not the same as for a probabilistic algorithm (Chapter 8): we are content even if there exist very few q (for some composite n) such that *VerifyCOMP* (n, q) is *true* and if our chance of hitting one at random would be staggeringly low. ☐

Problem 10.3.8. Let X be a decision problem for which there exists a polynomial-time *true*-biased Monte Carlo algorithm (section 8.6). Prove that $X \in$ **NP**. (*Hint*: the proof space is the set of all sequences of random decisions possibly taken by the Monte Carlo algorithm.) ☐

Problem 10.3.9. Prove that **P** $\subseteq$ **NP**. (*Hint*: Let X be a decision problem in **P**. It suffices to take $Q = \{0\}$ and $F = \{<x, 0> \mid x \in X\}$ to obtain a system of "proofs" that are succinct and easy to check. This example provides an extreme illustration of the fact that the same proof may serve for more than one instance of the same problem.) ☐

Example 10.3.6. The problems COLD and CLIQUE considered in example 10.3.4 and Problem 10.3.7 are in **NP**. ☐

Although COLD is in **NP** and COLO $\equiv_T^p$ COLD, it does not appear that **NP** contains the problem of deciding, given a graph G and an integer k, whether k is the chromatic number of G. Indeed, although it suffices to exhibit a valid colouring to prove that a graph can be coloured with a given number of colours (Example 10.3.4), no one has yet been able to invent an efficient proof system to demonstrate that a graph *cannot* be coloured with less than k colours.

Definition 10.3.6. Let $X \subseteq I$ be a decision problem. Its *complementary* problem consists of answering "Yes" for an instance $x \in I$ if and only if $x \notin X$. The class co-**NP** is the class of decision problems whose complementary problem is in **NP**. For instance, the preceding remark indicates that we do not know whether COLD $\in$ co-**NP**. Nonetheless, we know that COLD $\in$ co-**NP** if and only if **NP** = co-**NP** (Problems 10.3.27 and 10.3.16). The current conjecture is that **NP** $\neq$ co-**NP**, and therefore that COLD $\notin$ co-**NP**. ☐

Problem 10.3.10. Let A and B be two decision problems. Prove that if $A \leq_m^p B$ and $B \in$ **NP**, then $A \in$ **NP**. ☐

Problem 10.3.11. Let A and B be two decision problems. Do you believe that if $A \leq_T^p B$ and $B \in$ **NP**, then $A \in$ **NP**? Justify your answer. ☐

Problem 10.3.12.   Show that HAM, the Hamiltonian circuit problem defined in Example 10.3.2, is in **NP**.   □

Example 10.3.7.   In 1903, two centuries after Mersenne claimed without proof that $2^{67} - 1$ is a prime number, Frank Cole showed that

$$2^{67} - 1 = 193,707,721 \times 761,838,257,287 \;.$$

It took him "three years of Sundays" to discover this factorization. He was lucky that the number he chose to attack is indeed composite, since this enabled him to offer a proof of his result that is both short and easy to check. (This was not all luck: Lucas had already shown in the nineteenth century that $2^{67}-1$ is composite, but without finding the factors.)

The story would have had quite a different ending if this number had been prime. In this case the only "proof" of his discovery that Cole would have been able to produce would have been a thick bundle of papers covered in calculations. The proof would be far too long to have any practical value, since it would take just as long to check as it did to produce in the first place. (A similar argument may be advanced concerning the "proof" by computer of the famous four colour theorem.) This results from a phenomenon like the one mentioned in connection with the chromatic number of a graph: the problem of recognizing composite numbers is in **NP** (Example 10.3.3), but it seems certain at first sight not to be in co-**NP**, that is, the complementary problem of recognizing prime numbers seems not to be in **NP**.

However, nothing is certain in this world except death and taxes: this problem too is in **NP**, although the notion of a proof (or certificate) of primality is rather more subtle than that of a proof of nonprimality. A result from the theory of numbers shows that n, an odd integer greater than 2, is prime if and only if there exists an integer x such that

$$\begin{cases} 0 < x < n \\ x^{n-1} \equiv 1 \pmod{n}, \text{ and} \\ x^{(n-1)/p} \not\equiv 1 \pmod{n} \text{ for each prime factor } p \text{ of } n-1 \;. \end{cases}$$

A proof of primality for n therefore consists of a suitable x, the decomposition of $n-1$ into prime factors, and a collection of (recursive) proofs that each of these factors is indeed prime. (More succinct proof systems are known.)   □

*Problem 10.3.13.   Complete the proof sketched in Example 10.3.7 that the problem of primality is in **NP**. It remains to show that the length of a recursive proof of primality is bounded above by a polynomial in the size (that is, the logarithm) of the integer n concerned, and that the validity of such a proof can be checked in polynomial time.   □

Problem 10.3.14.   Let $F = \{<x,y> \mid x,y \in \mathbb{N}$ and x has a prime factor less than $y\}$. Let FACT be the problem of decomposing an integer into prime factors. Prove that

  i. $F \in \mathbf{NP} \cap \text{co-}\mathbf{NP}$; and
 ii. $F \equiv_T^p$ FACT.

If we accept the conjecture that no polynomial-time factorization algorithm exists, we can therefore conclude that $F \in (\mathbf{NP} \cap \text{co-}\mathbf{NP}) \setminus \mathbf{P}$.   □

10.3.1 NP-Complete Problems

The fundamental question concerning the classes **P** and **NP** is whether the inclusion $\mathbf{P} \subseteq \mathbf{NP}$ is strict. Does there exist a problem that allows an efficient proof system but for which it is inherently difficult to discover such proofs in the worst case? Our intuition and experience lead us to believe that it is generally more difficult to discover a proof than to check it: progress in mathematics would be much faster were this not so. In our context this intuition translates into the hypothesis that $\mathbf{P} \neq \mathbf{NP}$. It is a cause of considerable chagrin to workers in the theory of complexity that they can neither prove nor disprove this hypothesis. If indeed there exists a simple proof that $\mathbf{P} \neq \mathbf{NP}$, it has certainly not been easy to find!

On the other hand, one of the great successes of this theory is the demonstration that there exist a large number of practical problems in **NP** such that if any one of them were in **P** then **NP** would be equal to **P**. The evidence that supports the hypothesis $\mathbf{P} \neq \mathbf{NP}$ therefore also lends credence to the view that none of these problems can be solved by a polynomial-time algorithm in the worst case. Such problems are called **NP**-complete.

Definition 10.3.7.   A decision problem X is **NP**-complete if

  i. $X \in \mathbf{NP}$; and
 ii. for every problem $Y \in \mathbf{NP}$, $Y \leq_T^p X$.

Some authors replace the second condition by $Y \leq_m^p X$ or by other (usually stronger) kinds of reduction.   □

Problem 10.3.15.   Prove that there exists an **NP**-complete problem X such that $X \in \mathbf{P}$ if and only if $\mathbf{P} = \mathbf{NP}$.   □

*Problem 10.3.16.   Prove that if there exists an **NP**-complete problem X such that $X \in \text{co-}\mathbf{NP}$, then $\mathbf{NP} = \text{co-}\mathbf{NP}$.   □

Problem 10.3.17.   Prove that if the problem X is **NP**-complete and the problem $Z \in$ **NP**, then

  i. Z is **NP**-complete if and only if $X \leq_T^p Z$;
  ii. if $X \leq_m^p Z$, then Z is **NP**-complete.   □

Be sure to work this important problem. It provides the fundamental tool for proving **NP**-completeness. Suppose we have a pool of problems that have already been shown to be **NP**-complete. To prove that Z is **NP**-complete, we can choose an appropriate problem X from the pool and show that X is polynomially reducible to Z (either many-one or in the sense of Turing). We must also show that $Z \in$ **NP** by exhibiting an efficient proof system for Z. Several thousand **NP**-complete problems have been enumerated in this way.

This is all well and good once the process is under way, since the more problems there are in the pool, the more likely it is that we can find one that can be reduced without too much difficulty to some new problem. The trick, of course, is to get the ball rolling. What should we do at the outset when the pool is empty to prove for the very first time that some particular problem is **NP**-complete? (Problem 10.3.17 is then powerless.) This is the *tour de force* that Steve Cook managed to perform in 1971, opening the way to the whole theory of **NP**-completeness. (A similar theorem was discovered independently by Leonid Levin.)

10.3.2 Cook's Theorem

Definition 10.3.8.   A *Boolean variable* takes its values in the set $\mathbb{B} = \{ true, false \}$. Boolean variables are combined using logical operators (**not, and, or,** $\Leftrightarrow$, $\Rightarrow$, and so on) and parentheses to form *Boolean expressions*. It is customary to represent disjunction (**or**) in such expressions by the symbol "+" and conjunction (**and**) by simply juxtaposing the operands (as for arithmetic multiplication). Negation is often denoted by a horizontal bar above the variable or the expression concerned. A Boolean expression is *satisfiable* if there exists at least one way of assigning values to its variables so as to make the expression true. A Boolean expression is a *tautology* if it remains true whatever values are assigned to its variables. A Boolean expression is a *contradiction* if it is not satisfiable, that is, if its negation is a tautology. We denote by SAT, TAUT and CONT, respectively, the problems of deciding, given a Boolean expression, whether it is satisfiable, whether it is a tautology, and whether it is a contradiction.   □

Example 10.3.8.   Here are three Boolean expressions using the Boolean variables p and q .

  i. $(p + q) \Rightarrow pq$
  ii. $(p \Leftrightarrow q) \Leftrightarrow (\bar{p} + q)(p + \bar{q})$
  iii. $\bar{p} (p + q)\bar{q}$

Expression (i) is satisfiable because it is true if $p = true$ and $q = true$, but it is not a tautology because it is false if $p = true$ and $q = false$. Verify that expression (ii) is a tautology and that expression (iii) is a contradiction.   □

To prove that a Boolean expression is satisfiable, it suffices to produce an assignment that satisfies it. Moreover, such a proof is easy to check. This shows that $SAT \in NP$. It is not apparent that the same is true of the two other problems: what short and easy proof can one give in general of the fact that a Boolean expression is a tautology or a contradiction? These three problems are, nevertheless, polynomially equivalent in the sense of Turing.

Problem 10.3.18.   Prove that

  i. $SAT \equiv_T^p TAUT \equiv_T^p CONT$; and even
  ii. $TAUT \equiv_m^p CONT$.   □

It is possible in principle to decide whether a Boolean expression is satisfiable by working out its value for every possible assignment to its Boolean variables. However this approach is impractical when the number n of Boolean variables involved is large, since there are 2^n possible assignments. No efficient algorithm to solve this problem is known.

Definition 10.3.9.   A *literal* is either a Boolean variable or its negation. A *clause* is a literal or a disjunction of literals. A Boolean expression is in *conjunctive normal form* (CNF) if it is a clause or a conjunction of clauses. It is in k-CNF for some positive integer k if it is composed of clauses, each of which contains at most k literals (some authors say: exactly k literals).   □

Example 10.3.9.   Consider the following expressions.

  i. $(p + \bar{q} + r)(\bar{p} + q + r)q\,\bar{r}$
  ii. $(p + qr)(\bar{p} + q\,(p + r))$
  iii. $(p \Rightarrow q) \Leftrightarrow (\bar{p} + q)$

Expression (i) is composed of four clauses. It is in 3-CNF (and therefore in CNF), but not in 2-CNF. Expression (ii) is not in CNF since neither $p + qr$ nor $\bar{p} + q\,(p + r)$ is a clause. Expression (iii) is also not in CNF since it contains operators other than conjunction, disjunction and negation.   □

*Problem 10.3.19.

  i. Show that to every Boolean expression there corresponds an equivalent expression in CNF.
  ii. Show on the other hand that the shortest equivalent expression in CNF can be exponentially longer than the original Boolean expression.   □

Definition 10.3.10.   SAT-CNF is the restriction of the SAT problem to Boolean expressions in CNF. For any positive integer k, SAT-k-CNF is the restriction of SAT-CNF to Boolean expressions in k-CNF. The problems TAUT-$(k$-$)$CNF and CONT-$(k$-$)$CNF are defined similarly.   □

Clearly, all these problems are in **NP** or in co-**NP**. Polynomial-time algorithms are known for a few of them.

*Problem 10.3.20.   Prove that SAT-2-CNF and TAUT-CNF can be solved in polynomial time.   □

The interest of Boolean expressions in the context of **NP**-completeness arises from their ability to simulate algorithms. Consider an arbitrary decision problem that can be solved by a polynomial-time algorithm A. Suppose that the size of the instances is measured in bits. For every integer n there exists a Boolean expression $\Psi_n(A)$ in CNF that can be obtained efficiently (in a time polynomial in n, where the polynomial may depend on the algorithm A; the *size* of $\Psi_n(A)$ is also polynomial in n). This Boolean expression contains a large number of variables, among which $x_1, x_2, \ldots, x_n$ correspond in a natural way to the bits of instances of size n for A. The Boolean expression is constructed so that there exists a way to satisfy it by choosing the values of its other Boolean variables if and only if algorithm A accepts the instance corresponding to the Boolean value of the x variables. For example, algorithm A accepts the instance 10010 if and only if the expression $x_1 \bar{x}_2 \bar{x}_3 x_4 \bar{x}_5 \Psi_5(A)$ is satisfiable. More interestingly, algorithm A accepts at least one instance of size 5 if and only if $\Psi_5(A)$ is satisfiable.

The proof that this Boolean expression exists and that it can be constructed efficiently poses difficult technical problems. It usually requires a formal model of computation beyond the scope of this book, such as the Turing machine. We content ourselves with mentioning that the expression $\Psi_n(A)$ contains among other things a distinct Boolean variable b_{it} for each bit i of memory that algorithm A may need to use when solving an instance of size n, and for each unit t of time taken by this computation. Once the variables $x_1, x_2, \ldots, x_n$ are fixed, the clauses of $\Psi_n(A)$ force these Boolean variables to simulate the step-by-step execution of the algorithm on the corresponding instance. (The number of additional Boolean variables is polynomial in the size of the instance because the algorithm runs in polynomial time and because we have assumed without loss of generality that none of its variables or arrays can ever occupy more than a polynomial number of bits of memory.)

We are finally in a position to state and prove the fundamental theorem of the theory of **NP**-completeness.

Theorem 10.3.1.   SAT-CNF is **NP**-complete.

Proof. We already know that SAT-CNF is in **NP**. Thus it remains to prove that $X \leq_T^p$ SAT-CNF for every problem $X \in$ **NP**. Let Q be a proof space and F an efficient proof system for X. Let $p(n)$ be the polynomial (given by the definition of **NP**) such that to every $x \in X$ there corresponds a $q \in Q$ whose length is bounded above by $p(|x|)$ such that $<x,q> \in F$. Let A be a polynomial-time algorithm able to decide, given $<x,q>$ as input, whether q is a proof that $x \in X$ (that is, whether $<x,q>$ belongs to F). That such an algorithm exists is ensured by the fact that $F \in \mathbf{P}$, which is part of the definition of $X \in \mathbf{NP}$. For each x, let A_x be an algorithm whose purpose is to verify whether a given $q \in Q$ is a valid proof that $x \in X$.

> **function** $A_x(q)$
>   **if** $A(<x,q>)$ **then return** *true*
>   **else return** *false*

Here finally is an algorithm to solve problem X in polynomial time, if we imagine that answers concerning the satisfiability of certain Boolean expressions in CNF can be obtained at no cost by a call on *DecideSATCNF*.

> **function** *DecideX*(x)
>   **let** n be the size of x (in bits)
>   **for** $i \leftarrow 0$ **to** $p(n)$ **do**
>   **if** *DecideSATCNF*$(\Psi_i(A_x))$ **then return** *true*
>   **return** *false*

Let $x \in X$ be of size n, and let $q \in Q$ be a proof of size i that $x \in X$, where $0 \le i \le p(n)$. The fact that algorithm A accepts $<x,q>$ implies that algorithm A_x accepts q. This is an instance of size i, hence the Boolean expression $\Psi_i(A_x)$ is satisfiable, as *DecideX*(x) will discover. Conversely, if $x \notin X$, then there exists no $q \in Q$ such that $<x,q> \in F$, and therefore A_x accepts no inputs, which implies that *DecideX*(x) will find no integer i such that $\Psi_i(A_x)$ is satisfiable. This completes the proof that $X \le_T^p$ SAT-CNF. (To be precise, one technical detail is worth mentioning: to each algorithm A, there corresponds a two variable polynomial r such that the Boolean formula $\Psi_i(A_x)$ can be constructed in a time in $O(r(|x|,i))$ and such that its size is bounded similarly.) □

Problem 10.3.21. Prove that in fact $X \le_m^p$ SAT-CNF for any decision problem $X \in \mathbf{NP}$. □

10.3.3 Some Reductions

We have just seen that SAT-CNF is **NP**-complete. Let $X \in \mathbf{NP}$ be some other decision problem. To show that X too is **NP**-complete, we need only prove that SAT-CNF $\le_T^p X$ (Problem 10.3.17). Thereafter, to show that $Y \in \mathbf{NP}$ is **NP**-complete, we have the choice of proving SAT-CNF $\le_T^p Y$ or $X \le_T^p Y$. We illustrate this principle with several examples.

Example 10.3.10. SAT is **NP**-complete.

We already know that SAT $\in \mathbf{NP}$. It therefore remains to show that SAT-CNF $\le_T^p$ SAT. The reduction is immediate because Boolean expressions in CNF are simply a special case of general Boolean expressions. More precisely, if we imagine that the satisfiability of Boolean expressions can be decided at no cost by a call on *DecideSAT*, here is a polynomial-time algorithm for solving SAT-CNF.

> **function** *DecideSATCNF* (Ψ)
> **if** Ψ is not in CNF **then return** *false*
> **if** *DecideSAT* (Ψ) **then return** *true*
> > **else return** *false*   □

Problem 10.3.22.   Prove that SAT $\leq_T^p$ SAT-CNF. Using Example 10.3.10, conclude that SAT $\equiv_T^p$ SAT-CNF. (*Hint:* This problem has a very simple solution. However, resist the temptation to use Problem 10.3.19(i) to obtain the algorithm

> **function** *DecideSAT* (Ψ)
> let ξ be a Boolean expression in CNF equivalent to Ψ
> **if** *DecideSATCNF* (ξ) **then return** *true*
> > **else return** *false*

because Problem 10.3.19(ii) shows that expression ξ can be exponentially longer than expression Ψ, so it cannot be computed in polynomial time in the worst case.)   □

Example 10.3.11.   SAT-3-CNF is **NP**-complete.

We have already seen that SAT-3-CNF is in **NP**. It remains to show that SAT-CNF $\leq_T^p$ SAT-3-CNF. This time let us show that SAT-CNF $\leq_m^p$ SAT-3-CNF. Let Ψ be a Boolean expression in CNF. Our problem is to construct efficiently a Boolean expression ξ in 3-CNF that is satisfiable if and only if Ψ is satisfiable. Consider first how to proceed if Ψ contains only one clause, which is therefore a disjunction of k literals.

- **i.** If $k \leq 3$, set $\xi = \Psi$, which is already in 3-CNF.
- **ii.** If $k = 4$, let l_1, l_2, l_3, and l_4 be the literals such that Ψ is $l_1 + l_2 + l_3 + l_4$. Let u be a new Boolean variable. Take $\xi = (l_1 + l_2 + u)(\bar{u} + l_3 + l_4)$.
- **iii.** More generally, if $k \geq 4$, let $l_1, l_2, \ldots, l_k$ be the literals such that Ψ is $l_1 + l_2 + \cdots + l_k$. Let $u_1, u_2, \ldots, u_{k-3}$ be new Boolean variables. Take

$$\xi = (l_1 + l_2 + u_1)(\bar{u}_1 + l_3 + u_2)(\bar{u}_2 + l_4 + u_3) \cdots (\bar{u}_{k-3} + l_{k-1} + l_k).$$

If the expression Ψ consists of several clauses, treat each of them independently (using different new variables for each clause) and form the conjunction of all the expressions in 3-CNF thus obtained.   □

Example 10.3.12.   (Continuation of Example 10.3.11)   If

$$\Psi = (p + \bar{q} + r + s)(\bar{r} + s)(\bar{p} + s + \bar{x} + v + \bar{w})$$

we obtain

$$\xi = (p + \bar{q} + u_1)(\bar{u}_1 + r + s)(\bar{r} + s)(\bar{p} + s + u_2)(\bar{u}_2 + \bar{x} + u_3)(\bar{u}_3 + v + \bar{w}).$$ □

Problem 10.3.23. Prove that ξ is satisfiable if and only if Ψ is satisfiable in the construction of Example 10.3.11. ☐

Problem 10.3.24. Prove that SAT-CNF $\leq_m^p$ SAT-3-CNF still holds even if in the definition of SAT-3-CNF we insist that each clause should contain *exactly* three literals. ☐

Example 10.3.13. 3-COL is **NP**-complete.

Let G be an undirected graph and k an integer constant. The problem k-COL consists of determining whether G can be coloured with k colours (see Problem 10.3.6). It is easy to see that 3-COL $\in$ **NP**. To show that 3-COL is **NP**-complete, we shall prove this time that SAT-3-CNF $\leq_m^p$ 3-COL. Given a Boolean expression Ψ in 3-CNF, we have to construct efficiently a graph G that can be coloured in three colours if and only if Ψ is satisfiable. This reduction is considerably more complex than those we have seen so far.

Suppose for simplicity that every clause of the expression Ψ contains exactly three literals (see Problem 10.3.24). Let k be the number of clauses in Ψ. Suppose further without loss of generality that the Boolean variables appearing in Ψ are x_1, $x_2, \ldots$, x_t. The graph G that we are about to build contains $3 + 2t + 6k$ nodes and $3 + 3t + 12k$ edges. Three distinguished nodes of this graph are linked in a triangle: call them T, F, and C. When the time comes to colour G in three colours, imagine that the colours assigned to T and F represent the Boolean values *true* and *false*, respectively. The colour used for node C will be a control colour. See Figure 10.3.1.

For each Boolean variable x_i of Ψ the graph contains two nodes y_i and z_i that are linked to each other and to the control node C. In any colouring of G in three colours, this forces y_i to be the same colour as either T or F and z_i to be the complementary colour. If y_i is the same colour as T, think of this intuitively as an assignment of the value *true* to the Boolean variable x_i. Contrariwise, x_i is *false* if it is z_i that is the same colour as T. In every case node y_i corresponds to the literal x_i and node z_i corresponds to the literal $\bar{x}_i$. If $t = 3$, for example, Figure 10.3.2 shows the part of the graph that we have constructed up to now.

We still have to add 6 nodes and 12 edges for each clause in Ψ. These are added in such a way that the graph will be colourable with three colours if and only if the choice of colours for $y_1, y_2, \ldots$, y_t corresponds to an assignment of Boolean values to $x_1, x_2, \ldots$, x_t that satisfies every clause. This can be accomplished thanks to the *widget* illustrated in Figure 10.3.3. One copy of this widget is added to the graph for each clause in Ψ.

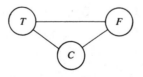

Figure 10.3.1. The control triangle.

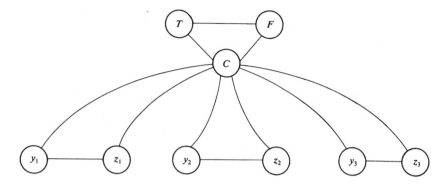

Figure 10.3.2.   Graph representation of three Boolean variables.

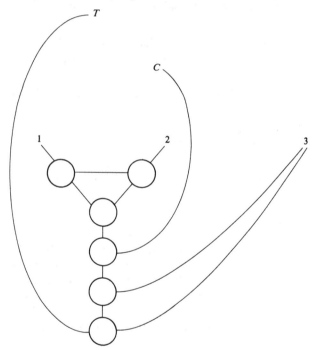

Figure 10.3.3.   A widget.

Each widget is linked to five other nodes of the graph: nodes C and T of the control triangle, and three nodes chosen from the y_i and z_i so as to correspond to the three literals of the clause concerned. Because these *input nodes* 1, 2 and 3 cannot be the same colour as C, Problem 10.3.25 shows that the widget can be coloured with the colours assigned to C, T, and F if and only if at least one of the nodes 1, 2, and 3 is coloured with the same colour as node T. In other words, since the colour assigned to node T represents *true*, the widget simulates the disjunction of the three literals represented by the nodes to which it is joined.

This ends the description of the graph G, which can be coloured with three colours if and only if Ψ is satisfiable. It is clear that the graph can be constructed efficiently starting from the Boolean expression Ψ in 3-CNF. We conclude that SAT-3-CNF $\leq_m^p$ 3-COL, and therefore that 3-COL is **NP**-complete.   □

Problem 10.3.25.   Verify that the colours attributed to nodes C, T, and F suffice to colour the widget of Figure 10.3.3 if and only if at least one of the input nodes is coloured with the same colour as node T (knowing that the input nodes cannot be coloured with the same colour as node C).   □

Problem 10.3.26.   Give a simple proof that 4-COL is **NP**-complete.   □

Problem 10.3.27.   Prove that COLD (Problem 10.3.6) is **NP**-complete.   □

** Problem 10.3.28.   Prove that 3-COL is still **NP**-complete even if we restrict ourselves to planar graphs of degree not greater than 4.   □

* Problem 10.3.29.   Show, on the other hand, that 2-COL is in **P**.   □

* Problem 10.3.30.   Prove that CLIQUE (Problem 10.3.7) is **NP**-complete. (*Hint :* prove that SAT-3-CNF $\leq_m^p$ CLIQUE.)   □

** Problem 10.3.31.   Prove that the problem of the Hamiltonian circuit (Example 10.3.2 and Problem 10.3.12) is **NP**-complete.   □

* Problem 10.3.32.   The *halting problem* consists of deciding, given any program as instance, whether the latter will ever halt when started. The notion of reducibility extends in a natural way to *unsolvable* problems such as the halting problem (although it is usual to drop the polynomial-time aspect of the reductions — which we do *not* do here). A function $f : \mathbb{N} \to \mathbb{N}$ is *polynomially bounded* if the size of its value on any argument is polynomially bounded by the size of its argument. Prove that the problem of computing *any* polynomially bounded computable function is polynomially reducible to the halting problem in the sense of Turing. Prove however that there exist decision problems that are not polynomially reducible to the halting problem in the sense of Turing.   □

10.3.4 Non-determinism

The class **NP** is usually defined quite differently, although the definitions are equivalent. The classic definition involves the notion of non-deterministic algorithms, which we only sketch here. The name **NP** arose from this other definition: it represents the class of problems that can be solved by a **N**on-deterministic algorithm in **P**olynomial time.

On the surface, a non-deterministic algorithm resembles a Las Vegas probabilistic algorithm (Section 8.5). To solve instance x of some problem X, we call $ND(x, y, success)$, where y and *success* are return parameters. If the algorithm sets *success* to *true*, then y is a correct solution to instance x; otherwise no solution is obtained. The difference between non-deterministic and Las Vegas algorithms is that for the former we do not set any bound on the probability of success. It is even allowable that for some instances the algorithm will never set *success* to *true*. For this reason non-deterministic algorithms are only a mathematical abstraction that cannot be used directly in practice: we would not program such an algorithm in the hope of running it successfully and efficiently on a computer. (This explains why, although non-deterministic algorithms can solve **NP**-complete problems in polynomial time, this does not imply that **P = NP**.)

To avoid confusion with probabilistic algorithms, we do not denote non-deterministic choices by calls on *uniform* $(i .. j)$ as in Chapter 8, but instead, we use a special instruction

> **choose** n **between** i **and** j ,

whose effect is to set n to some value between i and j inclusive. The actual value assigned to n is not specified by the algorithm, nor is it subject to the laws of probability. The effect of the algorithm is determined by the existence or the nonexistence of sequences of non-deterministic choices that lead to the production of a result. We are *not* concerned with how such sequences could be determined efficiently or how their nonexistence could be established. For simplicity, we write

> **return** *success* $\leftarrow$ *bool*

as an abbreviation for

> *success* $\leftarrow$ *bool*
> **return** .

Definition 10.3.11.   A *computation* of the algorithm is a sequence of non-deterministic choices that leads to setting *success* to *true*, hence to returning a solution. The *domain* of the algorithm is the set of instances on which it has at least one computation. The algorithm is *total* if its domain is the set of all possible instances. The algorithm is *consistent* if two different computations on any given instance always lead to the same result. When a non-deterministic algorithm is consistent, it computes a well-defined *function* on the instances of its domain.

The *time* taken by a non-deterministic algorithm on a given instance of its domain is defined as the shortest possible time that any computation can cause it to run on this instance; the time is undefined on instances that are not in the domain of the algorithm. A non-deterministic algorithm *runs in polynomial time* if the time it takes is bounded by some polynomial in the size of its instance provided that the instance is in the domain of the algorithm. Notice that there is no limit on how long a polynomial-time non-deterministic algorithm can run if the "wrong" non-deterministic choices are made. It is even possible for a computation to be arbitrarily long, provided that the same instance also admits at least one polynomially bounded computation.   □

Example 10.3.14.   Consider the following total consistent non-deterministic primality testing algorithm (recall from Section 4.8 that $dexpo(x, i, n)$ computes $x^i \bmod n$ efficiently).

```
procedure primeND (n , var prime, var success )
    { non-deterministically determines whether n is prime }
    if n ≤ 3 or n is even then prime ← (2 ≤ n ≤ 3)
                                    return success ← true
    choose guess between 0 and 1
    if guess = 0
    then { the guess is that n is composite — let's guess a proof! }
        prime ← false
        choose m between 2 and n − 1
        if m divides n then success ← true
                        else success ← false
    else { the guess is that n is prime — let's guess a proof! }
        prime ← true
        choose x between 1 and n − 1
        { the guess is that x is as in Example 10.3.7 }
        if dexpo (x , n − 1, n) ≠ 1 then return success ← false
        m ← n − 1
        while m > 1 do
            choose p between 2 and m
            { the guess is that p is a new prime divisor of n − 1 }
            primeND ( p , pr , suc )
            if suc and pr and p divides m and dexpo (x , (n − 1)/p , n) ≠ 1
            then while p divides m do m ← m /p
            else return success ← false
        success ← true
```

When n is an odd composite number, $n \geq 9$, the algorithm has a computation that consists of choosing $guess = 0$ and non-deterministically setting m to some nontrivial factor of n. When n is prime, $n \geq 5$, the algorithm also has a computation that consists of choosing $guess = 1$, choosing x in accordance with the theorem mentioned in example 10.3.7, and guessing successively each prime divisor of $n - 1$. Clearly, the algorithm also has a computation when $n \leq 3$ or when n is even. The algorithm *primeND* is therefore total. It is also consistent by the same theorem from Example 10.3.7. Notice again that it would be pointless to attempt implementing this algorithm on the computer by replacing the **choose** instructions by random choices in the same interval: the probability of success would be infinitesimal.   □

*Problem 10.3.33.   Prove that the non-deterministic algorithm *primeND* of Example 10.3.14 runs in a time polynomial in the size of its instance. Show, however, that there are sequences of non-deterministic choices that can cause the algorithm to run for a time exponential in the size of its instance.   □

*Problem 10.3.34.   Prove that a decision problem can be solved by a total consistent polynomial-time non-deterministic algorithm if and only if it belongs to **NP** ∩ co-**NP**.   □

Theorem 10.3.2   Every decision problem in **NP** is the domain of some polynomial-time non-deterministic algorithm. (The algorithm can easily be made consistent but it will not in general be total.)

Proof.   Let $X \in$ **NP** be a decision problem, let Q be its proof space, let $F \subseteq X \times Q$ be its efficient proof system, and let p be the polynomial that bounds the length of a proof as a function of the length of the corresponding positive instance. Assume for simplicity that Q is the set of binary strings. The following polynomial-time non-deterministic algorithm has a computation on instance x if and only if $x \in X$.

> **procedure** *XND* $(x$, **var** *ans*, **var** *success*)
> $n \leftarrow$ size of x
>   **choose** l **between** 0 **and** $p(n)$
> $q \leftarrow$ empty binary string
>   **for** $i \leftarrow 1$ **to** l **do**
>   **choose** b **between** 0 **and** 1
>   append bit b to the right of q
>   **if** $< x, q > \in F$ **then** *ans* $\leftarrow$ *true*, *success* $\leftarrow$ *true*
>   **else** *success* $\leftarrow$ *false*   □

Problem 10.3.35   Prove the converse of Theorem 10.3.2: whenever a decision problem is the domain of some polynomial-time non-deterministic algorithm, this problem belongs to **NP**. (*Hint:* use the set of all possible computations as proof space.)   □

The preceding theorem and problem suggest the alternative (and more usual) definition for **NP**: it is the class of decision problems that are the domain of some polynomial-time non-deterministic algorithm. In this case, we are only concerned with the existence or nonexistence of computations (usually called *accepting* computations in this context); the actual result returned by the algorithm in case of success is irrelevant, and the corresponding parameter may be ignored altogether (there is no point in algorithm *XND* setting *ans* to *true* when it finds a q such that $< x, q > \in F$). Although the authors prefer the definition based on proof systems, it is sometimes easier to show that a problem belongs to **NP** with this other definition. For instance, it makes Problems 10.3.8 and 10.3.9 completely obvious.

10.4 REFERENCES AND FURTHER READING

For an overview of algorithms for sorting by transformation consult Aho, Hopcroft, and Ullman (1974). In particular, the lexicographic sorting algorithm can sort n elements in a time in $\Theta(n+m)$, where m is the sum of the sizes of the elements to be sorted. For modifications of heapsort that come close to being optimal for the worst-case number of comparisons, consult Gonnet and Munro (1986) and Carlsson (1986, 1987). In his tutorial article, Pippenger (1978) describes a method similar to decision trees for determining a lower bound on the size of logic circuits such as those presented in Problems 4.11.8 to 4.11.12. In particular, this technique shows that the required solutions for Problems 4.11.9 and 4.11.12 are optimal.

The reduction IQ $\leq^l$ MQ (Problem 10.2.9) comes from Bunch and Hopcroft (1974). If f and g are two cost functions as in section 10.2.2, then an algorithm that is asymptotically more efficient than the naive algorithm for calculating fg is given in Fredman (1976). Theorem 10.2.5 is due to Fischer and Meyer (1971) and Theorem 10.2.6 is due to Furman (1970). In the case of cost functions whose range is restricted to $\{0,+\infty\}$, Arlazarov, Dinic, Kronrod, and Faradzev (1970) present an algorithm to calculate fg using a number of *Boolean* operations in $O(n^3/\log n)$. Problem 10.2.12 is solved in Fischer and Meyer (1971). The reduction INV $\leq^l$ MLT (Problem 10.2.22), which is crucial in the proof of Theorem 10.2.7, comes from Cook and Aanderaa (1969). For further information concerning the topics of Section 10.2, consult Aho, Hopcroft, and Ullman (1974).

The theory of **NP**-completeness originated with two fundamental papers: Cook (1971) proves that SAT-CNF is **NP**-complete, and Karp (1972) underlines the importance of this notion by presenting a large number of **NP**-complete problems. To be historically exact, the original statement from Cook (1971) is that $X \leq^p_T$ TAUT-DNF for every $X \in$ **NP**, where TAUT-DNF is concerned with tautologies in *disjunctive normal form*; however, it should be noted that TAUT-DNF is probably not **NP**-complete since otherwise **NP** = co-**NP** (Problem 10.3.16). A similar theory was developed independently by Levin (1973). The uncontested authority in matters of **NP**-completeness is Garey and Johnson (1979). A good introduction can also be found in Hopcroft and Ullman (1979). The fact that the set of prime numbers is in **NP** (Examples 10.3.7 and 10.3.14, and Problems 10.3.13, 10.3.33, and 10.3.34) was discovered by Pratt (1975); more succinct primality certificates can be found in Pomerance (1987). Problems 10.3.14 and 10.3.16 are from Brassard (1979). Part of the solution to Problem 10.3.28 can be found in Stockmeyer (1973). In practice, the fact that a problem is **NP**-complete does not make it go away. However in this case we have to be content with heuristics and approximations as described in Garey and Johnson (1976), Sahni and Horowitz (1978), and Horowitz and Sahni (1978). To find out more about non-determinism, see Hopcroft and Ullman (1979).

This chapter has done little more than sketch the theory of computational complexity. Several important techniques have gone unmentioned. An algebraic approach to lower bounds is described in Aho, Hopcroft, and Ullman (1974), Borodin and

Munro (1975), and Winograd (1980). For an introduction to adversary arguments (Problem 8.4.4), consult Horowitz and Sahni (1978). Although we do not know how to prove that there are no efficient algorithms for **NP**-complete problems, there exist problems that are intrinsically difficult, as described in Aho, Hopcroft, and Ullman (1974). These can be solved in theory, but it can be proved that no algorithm can solve them in practice when the instances are of moderate size, even if it is allowed to take a time comparable to the age of the universe and as many bits of memory as there are elementary particles in the known universe (Stockmeyer and Chandra 1979). As mentioned in Problem 10.3.32, there also exist problems that cannot be solved by any algorithm, whatever the resources available; read Turing (1936), Gardner and Bennett (1979), Hopcroft and Ullman (1979), and Brassard and Monet (1982).

Table of Notation

$\#T$	number of elements in array T ; cardinality of set T.
$i \mathrel{..} j$	interval of integers: $\{\, k \in \mathbb{N} \mid i \le k \le j \,\}$
div, mod	arithmetic quotient and modulo ; extended to polynomials in Section 10.2.3
$\times$	arithmetic and matrix multiplication ; Cartesian product
$\uparrow$	pointer
$\leftarrow$	assignment
var x	return parameter of a procedure or function
return	dynamic end of a **procedure**
return v	dynamic end of a **function** with value v returned
$\lvert x \rvert$	size of instance x ; $\lceil \lg(1+x) \rceil$ if x is an integer ; absolute value of x
$\log, \lg, \ln, \log_b$	logarithm in basis 10, 2, e, and b, respectively
e	basis of the natural logarithm: 2.7182818...
$n!$	n factorial ($0! = 1$ and $n! = n \times (n-1)!$, $n! \ge 1$)
$\dbinom{n}{k}$	number of combinations of k elements chosen among n

$<a, b>$	ordered pair consisting of elements a and b
(a, b)	same as $<a, b>$ (in particular: edge of a directed graph); open interval $\{x \in \mathbb{R} \mid a < x < b\}$
$[a, b]$	closed interval $\{x \in \mathbb{R} \mid a \leq x \leq b\}$
$[a, +\infty)$	set of real numbers larger than or equal to a
$\{ \cdots \}$	denotes comments in algorithms; set of elements " $\cdots$ "
$\mid$	such that
$\subseteq$	set inclusion (allowing equality)
$\subset$	strict set inclusion
$\cup$	set union: $A \cup B = \{x \mid x \in A$ **or** $x \in B\}$
$\cap$	set intersection: $A \cap B = \{x \mid x \in A$ **and** $x \in B\}$
$\in$	set membership
$\notin$	set nonmembership
$\backslash$	set difference: $A \backslash B = \{x \mid x \in A$ **and** $x \notin B\}$
$\varnothing$	empty set
$\mathbb{N}, \mathbb{R}, \mathbb{N}^+, \mathbb{R}^+, \mathbb{R}^*, \mathbb{B}$	sets of integers, reals, and Booleans (see Section 2.1.1)
$f: A \rightarrow B$	f is a function from A to B
$(\exists x)[P(x)]$	there exists an x such that $P(x)$
$\exists!$	there exists one and only one
$(\forall x)[P(x)]$	for all x, $P(x)$
O, Ω, Θ	asymptotic notation (see Section 2.1)
$\Rightarrow$	implies
$\Leftrightarrow$	if and only if
Σ	summation
$\int$	integral
$\pm$	plus or minus
$f'(x)$	derivative of the function $f(x)$
∞	infinity
$\lim\limits_{n \to \infty} f(x)$	limit of $f(x)$ when x goes to infinity
$\lfloor x \rfloor$	floor of x: largest integer less than or equal to x; extended to polynomials in Section 10.2.3
$\lceil x \rceil$	ceiling of x: smallest integer larger than or equal to x
$\lg^*$	iterated logarithm (see page 63)
$uniform(i .. j)$	randomly and uniformly selected integer between i and j
$x \equiv y \pmod{n}$	x is congruent to y modulo n (n divides $x - y$ exactly)
$x \oplus y$	exclusive-or of x and y for bits or Booleans; bit-by-bit exclusive-or for bit strings

and, or	Boolean conjunction and disjunction
$\bar{x}$	Boolean complement of x
$F_\omega(a)$	Fourier transform of vector a with respect to ω
$F_\omega^{-1}(a)$	inverse Fourier transform
$A \leq^l B$	A is linearly reducible to B
$A \equiv^l B$	A is linearly equivalent to B
$A \leq^p_T B$	A is polynomially reducible to B in the sense of Turing
$A \equiv^p_T B$	A is polynomially equivalent to B in the sense of Turing
$A \leq^p_m B$	A is many-one polynomially reducible to B
$A \equiv^p_m B$	A is many-one polynomially equivalent to B
P	class of decision problems that can be solved in polynomial time
NP	class of decision problems that have an efficient proof system
co-**NP**	class of decision problems whose complementary problem is in **NP**
choose	instruction for non-deterministic choice

Bibliography

ABADI, M., J. FEIGENBAUM, and J. KILIAN (1987), "On hiding information from an oracle", *Proceedings of 19th Annual ACM Symposium on the Theory of Computing*, pp. 195–203.

ACKERMANN, W. (1928), "Zum Hilbertschen Aufbau der reellen Zahlen", *Mathematische Annalen*, 99, 118–133.

ADEL'SON-VEL'SKII, G. M. and E. M. LANDIS (1962), "An algorithm for the organization of information" (in Russian), *Doklady Akademii Nauk SSSR*, 146, 263–266.

ADLEMAN, L. M. and M.-D. A. HUANG (1987), "Recognizing primes in random polynomial time", *Proceedings of 19th Annual ACM Symposium on the Theory of Computing*, pp. 462–469.

ADLEMAN, L. M., K. MANDERS, and G. MILLER (1977), "On taking roots in finite fields", *Proceedings of 18th Annual IEEE Symposium on the Foundations of Computer Science*, pp. 175–178.

ADLEMAN, L. M., C. POMERANCE, and R. S. RUMELY (1983), "On distinguishing prime numbers from composite numbers", *Annals of Mathematics*, 117, 173–206.

AHO, A. V. and M. J. CORASICK (1975), "Efficient string matching: An aid to bibliographic search", *Communications of the ACM*, 18(6), 333–340.

AHO, A. V., J. E. HOPCROFT, and J. D. ULLMAN (1974), *The Design and Analysis of Computer Algorithms*, Addison-Wesley, Reading, MA.

AHO, A. V., J. E. HOPCROFT, and J. D. ULLMAN (1976), "On finding lowest common ancestors in trees", *SIAM Journal on Computing*, 5(1), 115–132.

AHO, A. V., J. E. HOPCROFT, and J. D. ULLMAN (1983), *Data Structures and Algorithms*, Addison-Wesley, Reading, MA.

AJTAI, M., J. KOMLÓS, and E. SZEMERÉDI (1983), "An $O(n \log n)$ sorting network", *Proceedings of 15th Annual ACM Symposium on the Theory of Computing*, pp. 1–9.

ANON. (*c.* 1495) *Lytell Geste of Robyn Hode*, Wynkyn de Worde, London.

ARLAZAROV, V. L., E. A. DINIC, M. A. KRONROD, and I. A. FARADZEV (1970), "On economical construction of the transitive closure of a directed graph" (in Russian), *Doklady Akademii Nauk SSSR*, 194, 487–488.

BAASE, S. (1978), *Computer Algorithms: Introduction to Design and Analysis*, Addison-Wesley, Reading, MA; second edition, 1987.

BABAÏ, L. (1979), "Monte Carlo algorithms in graph isomorphism techniques", Research report, Département de mathématiques et de statistique, Université de Montréal, D.M.S. no. 79-10.

BACH, E., G. MILLER, and J. SHALLIT (1986), "Sums of divisors, perfect numbers and factoring", *SIAM Journal on Computing*, 15(4), 1143–1154.

BACHMANN, P. G. H. (1894), *Zahlentheorie*, vol. 2: *Die Analytische Zahlentheorie*, B. G. Teubner, Leipzig.

BATCHER, K. (1968), "Sorting networks and their applications", *Proceedings of AFIPS 32nd Spring Joint Computer Conference*, pp. 307–314.

BEAUCHEMIN, P., G. BRASSARD, C. CRÉPEAU, C. GOUTIER, and C. POMERANCE (1988), "The generation of random numbers that are probably prime", *Journal of Cryptology,* 1(1), in press.

BELAGA, E. C. (1961), "On computing polynomials in one variable with initial preconditioning of the coefficients", *Problemi Kibernetiki*, 5, 7–15.

BELLMAN, R. E. (1957), *Dynamic Programming*, Princeton University Press, Princeton, NJ.

BELLMAN, R. E. and S. E. DREYFUS (1962), *Applied Dynamic Programming*, Princeton University Press, Princeton, NJ.

BELLMORE, M. and G. NEMHAUSER (1968), "The traveling salesman problem: A survey", *Operations Research*, 16(3), 538–558.

BENNETT, C. H., G. BRASSARD, and J.-M. ROBERT (1988), "Privacy amplification through public discussion", *SIAM Journal on Computing*, in press.

BENTLEY, J. L. (1984), "Programming pearls: Algorithm design techniques", *Communications of the ACM*, 27(9), 865–871.

BENTLEY, J. L., D. HAKEN, and J. B. SAXE (1980), "A general method for solving divide-and-conquer recurrences", *SIGACT News*, ACM, 12(3), 36–44.

BENTLEY, J. L. and I. SHAMOS (1976), "Divide-and-conquer in multidimensional space", *Proceedings of 8th Annual ACM Symposium on the Theory of Computing*, pp. 220–230.

BENTLEY, J. L., D. F. STANAT, and J. M. STEELE (1981), "Analysis of a randomized data structure for representing ordered sets", *Proceedings of 19th Annual Allerton Conference on Communication, Control, and Computing*, pp. 364–372.

BERGE, C. (1958), *Théorie des graphes et ses applications,* Dunod, Paris; second edition, 1967; translated as: *The Theory of Graphs and Its Applications* (1962), Methuen & Co., London.

BERGE, C. (1970), *Graphes et hypergraphes*, Dunod, Paris; translated as: *Graphs and Hypergraphs* (1973), North Holland, Amsterdam.

BERLEKAMP, E. R. (1970), "Factoring polynomials over large finite fields", *Mathematics of Computation*, 24(111), 713–735.

BERLINER, H. J. (1980), "Backgammon computer program beats world champion", *Artificial Intelligence*, 14, 205–220.

BLUM, L., M. BLUM, and M. SHUB (1986), "A simple unpredictable pseudo-random number generator", *SIAM Journal on Computing*, 15(2), 364–383.

BLUM, M., R. W. FLOYD, V. R. PRATT, R. L. RIVEST, and R. E. TARJAN (1972), "Time bounds for selection", *Journal of Computer and System Sciences*, 7(4), 448–461.

BLUM, M. and S. MICALI (1984), "How to generate cryptographically strong sequences of pseudo-random bits", *SIAM Journal on Computing*, 13(4), 850–864.

BORODIN, A. B. and J. I. MUNRO (1971), "Evaluating polynomials at many points", *Information Processing Letters*, 1(2), 66–68.

BORODIN, A. B. and J. I. MUNRO (1975), *The Computational Complexity of Algebraic and Numeric Problems*, American Elsevier, New York, NY.

BORŮVKA, O. (1926), "O jistém problému minimálnim", *Praca Moravské Prirodovedecké Spolecnosti*, 3, 37–58.

BOYER, R. S. and J. S. MOORE (1977), "A fast string searching algorithm", *Communications of the ACM*, 20(10), 762–772.

BRASSARD, G. (1979), "A note on the complexity of cryptography", *IEEE Transactions on Information Theory*, IT-25(2), 232–233.

BRASSARD, G. (1985), "Crusade for a better notation", *SIGACT News*, ACM, 17(1), 60–64.

BRASSARD, G. (1988), *Modern Cryptology: A Tutorial*, Lecture Notes in Computer Science, Springer-Verlag, New York, NY.

BRASSARD, G. and S. KANNAN (1988), "The generation of random permutations on the fly", *Information Processing Letters* (in press).

BRASSARD, G. and S. MONET (1982), "L'indécidabilité sans larme (ni diagonalisation)", Publication no. 445, Département d'informatique et de recherche opérationnelle, Université de Montréal.

BRASSARD, G., S. MONET, and D. ZUFFELLATO (1986), "L'arithmétique des très grands entiers", *TSI : Technique et Science Informatiques*, 5(2), 89–102.

BRATLEY, P., B. L. FOX, and L. E. SCHRAGE (1983), *A Guide to Simulation*, Springer-Verlag, New York, NY; second edition, 1987.

BRIGHAM, E. O. (1974), *The Fast Fourier Transform*, Prentice–Hall, Englewood Cliffs, NJ.

BUNCH, J. and J. E. HOPCROFT (1974), "Triangular factorization and inversion by fast matrix multiplication", *Mathematics of Computation*, 28(125), 231–236.

BUNEMAN, P. and L. LEVY (1980), "The towers of Hanoi problem", *Information Processing Letters*, 10(4,5), 243–244.

CARASSO, C. (1971), *Analyse numérique*, Lidec, Montréal, Québec, Canada.

CARLSSON, S. (1986), *Heaps*, Doctoral dissertation, Department of Computer Science, Lund University, Lund, Sweden, CODEN : LUNFD6/(NFCS-1003)/(1-70)/(1986).

CARLSSON, S. (1987), "Average case results on heapsort", *Bit*, 27, 2–17.

CARTER, J. L. and M. N. WEGMAN (1979), "Universal classes of hash functions", *Journal of Computer and System Sciences*, 18(2), 143–154.

CHANG, L. and J. KORSH (1976), "Canonical coin changing and greedy solutions", *Journal of the ACM*, 23(3), 418–422.

CHERITON, D. and R. E. TARJAN (1976), "Finding minimum spanning trees", *SIAM Journal on Computing*, 5(4), 724–742.

CHRISTOFIDES, N. (1975), *Graph Theory: An Algorithmic Approach*, Academic Press, New York, NY.

CHRISTOFIDES, N. (1976), "Worst-case analysis of a new heuristic for the traveling salesman problem", Management Sciences Research Report no. 388, Carnegie-Mellon University, Pittsburgh, PA.

COHEN, H. and A. K. LENSTRA, (1987), "Implementation of a new primality test", *Mathematics of Computation*, 48(177), 103–121.

COOK, S. A. (1971), "The complexity of theorem-proving procedures", *Proceedings of 3rd Annual ACM Symposium on the Theory of Computing*, pp. 151–158.

COOK, S. A. and S. O. AANDERAA (1969), "On the minimum complexity of functions", *Transactions of the American Mathematical Society*, 142, 291–314.

COOLEY, J. M., P. A. LEWIS, and P. D. WELCH (1967), "History of the fast Fourier transform", *Proceedings of the IEEE*, 55, 1675–1679.

COOLEY, J. M. and J. W. TUKEY (1965), "An algorithm for the machine calculation of complex Fourier series", *Mathematics of Computation*, 19(90), 297–301.

COPPERSMITH, D. and S. WINOGRAD (1987), "Matrix multiplication via arithmetic progressions", *Proceedings of 19th Annual ACM Symposium on the Theory of Computing*, pp. 1–6.

CRAY RESEARCH (1986), "CRAY-2 computer system takes a slice out of pi", *Cray Channels*, 8(2), 39.

CURTISS, J. H. (1956), "A theoretical comparison of the efficiencies of two classical methods and a Monte Carlo method for computing one component of the solution of a set of linear algebraic equations", in *Symposium on Monte Carlo Methods*, H. A. Meyer, ed., John Wiley & Sons, New York, NY, pp. 191–233.

DANIELSON, G. C. and C. LANCZOS (1942), "Some improvements in practical Fourier analysis and their application to X-ray scattering from liquids", *Journal of the Franklin Institute*, 233, 365–380, 435–452.

DE BRUIJN, N. G. (1961), *Asymptotic Methods in Analysis*, North Holland, Amsterdam.

DEMARS, C. (1981), "Transformée de Fourier rapide", *Micro-Systèmes"*, 155–159.

DENNING, D.E.R. (1983), *Cryptography and Data Security*, Addison-Wesley, Reading, MA.

DEVROYE, L. (1986), *Non-Uniform Random Variate Generation*, Springer-Verlag, New York, NY.

DEWDNEY, A. K. (1984), "Computer recreations: Yin and yang: recursion and iteration, the tower of Hanoi and the Chinese rings", *Scientific American*, 251(5), 19–28.

DEYONG, L. (1977), *Playboy's Book of Backgammon*, Playboy Press, Chicago, IL.

DIFFIE, W. and M. E. HELLMAN (1976), "New directions in cryptography", *IEEE Transactions on Information Theory*, IT-22(6), 644–654.

DIJKSTRA, E. W. (1959), "A note on two problems in connexion with graphs", *Numerische Mathematik*, 1, 269–271.

DIXON, J. D. (1981), "Asymptotically fast factorization of integers", *Mathematics of Computation*, 36(153), 255–260.

DROMEY, R. G. (1982), *How to Solve It by Computer*, Prentice-Hall, Englewood Cliffs, NJ.

ERDŐS, P. and C. POMERANCE (1986), "On the number of false witnesses for a composite number", *Mathematics of Computation*, 46(173), 259–279.

EVEN, S. (1980), *Graph Algorithms*, Computer Science Press, Rockville, MD.

FEIGENBAUM, J. (1986), "Encrypting problem instances, or ..., can you take advantage of someone without having to trust him?", *Proceedings of CRYPTO 85*, Springer-Verlag, Berlin, pp. 477–488.

FISCHER, M. J. and A. R. MEYER (1971), "Boolean matrix multiplication and transitive closure", *Proceedings of IEEE 12th Annual Symposium on Switching and Automata Theory*, pp. 129–131.

FLAJOLET, P. and G. N. MARTIN (1985), "Probabilistic counting algorithms for data base applications", *Journal of Computer and System Sciences*, 31(2), 182–209.

FLOYD, R. W. (1962), "Algorithm 97: Shortest path", *Communications of the ACM*, 5(6), 345.

FOX, B. L. (1986), "Algorithm 647: Implementation and relative efficiency of quasirandom sequence generators", *ACM Transactions on Mathematical Software*, 12(4), 362–376.

FREDMAN, M. L. (1976), "New bounds on the complexity of the shortest path problem", *SIAM Journal on Computing*, 5(1), 83–89.

FREDMAN, M. L. and R. E. TARJAN (1984), "Fibonacci heaps and their uses in improved network optimization algorithms", *Proceedings of 25th Annual IEEE Symposium on the Foundations of Computer Science*, pp. 338–346.

FREIVALDS, R. (1977), "Probabilistic machines can use less running time", *Proceedings of Information Processing 77*, pp. 839–842.

FREIVALDS, R. (1979), "Fast probabilistic algorithms", *Proceedings of 8th Symposium on the Mathematical Foundations of Computer Science*, Lecture Notes in Computer Science, 74, Springer-Verlag, Berlin, pp. 57–69.

FURMAN, M. E. (1970), "Application of a method of fast multiplication of matrices in the problem of finding the transitive closure of a graph" (in Russian), *Doklady Akademii Nauk SSSR*, 194, 524.

GARDNER, M. (1977), "Mathematical games: A new kind of cipher that would take millions of years to break", *Scientific American*, 237(2), 120–124.

GARDNER, M. and C. H. BENNETT (1979), "Mathematical games: The random number omega bids fair to hold the mysteries of the universe", *Scientific American*, 241(5), 20–34.

GAREY, M. R. and D. S. JOHNSON (1976), "Approximation algorithms for combinatorial problems: An annotated bibliography", in Traub (1976), pp. 41–52.

GAREY, M. R. and D. S. JOHNSON (1979), *Computers and Intractability: A Guide to the Theory of NP-Completeness*, W. H. Freeman and Co., San Francisco, CA.

GENTLEMAN, W. M. and G. SANDE (1966), "Fast Fourier transforms—for fun and profit", *Proceedings of AFIPS Fall Joint Computer Conference*, 29, Spartan, Washington, DC, pp. 563–578.

GILBERT, E. N. and E. F. MOORE (1959), "Variable length encodings", *Bell System Technical Journal*, 38(4), 933–968.

GLEICK, J. (1987), "Calculating pi to 134 million digits hailed as great test for computer", *New York Times*, c. March 14.

GODBOLE, S. (1973), "On efficient computation of matrix chain products", *IEEE Transactions on Computers*, C-22(9), 864–866.

GOLDWASSER, S. and J. KILIAN (1986), "Almost all primes can be quickly certified", *Proceedings of 18th Annual ACM Symposium on the Theory of Computing*, pp. 316–329.

GOLDWASSER, S. and S. MICALI (1984), "Probabilistic encryption", *Journal of Computer and System Sciences*, 28(2), 270–299.

GOLOMB, S. and L. BAUMERT (1965), "Backtrack programming", *Journal of the ACM*, 12(4), 516–524.

GONDRAN, M. and M. MINOUX (1979), *Graphes et algorithmes*, Eyrolles, Paris; translated as: *Graphs and Algorithms* (1984), John Wiley & Sons, New York, NY.

GONNET, G. H. (1984), *Handbook of Algorithms and Data Structures*, Addison-Wesley, Reading, MA.

GONNET, G. H. and J. I. MUNRO (1986), "Heaps on heaps", *SIAM Journal on Computing*, 15(4), 964–971.

GOOD, I. J. (1968), "A five-year plan for automatic chess", in *Machine Intelligence*, vol. 2, E. Dale and D. Michie, eds., American Elsevier, New York, NY, pp. 89–118.

GRAHAM, R. L. and P. HELL (1985), "On the history of the minimum spanning tree problem", *Annals of the History of Computing*, 7(1), 43–57.

GREENE, D. H. and D. E. KNUTH (1981), *Mathematics for the Analysis of Algorithms*, Birkhauser, Boston, MA.

GRIES, D. (1981), *The Science of Programming*, Springer-Verlag, New York, NY.

GRIES, D. and G. LEVIN (1980), "Computing Fibonacci numbers (and similarly defined functions) in log time", *Information Processing Letters*, 11(2), 68–69.

HALL, A. (1873), "On an experimental determination of π", *Messenger of Mathematics*, 2, 113–114.

HAMMERSLEY, J. M. and D. C. HANDSCOMB (1965), *Monte Carlo Methods*; reprinted in 1979 by Chapman and Hall, London.

HARDY, G. H. and E. M. WRIGHT (1938), *An Introduction to the Theory of Numbers*, Oxford Science Publications, Oxford, England; fifth edition, 1979.

HAREL, D. (1987), *Algorithmics: The Spirit of Computing*, Addison-Wesley, Reading, MA.

HARRISON, M. C. (1971), "Implementation of the substring test by hashing", *Communications of the ACM*, 14(12), 777–779.

HELD, M. and R. KARP (1962), "A dynamic programming approach to sequencing problems", *SIAM Journal on Applied Mathematics*, 10(1), 196–210.

HELLMAN, M. E. (1980), "The mathematics of public-key cryptography", *Scientific American*, 241(2), 146–157.

HOARE, C. A. R. (1962), "Quicksort", *Computer Journal*, 5(1), 10–15.

HOPCROFT, J. E. and R. KARP (1971), "An algorithm for testing the equivalence of finite automata", Technical report TR-71-114, Department of Computer Science, Cornell University, Ithaca, NY.

HOPCROFT, J. E. and L. R. KERR (1971), "On minimizing the number of multiplications necessary for matrix multiplication", *SIAM Journal on Applied Mathematics*, 20(1), 30–36.

HOPCROFT, J. E. and R. E. TARJAN (1973), "Efficient algorithms for graph manipulation", *Communications of the ACM*, 16(6), 372–378.

HOPCROFT, J. E. and R. E. TARJAN (1974), "Efficient planarity testing", *Journal of the ACM*, 21(4), 549–568.

HOPCROFT, J. E. and J. D. ULLMAN (1973), "Set merging algorithms", *SIAM Journal on Computing*, 2(4), 294–303.

HOPCROFT, J. E. and J. D. ULLMAN (1979), *Introduction to Automata Theory, Languages, and Computation*, Addison-Wesley, Reading, MA.

HOROWITZ, E. and S. SAHNI (1976), *Fundamentals of Data Structures*, Computer Science Press, Rockville, MD.

HOROWITZ, E. and S. SAHNI (1978), *Fundamentals of Computer Algorithms*, Computer Science Press, Rockville, MD.

HU, T. C. and M. R. SHING (1982), "Computations of matrix chain products", Part I, *SIAM Journal on Computing*, 11(2), 362–373.

HU, T. C. and M. R. SHING (1984), "Computations of matrix chain products", Part II, *SIAM Journal on Computing*, 13(2), 228–251.

ITAI, A. and M. RODEH (1981), "Symmetry breaking in distributive networks", *Proceedings of 22nd Annual IEEE Symposium on the Foundations of Computer Science*, pp. 150–158.

JANKO, W. (1976), "A list insertion sort for keys with arbitrary key distribution", *ACM Transactions on Mathematical Software*, 2(2), 143–153.

JARNÍK, V. (1930), "O jistém problému minimálnim", *Praca Moravské Prirodovedecké Spolecnosti*, 6, 57–63.

JENSEN, K. and N. WIRTH (1985), *Pascal User Manual and Report*, third edition revised by A. B. Michel and J. F. Miner, Springer-Verlag, New York, NY.

JOHNSON, D. B. (1975), "Priority queues with update and finding minimum spanning trees", *Information Processing Letters*, 4(3), 53–57.

JOHNSON, D. B. (1977), "Efficient algorithms for shortest paths in sparse networks", *Journal of the ACM*, 24(1), 1–13.

KAHN, D. (1967), *The Codebreakers: The Story of Secret Writing*, Macmillan, New York, NY.

KALISKI, B. S., R. L. RIVEST, and A. T. SHERMAN (1988), "Is the Data Encryption Standard a group?", *Journal of Cryptology*, (1), in press.

KANADA, Y, Y. TAMURA, S. YOSHINO, and Y. USHIRO (1986), "Calculation of π to 10,013,395 decimal places based on the Gauss-Legendre algorithm and Gauss arctangent relation", manuscript.

KARATSUBA, A. and Y. OFMAN (1962), "Multiplication of multidigit numbers on automata" (in Russian), *Doklady Akademii Nauk SSSR*, 145, 293–294.

KARP, R. (1972), "Reducibility among combinatorial problems", in *Complexity of Computer Computations*, R. E. Miller and J. W. Thatcher, eds., Plenum Press, New York, NY, pp. 85–104.

KARP, R. and M. O. RABIN (1987), "Efficient randomized pattern-matching algorithms", *IBM Journal of Research and Development*, 31(2), 249–260.

KASIMI, T. (1965), "An efficient recognition and syntax algorithm for context-free languages", Scientific Report AFCRL-65-758, Air Force Cambridge Research Laboratory, Bedford, MA.

KLAMKIN, M. S. and D. J. NEWMAN (1967), "Extensions of the birthday surprise", *Journal of Combinatorial Theory*, 3(3), 279–282.

KLEENE, S. C. (1956), "Representation of events in nerve nets and finite automata", in *Automata Studies*, C.E. Shannon and J. McCarthy, eds., Princeton University Press, Princeton, NJ, pp. 3–40.

KNUTH, D. E. (1968), *The Art of Computer Programming*, 1: *Fundamental Algorithms*, Addison-Wesley, Reading, MA; second edition, 1973.

KNUTH, D. E. (1969), *The Art of Computer Programming*, 2: *Seminumerical Algorithms*, Addison-Wesley, Reading, MA; second edition, 1981.

KNUTH, D. E. (1971), "Optimal binary search trees", *Acta Informatica*, 1, 14–25.

KNUTH, D. E. (1973), *The Art of Computer Programming*, 3: *Sorting and Searching*, Addison-Wesley, Reading, MA.

KNUTH, D. E. (1975a), "Estimating the efficiency of backtrack programs", *Mathematics of Computation*, 29, 121–136.

KNUTH, D. E. (1975b), "An analysis of alpha-beta cutoffs", *Artificial Intelligence*, 6, 293–326.

KNUTH, D. E. (1976), "Big Omicron and big Omega and big Theta", *SIGACT News*, ACM, 8(2), 18–24.

KNUTH, D. E. (1977), "Algorithms", *Scientific American*, 236(4), 63–80.

KNUTH, D. E., J. H. MORRIS, and V. R. PRATT (1977), "Fast pattern matching in strings", *SIAM Journal on Computing*, 6(2), 240–267.

KRANAKIS, E. (1986), *Primality and Cryptography*, Wiley-Teubner Series in Computer Science.

KRUSKAL, J. B., Jr. (1956), "On the shortest spanning subtree of a graph and the traveling salesman problem", *Proceedings of the American Mathematical Society*, 7(1), 48–50.

LAURIÈRE , J.-L. (1979), *Eléments de programmation dynamique*, Bordas, Paris.

LAWLER, E. L. (1976), *Combinatorial Optimization: Networks and Matroids*, Holt, Rinehart and Winston, New York, NY.

LAWLER, E. L. and D. W. WOOD (1966), "Branch–and–bound methods: A survey", *Operations Research*, 14(4), 699–719.

LECARME, O. and J.-L. NEBUT (1985), *Pascal pour programmeurs*, McGraw-Hill, Paris.

LECLERC, G. L. (1777), *Essai d'arithmétique morale*.

LEHMER, D. H. (1969), "Computer technology applied to the theory of numbers", in *Studies in Number Theory*, W. J. LeVeque, ed., Mathematical Association of America, Providence, RI, p. 117.

LENSTRA, H. W., Jr. (1982), "Primality testing", in Lenstra and Tijdeman (1982), pp. 55–97.

LENSTRA, H. W., Jr. (1986), "Factoring integers with elliptic curves", report 86-18, Mathematisch Instituut, Universiteit van Amsterdam; to appear in *Annals of Mathematics*.

LENSTRA, H. W., Jr. and R. TIJDEMAN, eds. (1982), *Computational Methods in Number Theory*, Part I, Mathematical Centre Tracts 154, Mathematisch Centrum, Amsterdam.

LEVIN, L. (1973), "Universal search problems" (in Russian), *Problemy Peredaci Informacii*, 9, 115–116.

LEWIS, H. R. and C. H. Papadimitriou (1978), "The efficiency of algorithms", *Scientific American*, 238(1), 96–109.

LUEKER, G. S. (1980), "Some techniques for solving recurrences", *Computing Surveys*, 12(4), 419–436.

MARSH, D. (1970), "Memo functions, the Graph Traverser, and a simple control situation", in *Machine Intelligence*, 5, B. Meltzer and D. Michie, eds., American Elsevier, New York, NY, and Edinburgh University Press, pp. 281–300.

MCDIARMID, C. J. H. and B. A. REED (1987), "Building heaps fast", submitted to the *Journal of Algorithms*.

MELHORN, K. (1984a), *Data Structures and Algorithms*, 1: *Sorting and Searching*, Springer-Verlag, Berlin.

MELHORN, K. (1984b), *Data Structures and Algorithms*, 2: *Graph Algorithms and* **NP**-*Completeness*, Springer-Verlag, Berlin.

MELHORN, K. (1984c), *Data Structures and Algorithms,* 3 : *Multi-Dimensional Searching and Computational Geometry,* Springer-Verlag, Berlin.

METROPOLIS, I. N. and S. ULAM (1949), "The Monte Carlo method", *Journal of the American Statistical Association,* 44(247), 335–341.

MICHIE, D. (1968), " 'Memo' functions and machine learning", *Nature,* 218, 19–22.

MONIER, L. (1980), "Evaluation and comparison of two efficient probabilistic primality testing algorithms", *Theoretical Computer Science,* 12, 97–108.

MONTGOMERY, P. L. (1987), "Speeding the Pollard and elliptic curve methods of factorization", *Mathematics of Computation,* 48(177), 243–264.

NEMHAUSER, G. (1966), *Introduction to Dynamic Programming,* John Wiley & Sons, Inc., New York, NY.

NILSSON, N. (1971), *Problem Solving Methods in Artificial Intelligence,* McGraw-Hill, New York, NY.

PAN, V. (1978), "Strassen's algorithm is not optimal", *Proceedings of 19th Annual IEEE Symposium on the Foundations of Computer Science,* pp. 166–176.

PAPADIMITRIOU, C. H. and K. STEIGLITZ (1982), *Combinatorial Optimization : Algorithms and Complexity,* Prentice-Hall, Englewood Cliffs, NJ.

PERALTA, R. C. (1986), "A simple and fast probabilistic algorithm for computing square roots modulo a prime number", *IEEE Transactions on Information Theory,* IT-32(6), 846–847.

PIPPENGER, N. (1978), "Complexity theory", *Scientific American,* 238(6), 114–124.

POHL, I. (1972), "A sorting problem and its complexity", *Communications of the ACM,* 15(6), 462–463.

POLLARD, J. M. (1971), "The fast Fourier transform in a finite field", *Mathematics of Computation,* 25(114), 365–374.

POLLARD, J. M. (1975), "A Monte Carlo method of factorization", *Bit,* 15, 331–334.

POMERANCE, C. (1982), "Analysis and comparison of some integer factoring algorithms", in Lenstra and Tijdeman (1982), pp. 89–139.

POMERANCE, C. (1987), "Very short primality proofs", *Mathematics of Computation,* 48(177), 315–322.

PRATT, V. R. (1975), "Every prime has a succinct certificate", *SIAM Journal on Computing,* 4(3), 214–220.

PRIM, R. C. (1957), "Shortest connection networks and some generalizations", *Bell System Technical Journal,* 36, 1389–1401.

PURDOM, P. W., Jr. and C. A. BROWN (1985), *The Analysis of Algorithms,* Holt, Rinehart and Winston, New York, NY.

RABIN, M. O. (1976), "Probabilistic algorithms", in Traub (1976), pp. 21–39.

RABIN, M. O. (1980a), "Probabilistic algorithms in finite fields", *SIAM Journal on Computing,* 9(2), 273–280.

RABIN, M. O. (1980b), "Probabilistic algorithm for primality testing", *Journal of Number Theory,* 12, 128–138.

RABINER, L. R. and B. GOLD (1974), *Digital Signal Processing,* Prentice-Hall, Englewood Cliffs, NJ.

REINGOLD, E. M., J. NIEVERGELT, and N. DEO (1977), *Combinatorial Algorithms : Theory and Practice,* Prentice-Hall, Englewood Cliffs, NJ.

RIVEST, R. L. and R. W. FLOYD (1973), "Bounds on the expected time for median computations", in *Combinatorial Algorithms*, R. Rustin, ed., Algorithmics Press, New York, NY, pp. 69–76.

RIVEST, R. L., A. SHAMIR, and L.M. ADLEMAN, (1978), "A method for obtaining digital signatures and public-key cryptosystems", *Communications of the ACM*, 21(2), 120–126.

ROBSON, J.M. (1973), "An improved algorithm for traversing binary trees without auxiliary stack", *Information Processing Letters*, 2(1), 12–14.

ROSENTHAL, A. and A. GOLDNER (1977), "Smallest augmentation to biconnect a graph", *SIAM Journal on Computing*, 6(1), 55–66.

RUDIN, W. (1953), *Principles of Mathematical Analysis*, McGraw-Hill, New York, NY.

RUNGE, C. and H. KÖNIG (1924), *Die Grundlehren der Mathematischen Wissenschaften*, 11, Springer, Berlin.

RYTTER, W. (1980), "A correct preprocessing algorithm for Boyer-Moore string searching", *SIAM Journal on Computing*, 9(3), 509–512.

SAHNI, S. and E. HOROWITZ (1978), "Combinatorial problems: reducibility and approximation", *Operations Research*, 26(4), 718–759.

SCHÖNHAGE, A. and V. STRASSEN (1971), "Schnelle Multiplikation grosser Zahlen", *Computing*, 7, 281–292.

SCHWARTZ, E. S. (1964), "An optimal encoding with minimum longest code and total number of digits", *Information and Control*, 7(1), 37–44.

SCHWARTZ, J. (1978), "Probabilistic algorithms for verification of polynomial identities", Computer Science Department, Courant Institute, New York University, Technical Report no. 604.

SEDGEWICK, R. (1983), *Algorithms*, Addison-Wesley, Reading, MA.

SHAMIR, A. (1979), "Factoring numbers in $O(\log n)$ arithmetic steps", *Information Processing Letters*, 8(1), 28–31.

SHANKS, D. (1972), "Five number-theoretic algorithms", *Proceedings of the Second Manitoba Conference on Numerical Mathematics*, pp. 51–70.

SLOANE, N. J. A. (1973), *A Handbook of Integer Sequences*, Academic Press, New York, NY.

SOBOL', I. M. (1974), *The Monte Carlo Method*, second edition, University of Chicago Press, Chicago, IL.

SOLOVAY, R. and V. STRASSEN (1977), "A fast Monte-Carlo test for primality", *SIAM Journal on Computing*, 6(1), 84–85; erratum (1978), *ibid*, 7, 118.

STANDISH, T. A. (1980), *Data Structure Techniques*, Addison-Wesley, Reading, MA.

STINSON, D. R. (1985), *An Introduction to the Design and Analysis of Algorithms*, The Charles Babbage Research Centre, St. Pierre, Manitoba.

STOCKMEYER, L. J. (1973), "Planar 3-colorability is polynomial complete", *SIGACT News*, 5(3), 19–25.

STOCKMEYER, L. J. and A. K. CHANDRA (1979), "Intrinsically difficult problems", *Scientific American*, 240(5), 140–159.

STONE, H. S. (1972), *Introduction to Computer Organization and Data Structures*, McGraw-Hill, New York, NY.

STRASSEN, V. (1969), "Gaussian elimination is not optimal", *Numerische Mathematik*, 13, 354–356.

TARJAN, R. E. (1972), "Depth-first search and linear graph algorithms", *SIAM Journal on Computing*, 1(2), 146–160.

TARJAN, R. E. (1975), "On the efficiency of a good but not linear set merging algorithm", *Journal of the ACM*, 22(2), 215–225.

TARJAN, R. E. (1981), "A unified approach to path problems", *Journal of the ACM*, 28(3), 577–593.

TARJAN, R. E. (1983), *Data Structures and Network Algorithms*, SIAM, Philadelphia, PA.

TRAUB, J. F., ed. (1976), *Algorithms and Complexity: Recent Results and New Directions*, Academic Press, New York, NY.

TURING, A. M. (1936), "On computable numbers with an application to the Entscheidungsproblem", *Proceedings of the London Mathematical Society*, 2(42), 230–265.

TURK, J. W. M. (1982), "Fast arithmetic operations on numbers and polynomials", in Lenstra and Tijdeman (1982), pp. 43–54.

URBANEK, F. J. (1980), "An $O(\log n)$ algorithm for computing the nth element of the solution of a difference equation", *Information Processing Letters*, 11(2), 66–67.

VALOIS, D. (1987), *Algorithmes probabilistes: une anthologie*, Masters Thesis, Département d'informatique et de recherche opérationnelle, Université de Montréal.

VAZIRANI, U. V. (1986), *Randomness, Adversaries and Computation*, Doctoral dissertation, Computer Science, University of California, Berkeley, CA.

VAZIRANI, U. V. (1987), "Efficiency considerations in using semi-random sources", *Proceedings of 19th Annual ACM Symposium on the Theory of Computing*, pp. 160–168.

VICKERY, C. W. (1956), "Experimental determination of eigenvalues and dynamic influence coefficients for complex structures such as airplanes", *Symposium on Monte Carlo Methods*, H. A. Meyer, ed., John Wiley & Sons, New York, NY, pp. 145–146.

WAGNER, R. A. and M. J. FISCHER (1974), "The string-to-string correction problem", *Journal of the ACM*, 21(1), 168–173.

WARSHALL, S. (1962), "A theorem on Boolean matrices", *Journal of the ACM*, 9(1), 11–12.

WARUSFEL, A. (1961), *Les nombres et leurs mystères*, Editions du Seuil, Paris.

WEGMAN, M. N. and J. L. CARTER (1981), "New hash functions and their use in authentication and set equality", *Journal of Computer and System Sciences*, 22(3), 265–279.

WILLIAMS, H. (1978), "Primality testing on a computer", *Ars Combinatoria*, 5, 127–185.

WILLIAMS, J. W. J. (1964), "Algorithm 232: Heapsort", *Communications of the ACM*, 7(6), 347–348.

WINOGRAD, S. (1980), *Arithmetic Complexity of Computations*, SIAM, Philadelphia, PA.

WRIGHT, J. W. (1975), "The change-making problem", *Journal of the ACM*, 22(1), 125–128.

YAO, A. C. (1975), "An $O(|E|\log\log|V|)$ algorithm for finding minimum spanning trees", *Information Processing Letters*, 4(1), 21–23.

YAO, A. C. (1982), "Theory and applications of trapdoor functions", *Proceedings of 23rd Annual IEEE Symposium on the Foundations of Computer Science*, pp. 80–91.

YAO, F. F. (1980), "Efficient dynamic programming using quadrangle inequalities", *Proceedings of 12th Annual ACM Symposium on the Theory of Computing*, pp. 429–435.

YOUNGER, D. H. (1967), "Recognition of context-free languages in time n^3", *Information and Control*, 10(2), 189–208.

ZIPPEL, R. E. (1979), *Probabilistic Algorithms for Sparse Polynomials*, Doctoral dissertation, Massachusetts Institute of Technology, Cambridge, MA.

Index